Future of Economy in Asia

Evan Lau · Weng Marc Lim · Huey Fen Cheong
Editors

Future of Economy in Asia

Geopolitical Diplomacy, Technological Revolutions and Sustainable Pathways

Springer

Editors
Evan Lau
Faculty of Economics and Business
Universiti Malaysia Sarawak
Kota Samarahan, Malaysia

Weng Marc Lim
Sunway Business School
Sunway University
Selangor, Malaysia

Huey Fen Cheong
Universiti Malaya
Kuala Lumpur, Malaysia

ISBN 978-3-031-95499-3 ISBN 978-3-031-95500-6 (eBook)
https://doi.org/10.1007/978-3-031-95500-6

This Springer imprint is published by the registered company Springer Nature Switzerland AG
The registered company address is: Gewerbestrasse 11, 6330 Cham, Switzerland

Titles on the "Futures of Asia"

Future of Business in Asia: Exploring Pathways to Tomorrow's Business Landscape
Future of Cities in Asia: Future Direction of Urban Planning, Technology and Environmental Sustainability
Future of Economy in Asia: Geopolitical Diplomacy, Technological Revolutions and Sustainable Pathways
Future of Education in Asia: Sustainability, Technology and Internationalisation
Future of Healthcare in Asia: Advancing Equity, Ethics and Technology in Asian Healthcare
Future of Media in Asia: Artificial Intelligence, Digital Technology and Media Practice
Future of Work in Asia: Transforming Work Models, Workforce Dynamics and Workplace Technology

Preface

In the rapidly evolving global economy, Asia stands at the epicentre of dynamic transformation, shaped by unprecedented economic, technological, and geopolitical shifts. This book delves into these transformative forces through the lens of future studies, blending theoretical insights, empirical evidence, and conceptual frameworks to map Asia's economic trajectories. Drawing on Fergnani's (2019) six research clusters in future studies—corporate foresight, past and futures, humanity at the limen, environmental futures, post-normality and complexity, and technological trends—this book offers a holistic view of Asia's future.

This book is structured into three interconnected parts. Part I: *Economic Growth and Geopolitical Diplomacy* explores Asia's evolving economic landscape, examining historical patterns, policy decisions, and governance structures that influence regional growth and resilience. It highlights ASEAN's long-term economic trajectory, Vietnam's strategic ascent, Sri Lanka's crisis, and the post-pandemic recovery of India and Southeast Asia, showcasing how geopolitics and economic policies shape the region's future. Part II: *Technological and Transformative Future* investigates Asia's digital transformation, emphasising the role of disruptive innovation, digital payments, leadership, and ICT infrastructure in driving economic progress. Case studies from Malaysia, China, Japan, and other emerging economies illustrate how visionary leadership and technological adaptation are key to sustaining Asia's global competitiveness. Part III: *The Future of Sustainability* addresses the critical need for balancing economic expansion with environmental and social responsibility, focusing on themes such as ocean sustainability, clean energy transitions, trade policies, and the adoption of green hydrogen technology.

Part I: Economic Growth and Geopolitical Diplomacy

This part unravels the intricate relationship between economic development and geopolitics, exploring how strategic decisions shape Asia's economic destiny. It begins by tracing historical trends, identifying structural challenges, and examining the geopolitical strategies that have propelled Asia's rise. Chapter 1 projects the ASEAN economy through 2050 and 2100, emphasising its interconnectedness with China and India. By analysing macroeconomic trends, the chapter situates ASEAN within the broader global economic landscape, highlighting the region's evolving role as a geopolitical bridge. Chapter 2 explores Vietnam's strategic economic rise, demonstrating how a combination of geopolitical positioning, governance, and policy decisions contribute to sustainable growth. Chapter 3 provides an in-depth analysis of Sri Lanka's economic crisis, offering lessons on fiscal responsibility, governance, and risk management. It highlights how governance failures contribute to economic vulnerabilities. The key lesson for Asia is the importance of sound governance, strategic partnerships, and sustainable economic policies for resilience and growth. Chapter 4 shifts focus to India and Southeast Asia's post-pandemic recovery, exploring how these economies are navigating the dual challenges of crisis management and adaptation to new economic realities. Together, these chapters highlight the delicate balance between growth, governance, and geopolitical strategy in shaping Asia's future.

Part II: Technological and Transformative Future

Technology is reshaping Asia's economic landscape at an unprecedented pace. This part examines how innovation, digital infrastructure, and visionary leadership are driving the region's transformation. Chapter 5 analyses the impact of disruptive vision on firm performance, using evidence from Malaysia to highlight a broader principle relevant to Asia's diverse economies. Firms embracing disruptive visions drive innovation, efficiency, and growth in emerging sectors like digital technology and renewable energy. Empirical findings suggest that visionary leadership leads to superior financial outcomes. Policymakers can support this by fostering innovation-driven growth through incentives, regulations, and knowledge networks, positioning Asia's firms for sustained success in the global economy. Chapter 6 examines the evolution of cross-border digital payments in Southeast Asia, emphasising regulatory frameworks and technological infrastructure. Strengthening digital payments will be crucial for Asia's economic future while fostering financial inclusion and deeper global integration. Chapter 7 compares innovation strategies in Japan and

China, focusing on their responses to the COVID-19 pandemic. The chapter examines how different economies adapt to technological disruptions. Chapter 8 investigates digital banking adoption in Malaysia, incorporating financial literacy, security, and reputation into the technology acceptance model. It explores behavioural factors shaping digital banking adoption. Chapter 9 examines the role of leadership quality and ICT infrastructure in the finance-growth nexus across 50 emerging economies, highlighting the critical role of digital infrastructure in sustainable economic development. Policymakers must leverage these synergies for long-term prosperity. Strengthening leadership and innovation will be crucial for Asia's economic resilience and global competitiveness in the digital era. Together, these chapters paint a picture of a region on the cusp of a technological revolution, where innovation and leadership will determine Asia's economic resilience and global competitiveness.

Part III: The Future of Sustainability

As Asia continues to experience rapid economic growth, sustainability has emerged as a defining challenge for the region. This part delves into how Asia can strike a balance between economic expansion and environmental and social responsibility, with a focus on sustainable resource management, clean energy investments, and innovative policies. The interconnected themes of ocean sustainability, clean energy transitions, trade dynamics, and green hydrogen technology are explored across four chapters, each contributing to a holistic vision for Asia's sustainable future. Ocean sustainability is vital for Asia's economic growth and food security, yet ocean literacy remains low. Chapter 10 explores its evolution, impact on sustainable seafood consumption, and research gaps. Effective ocean literacy fosters marine conservation and sustainable fisheries, ensuringlong-term economic stability. To support the UN 2030 SDGs, biodiversity education, seafood sustainability policies, and innovative methodologies are essential. Institutional change is needed to promote "ocean literacy for all," securing Asia's blue economy and future economic resilience. Building on the theme of sustainability, Chapter 11 explores Asia's clean energy transition as a pivotal driver of sustainable growth in the face of rising carbon emissions. The chapter analyses the significant influence of monetary policy, urbanisation, and green growth on clean energy investments, while noting the limited impact of green technology innovation. To accelerate progress, the study advocates for effective policies such as green finance, smart cities, and regulatory incentives. Strengthening public-private partnerships and implementing robust risk mitigation strategies are identified as critical steps to ensure Asia's long-term energy sustainability and economic resilience.

Chapter 12 examines the interconnected dynamics of energy, trade, and economic growth in the Asia-Pacific region, emphasising the need for trade policies that balance economic expansion with environmental sustainability. The study highlights how trade can drive growth without compromising energy efficiency and conservation efforts. By reinforcing the trade-led growth hypothesis, the chapter underscores the importance of strategic policies that promote sustainable development while enhancing the region's economic resilience and global competitiveness. Chapter 13 explores green hydrogen technology and the industrial capabilities needed for its adoption in Asia's energy transition. While China leads production, lower-income nations face industrial and policy challenges, and high-income countries like Japan must maximise their potential. Strengthening technology transfers, value chains, and market development is key. Green hydrogen will enhance Asia's energy security, drive industrial modernisation, and support sustainable growth, positioning the region as a leader in clean energy.

Shaping Asia's Future

Asia's future will be shaped by the dynamic interplay of geopolitics, technology, and sustainability. As the region deepens its integration with the global economy, proactive policies and strengthened regional collaboration will be essential in navigating both emerging challenges and new opportunities. By addressing structural challenges, fostering innovation, and prioritising sustainable practices, the region can unlock its immense potential and lay the foundation for a resilient, inclusive, and prosperous future. This book offers a roadmap for navigating Asia's complex economic transformation. Through informed decision-making and strategic action, Asia can unlock its vast potential, fostering a resilient, inclusive, and prosperous global economy.

Kota Samarahan, Malaysia Evan Lau
lphevan@unimas.my

References

Fergnani, A. (2019). Mapping futures studies scholarship from 1968 to present: A bibliometric review of thematic clusters, research trends, and research gaps. Futures 105:104–123.

Fergnani, A. (2020). Mapping Fifty Years of Futures Studies Scholarship (1968–2017). In R. Slaughter & A. Hines (Eds.), *The Knowledge Base of Futures Studies 2020* (p. 24–47). US: Association of Professional Futurists & Foresight International.

Acknowledgements

The editors would like to express their sincerest gratitude to all those who contributed to the completion of this book and the *Futures of Asia* series, which comprises seven volumes, including this one.

First and foremost, we extend our deepest appreciation to the chapter authors for their hard work and kind cooperation throughout the process. We are also grateful to the individuals and organisations who helped disseminate the calls for papers (CFPs), such as the Regional Studies Association (RSA), various academic research groups on social media (particularly Facebook), and our respective institutions, including our faculties and international relations offices. Special thanks go to our assistant, Fatin Najihah, for her invaluable support in the early stages of this project.

We are also deeply grateful for the constructive feedback provided by the peer reviewers of this book, as well as the opportunities, guidance and support from Springer's commissioning editor, Doris Bleier.

Additionally, we would also like to acknowledge and thank the following scholars for their insightful feedback on individual chapter manuscripts, both published and unpublished in this volume.

Reviewed Two Manuscripts

1. Abdul Hayy Haziq Mohamad (University of Technology Sarawak, Malaysia)
2. Alfie Maria Custodio (Notre Dame of Dadiangas University, Philippines)
3. Junlin He (Universiti Malaya, Malaysia)
4. Muhammad Nouman Akhtar (Allama Iqbal College Satellite Town, Pakistan)
5. Tariq Mahmood Khan (Education Department, Government of Punjab, Pakistan)
6. Chang Yee Kwan (National Chengchi University, Taiwan).

Reviewed One Manuscript

1. Abdul Rahim Ridzuan (Universiti Teknologi MARA, Malaysia)
2. George Halkos (University of Thessaly, Greece)
3. Lim Thye Goh (Universiti Malaya, Malaysia)
4. Ka Shing Ko (City University, Cambodia)
5. Masom Jan Masomy (Afghanistan Academy of Sciences, Afghanistan)
6. Mst Sabina Tabasum (South Asian University, India)
7. Sharizal Hashim (Universiti Kebangsaan Malaysia, Malaysia)
8. Silmara Furtado da Silva (Pontifical Catholic University of Rio de Janeiro, Brazil)
9. Zahid Hussain (Khadim Hussain Shah Institute of Technology, Pakistan).

We also thank peer reviewers who opted for anonymity.

Once again, we extend our deepest gratitude for the support, time, and expertise of these individuals. Their contributions have been instrumental in making this book a reality. In case we have unintentionally overlooked any individual or organisation that has contributed to this project, we express our sincere appreciation and gratitude as well.

Contents

List of Figures

List of Tables

About the Editors

Evan Lau is a Professor at Universiti Malaysia Sarawak (UNIMAS), recognised for his academic leadership and research excellence. He has served in key roles, including Deputy Dean for Research and Commercialisation and Director of the Centre for Business, Economics, and Finance Forecasting. An internationally acclaimed scholar, Evan has held visiting positions at the University of Cambridge and several universities in Indonesia. His prolific research output includes numerous journal articles, conference papers, and influential publications, earning him multiple awards for research excellence and high citation impact. A sought-after speaker at international conferences, he is also a dedicated mentor to postgraduate students. Outside academia, Evan enjoys running and travelling.

Weng Marc Lim is a Distinguished Professor and the Dean of Sunway Business School at Sunway University, an ASU-Cintana Alliance Global Partner Affiliate Faculty at Arizona State University in the USA, and an Adjunct Professor at Swinburne University of Technology's home campus in Melbourne, Australia, and international branch campus in Sarawak, Malaysia, where he served as Dean and Head of School as well as on the University Council. Professor Lim is the Record Holder for "Youngest Professor" in the Malaysia Book of Records and features in the Academy of Sciences Malaysia's Top Research Scientists Malaysia and JCI's Ten Outstanding Young Malaysians. With over 100 publications in prestigious "A*" and "A" ranked journals, Prof. Lim's research has been cited more than 45,000 times. As a thought leader, Prof. Lim has led high-level discussions at AppliedHE, Association to Advance Collegiate Schools of Business (AACSB), Association of MBAs (AMBA), Business Graduates Association (BGA), Quacquarelli Symonds (QS), Times Higher Education (THE), United Nations Educational, Scientific and Cultural Organization (UNESCO) and Principles of Responsible Management Education (UNPRME), Nikkei Forum, and World Economic Forum (WEF). Professor Lim holds a Bachelor of Business and Commerce with First Class Honors and a Doctor of Philosophy from Monash University, and multiple certificates post-Ph.D., including a Graduate

Certificate in Learning and Teaching (Higher Education) with High Distinction from Swinburne University of Technology and a Certificate in School Management and Leadership from Harvard University.

Huey Fen Cheong is a Senior Lecturer at Universiti Malaya, Malaysia. Her interests have been multidisciplinary, mainly linguistics, sociology, education, marketing, and psychology. Driven by a cultural and postcolonial perspective, she collaborated with a group of established scholars to initiate this *Futures of Asia* series (which contains this book) and the earlier *COVID-19 in Malaysia* series. These initiatives aim to promote Global South research and publications, especially on significant events such as the COVID-19 pandemic, post-pandemic normality and transformations, and the rise of AI, which equally matter to non-Western regions and deserve equal scholarly attention (and acceptance) beyond Western-centric frameworks in academia.

Editors' Note

The clusters of future studies in each chapter are based on Fergnani (2019, 2020), who identified six (6) key research/thematic clusters from reviews of future studies since 1968: "corporate foresight; past and futures; humanity at the limen; environmental futures; post-normality and complexity, and technological trends". While the chapter topics present different aspects of economy in Asia, the clusters of future studies show different dimensions of futures and future studies.

References

Fergnani, A. (2019). Mapping futures studies scholarship from 1968 to present: A bibliometric review of thematic clusters, research trends, and research gaps. *Futures*, *105*: 104–23.

Fergnani, A. (2020). Mapping fifty years of futures studies scholarship (1968–2017). In R. Slaughter & A. Hines (Eds.), *The Knowledge Base of Futures Studies 2020* (pp. 24–47). US: Association of Professional Futurists and Foresight International.

Part I
Economic Growth and Geopolitical Diplomacy

Chapter 1
Foreseeing the ASEAN Economy in the Shadow of Dragon (China) and Elephant (India): 2050 and 2100

Cluster of Future Studies: Humanity at the Limen

Tuck Cheong Tang

Abstract The rise of China and India is observable recently. This future study of both empirical and conceptual types aims to foresees the ASEAN economy in association with China, and India in 2050 and 2100. The current data 1973 (1982)–2022 are utilised for generating the future data of selected core macroeconomic variables, namely real GDP, trade, and net financial account. The future trends of these variables are relatively different from the presently observed. Indeed, the connectedness of them among ASEAN, China, and India is supported by cointegration and Granger non-causality tests. Such connectedness is stronger toward the future of 2050 and 2100. This chapter forwards that Asia can help Asia. The future of ASEAN in 2050 remains as present, while 2010 remarks a take-off for ASEAN to integrate 'sufficiently' with China, and India.

Keywords ASEAN · China · India · Interconnectedness · Trendline

1.1 Introduction

China-India-ASEAN cooperation is expanding for a long especially in the fields of economic and scientific cooperation in which the rise of China and India could shift the political and economic center of the world to Asia in the next decade (Hoontrakul et al. 2014). Perhaps, India is recently witnessed her major strides in expanding its influence in Southeast Asia, which is inevitable that a move that will allow countries to counter China's dominance in the region. This chapter is about a future study of the Association of Southeast Asian Nations (ASEAN) which is a regional intergovernmental organisation focused on promoting economic and security cooperation among

T. C. Tang (✉)
Department of Economics, Faculty of Business and Economics, Universiti Malaya, Kuala Lumpur, Malaysia
e-mail: tangtuckcheong@um.edu.my

E. Lau et al. (eds.), *Future of Economy in Asia*,
https://doi.org/10.1007/978-3-031-95500-6_1

its ten members, viz. Brunei, Cambodia, Indonesia, Laos, Malaysia, Myanmar, the Philippines, Singapore, Thailand, and Vietnam, along the rises [growing influences] of China (symbolised as 'Dragon') and India (symbolised as 'Elephant'). According to Kuosa (2011), futures study focuses on long-term future spanning between 10 and 50 years where it considers on visionary decisions, aiming to delve deeply and more critically into the unpredictable and uncertain foundations of futures developments (see, Asian Development Bank, 2011, 2013). It seeks to define visions and innovative actions including creative ideas for ensuring survival and a prosperous future. Geographically, they are illustrated in Fig. 1.1 for visual convenience.

The ASEAN countries record a total population of 663 million with a combined gross domestic product (GDP) of $3.3 trillion in 2021. From 2012 to 2021, they achieved an average annual of 4.1%. ASEAN plays a crucial role in Asian economic

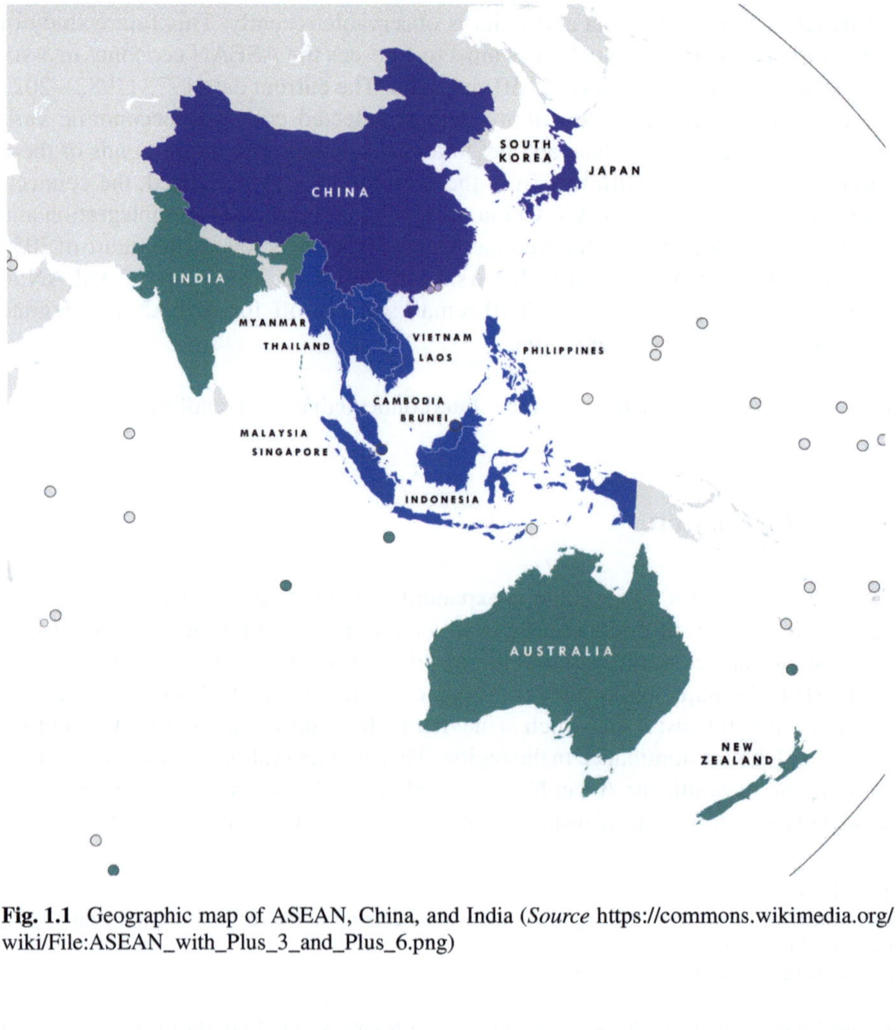

Fig. 1.1 Geographic map of ASEAN, China, and India (*Source* https://commons.wikimedia.org/wiki/File:ASEAN_with_Plus_3_and_Plus_6.png)

integration by participating in negotiations to established the largest free trade agreement in the world and signing six free trade deals with other economies in the region.[1] China's GDP has reported a $17.8 trillion in 2021 with a population of 1.4 billion, while India's GDP is $3.2 trillion and eventually same population size (1.4 billion).[2] China-ASEAN cooperation is expanding for a long especially in the fields of economic and scientific cooperation which is an significant and dynamic aspect of China-ASEAN relations. While, India is recently witnessed her major strides in expanding its influence in Southeast Asia, which is inevitable that a move that will allow countries to counter China's dominance in the region. In fact, ASEAN's consumer market has surpassed India's and approaching the size of China's market (ASEAN Secretariat 2006, p. 2).

According Zhang and Shen (2012), the rise of ASEAN, China and India signifies an important shift in global economic power and, they are foreseen to become the largest economic bloc in the world by 2030 (p. 93). Despite being nestled between Asia's giants of China and India, each with populations exceeding a billion, ASEAN stands out as an important regional player with a population of 622 million is nearly double the America's 320 million and surpasses the European Union's (EU) 506 million. Collectively, ASEAN is being considered an economic powerhouse, ranking as the world's seventh-biggest economy. Although ASEAN's economic growth hasn't matched the rapid pace of either China or India, it has been one of the fastest-growing markets globally, with an average annual growth of 5.1% between 2000 and2013. Perhaps, Indonesia with its 260 million people, is the largest economy within ASEAN which contributes for nearly 40% of ASEAN's GDP (West 2018, p. 301).

ASEAN Secretariat (2006, p. 1) acknowledges the dynamic performance of China and India as leading economic powers has substantially influenced the global and regional economic landscape, particularly in terms of export markets, resources allocations, and new sources of import demand. The economic influence of China and India on ASEAN as well as other developing and developed countries and regions, is most evident through international and regional trade (p. 2). Reddy (2020) probes the China factor in India-ASEAN relations. According to Chatterji (2021), historically, China viewed ASEAN as a model designed to encircle China. China positioned its approach to regionalism by promoting a "peaceful rise", committing itself to harmony, peace and stability. In 1991, ASEAN and China have initiated a dialogue process. In 1996, China has officially participated ASEAN as a full dialogue partner. The response of individual ASEAN member countries to China can be expressed as 'bandwagoners' (Cambodia, Laos, and Brunei), 'hesitant hedgers' (Thailand, Indonesia, the Philippines, Malaysia, and Myanmar), and 'active hedgers' (Singapore, and Vietnam). The study notes that ASEAN has committed to upholding a

[1] What is ASEAN? https://www.cfr.org/backgrounder/what-asean; ASEAN Statistical Yearbook 2022, the latest publication searched in 8th January, 2024. https://www.aseanstats.org/wp-content/uploads/2023/04/ASYB_2022_423.pdf.

[2] World Development Indicators, World Bank. https://databank.worldbank.org/source/world-development-indicators.

'rules-based regional architecture' in order to pursue cooperation, peace, stability and prosperity. Meanwhile, India became ASEAN Dialogue Partner in 1996. In 2019, India is being recorded as ASEAN's sixth largest trading partner and eight largest source of FDI among ASEAN Dialogue Partners. At the 2nd ASEAN-India Summit in 2003, the leaders signed ASEAN-India Framework Agreement on Comprehensive Economic Cooperation. And, India's relationship with ASEAN is a cornerstone of its foreign policy and the foundation of Act East Policy, which emphasises on connectivity, commerce and culture to enhance ASEAN-India integration.[3]

The next section looks at the methods of futures studies for the section are quantitatively based on time series methods, namely trendline analysis, cointegration tests, and Granger non-causality tests, respectively. Also, using visioning futures studies method, this chapter further reviews the available documents about ASEAN, China, and India, and foresee them for 2050, and 2100, respectively from the economic perspective. Two key issues to be considered here. First is about the current trend of real GDP, trade (as % of GDP), and net financial account till 2022, and foresees their future data (i.e. trend forecasting) to 2050 and 2100. The second issue is the "interconnectedness" of Asian Economies in a relation of ASEAN, China, and India with both observed data and future data (2023–2050 and–2100). Before concluding this chapter, a section to answer a future direction either Asia can help Asia or not?

1.2 Methodology

This future study incorporates both quantitative and qualitative methods. First, it provides a brief overview of three time-series econometric methods: trend forecasting, cointegration tests, and Granger non-causality tests. These methods are then applied to project future scenarios for ASEAN, China, and India for the years 2050 and 2100, using a visioning approach.

1.2.1 Polynomial Trendline

The trend analysis considers a polynomial trendline technique which is based on historical data i.e. time series 1973/1982–2022. Such trend extrapolation is one of the most commonly used for forecasting to 2050 and 2100, respectively. More technically, it is a curved line to capture fluctuating data. The order of the polynomial is suggested by the number of fluctuations in the data or the number of bends (i.e. hills and valleys) presented in the curve. A second-order (quadratic) polynomial trendline

---

[3] ASEAN-India Economic Relations, https://asean.org/our-communities/economic-community/integration-with-global-economy/asean-india-economic-relations/

India-ASEAN Relations, https://www.mea.gov.in/Portal/ForeignRelation/India-ASEAN-Relations-August-2018.pdf.

typically assumes has only one hill or valley which can be written as $y = a_2{*}x^2 + a_1{*}x + b$, and computed by Microsoft Excel.[9,4]

1.2.2 *Cointegration Test–Johansen's Method (Johansen and Juselius, 1990)*

Economic theory advocates that numerous time series are moving together over time, fluctuating around a long-run equilibrium. Such long-run equilibrium is captured and tested by the concept of cointegration. Technically, cointegration exists iftwo or more nonstationary time seriesshare a long-run equilibrium, that is move together in such that their linear combination produces a stationary property, and they exhibit a common stochastic trend. To consider a matrix, Y_t, which is composed of three time series variables: $y_1 = (y_{11}, y_{12}, \ldots, y_{1t})$, $y_2 = (y_{21}, y_{22}, \ldots, y_{2t})$, and $y_3 = (y_{31}, y_{32}, \ldots, y_{3t})$. All three series are nonstationary time series, at least $I(1)$. Cointegration implies that while y_1, y_2, and y_3 are independently nonstationary, given that their linear combination is stationary: $Y_t = \beta_1 y_{1t} + \beta_2 y_{2t} + \beta_3 y_{3t} \sim (0)$.[5] The Johansen test is one of the methods to test the existence of a cointegrating relation among several non-stationary time series. The Johansen test considers more than one cointegrating relation with two test statistics, i.e., trace tests and maximum eigenvalue test. Firstly, the trace test evaluates the number of linear combinations in a time series data, i.e., K to be equal to the value $K_{0,}$ and the hypothesis for the value K to be greater than K_0 that is H_0: $K = K_0$, and H_0: $K > K_0$. By setting K_0 to zero to test whether the null hypothesis will be rejected. If it is rejected, there exists a cointegration relationship between the time series. Therefore, the null hypothesis should be rejected to confirm the existence of a cointegrating relation. While, the maximum eigenvalue test is a non-zero vector which, when a linear transformation is applied to it, changes by a scalar factor. The key difference to trace test, is the null hypothesis. H_0: $K = K_0$, and H_0: $K = K_0 + 1$. If $K = K_0$ and the null hypothesis is rejected, there is only one possible outcome of the variable to produce a stationary process. However, when $K_0 = m-1$ and the null hypothesis is rejected, there are M possible linear combinations.[6]

---

[4] Microsoft. Choosing the Best Trendline for Your Data, https://support.microsoft.com/en-gb/office/choosing-the-best-trendline-for-your-data-1bb3c9e7-0280-45b5-9ab0-d0c93161daa8.

[5] A Guide to Conducting Cointegration Tests, https://www.aptech.com/blog/a-guide-to-conducting-cointegration-tests/.

[6] Cointegration. https://corporatefinanceinstitute.com/resources/data-science/cointegration/.

1.2.3 Granger Non-Causality Tests (Granger, 1980)

The concept of causality involves the relationship between cause and effect, or directed influence in time series data, defined by predictability and temporal precedence. One event leads to another, exemplified by 'chicken or egg' dilemma. Granger causality is a widely applied and conventional method for studying casual links between at least two random variables, helping to understand how different phenomena are interconnected. A variable **y** Granger-causes another variable **x** if the prediction of **x**'s values improves when the past values of **y**, when all other relevant information z is taken into account.[7] It assumes the data are stationary, $I(0)$ that is with a constant mean, constant variance, and no seasonal component. Stationary can be achieved by transforming the non-stationary, $I(d)$ data to stationary data by differencing it, either once or twice.[8] For example, in a trivariate framework of X_t, Y_t and Z_t, the following VAR equations can be specified:

$$\Delta x_t = \alpha + \sum_{j=1}^{p=3} \beta_{1,i}(\Delta x_{t-i}) + \sum_{j=1}^{p=3} \lambda_{1,i}(\Delta y_{t-i}) + \sum_{j=1}^{p=3} \psi_{1,i}(\Delta z_{t-i}) + \mu_t \quad (1)$$

$$\Delta y_t = \theta + \sum_{j=1}^{p=3} \lambda_{2,i}(\Delta y_{t-i}) + \sum_{j=1}^{p=3} \beta_{2,i}(\Delta x_{t-i}) + \sum_{j=1}^{p=3} \psi_{2,i}(\Delta z_{t-i}) + \varepsilon_t \quad (2)$$

$$\Delta z_t = \phi + \sum_{j=1}^{p=3} \psi_{3,i}(\Delta z_{t-i}) + \sum_{j=1}^{p=3} \beta_{3,i}(\Delta x_{t-i}) + \sum_{j=1}^{p=3} \lambda_{3,i}(\Delta y_{t-i}) + \tau_t \quad (3)$$

where Δ is first difference operator, p is the lag length to be included, μ_t, ε_t and τ_t are error terms for three equations respectively which is assumed to have zero mean, constant variances and uncorrelated. In VAR equations, the computed *Chi*-square test statistics are used to reject the null hypothesis if the computed p-value for the relevant VAR model equations is less than the level of significance, let's say 1%, 5% and 10%. For example, the first null hypothesis H_0: $\sum \lambda_{1,i} = 0$ reflects the y does not Granger cause the x. If the computed *Chi*-square test statistics is greater than its critical value let say at 10% level, the null hypothesis can be rejected and to conclude that the direction of Granger causality is from y to the x. The second null hypothesis H_0: $\sum \psi_{1,i} = 0$ reflects z does not Granger cause x. For Eqs. (2) and (3), similar procedure applies.

---

[7] Granger Causality, https://www.sciencedirect.com/topics/neuroscience/granger-causality#:~:text=Granger%20causality%20is%20defined%20as,values%20of%20another%20time%20series.

[8] Granger Causality in Time Series–Explained using Chicken and Egg Problem, https://www.analyticsvidhya.com/blog/2021/08/granger-causality-in-time-series-explained-using-chicken-and-egg-problem/#:~:text=A%20prerequisite%20for%20performing%20the,order%20or%20second%2Dorder%20differencing.

1.2.4 Visioning

Visioning is one of the most popular and important futures studies (qualitative) methods which helps futurist and economist to develop their vision of a desirable future for themselves, organisation, or community, in general. Indeed, visioning on larger scales such as national or global scales as this chapter, remains relatively undeveloped.[9] Hence, this futures studies method of visioning which identifies the economic related sources (via. Google search) in the past and present, employ their current projections/forecasts, and offer a sense of current drivers of change, so a range of alternative futures can be imagined, and facilitates a process of achieving some consensus of a preferred vision for the future of ASEAN, China, and India in 2050 and 2100, respectively.

1.3 Key Concepts and Issues

This section identifies and analyses three key issues related to the future economy of ASEAN, China, and India. The first issue examines future trends in three critical macroeconomic variables—real GDP, trade, and the financial account of the balance of payments (BoP)—and whether these variables can be forecasted for 2050 and 2100. The second issue evaluates the results of cointegration tests to determine whether ASEAN, China, and India have exhibited economic interconnectedness in the past and whether such connections are likely to persist or evolve by 2050 and 2100. Finally, the third issue explores future causality patterns among ASEAN, China, and India with respect to these three macroeconomic variables.

1.3.1 The Past Tells the Future

Is the future economy predicable from the current economy (i.e. the existing information such as trend)? Probably–"*We experience things 'now' and assume that the past is done and the future has yet to happen, but depending on speed and location (Einstein's Theory of General Relativity), your future can be someone else's past*".[10] Retrospectively, Aristotle expressed by saying "*if the 'now' were not different but one and the same, there would not have been time*" (Shoemaker 1969, p. 363). The future data are windowed by the periods of 2023 to 2050 and to 2100 benchmarking two perspectives. First is to benchmark the two available documents from Google

[9] See, Prallagon consulting Group, Methods and approaches of futures studies. https://prallagon.com/wp-content/uploads/2021/05/Methods-For-Futures-Studies.pdf.

[10] Why the Past, Present, and Future May Exist All at Once. https://www.grunge.com/731050/why-the-past-present-and-future-may-exist-all-at-once/.

search, namely "The World in 2050",[11] and "What Will the World Economy Look Like in 2100?"[12,13]

In line with these impressionistic views, this section offers a trend (forecast) analysis of three core macroeconomic variables, namely real GDP (gross domestic product, constant 2015 US$), Trade (i.e. sum of exports and imports, % of GDP), and NFA (Net Financial Account, BoP, current US$). The sample are ASEAN which is the sum of all 10 member countries with their data availability, China, and India. The data are collected from World Development Indicators (WDI), the World Bank[14] covering the sample period between 1973 (1982 for NFA) and 2022. The so-called 'future data' are forecasted ("trended") from 2023 to 2100 by a second-order polynomial trend (or quadratic polynomial trendline) trend estimator, which is the best fit of the observed data (1973/1982−2022) in a comparison of other methods such as exponential, linear, logarithmic, and power. If the data set is large enough and alternately increases and decreases, fluctuates, the best type of trendline is polynomial.[15] The time series plots with future data of RGDP, trade, and NFA are illustrated by Figs. 1.2, 1.3 and 1.4.

Figure 1.2 illustrates an upward trend of the observed real GDP of ASEAN, China, and India (1973–2022) including their forecasted behaviour to 2100. China's real GDP is escalating above ASEAN, and India all the time by 292% of 2050 is above 2022, and 305% of 2010 above 2050. By 2042, ASEAN's real GDP is above India,

---

[11] The World in 2050–The long view: how will the global economic order change by 2050? https://www.pwc.com/gx/en/research-insights/economy/the-world-in-2050.html.

[12] What Will the World Economy Look Like in 2100? https://www.ubss.edu.au/articles/2022/july/what-will-the-world-economy-look-like-in-2100/.

[13] Based on a Google search, this chapter can't ignore the inspiration by '*Tui bei tu*' (推背图), the 43rd prophecy as interpreted that, because the Belt and Road Initiative (BRI) will connect China with many countries, Chinese culture, politics and economy will surely influence the world, and Chinese culture will become the mainstream of the world after 2050. The 47th prophecy is predicted onset 2100 that, giving up force and ruling the world with civilisation is to practice martial arts and cultivate culture-it is the key for "Eastern Galaxy Civilisation" to defeat "Western Galaxy Civilisation"-everyone feels that they should make some contribution to society, which is everyone's responsibility. *Tui bei tu* is a Chinese prophecy book from the 7th-century Tang dynasty. The book is known for predicting the future of China, and is written by Li Chunfeng (李淳风) and Yuan Tiangang (袁天罡) (Wiki). The 43rd prophecy odes as "The black rabbit walks into the green dragon's cave; it can't express its desire; it can only be found outside on the tree; in thirty years, descendants will be born" (黑兔走入青龙穴。欲尽不尽不可说。惟有外边根树上。三十中子孙结). The 47rd prophecy odes as translated as "There is no king or emperor to determine the world; the first person from the field; so I can read more old books; a wise word can be seen once a righteous word is spoken" (无王无帝定乾坤。来自田间第一人。好把旧书多读到。义言一出见英明). Their interpretations are based on the available search bellows which relates to economics and Asia in 2050 and 2100, respectively that does not represent the author's view (interpretation). They are (1) What does the 43 prophecy in *Tui bei tu* mean? (https://ichingnfengshui.blogspot.com/2018/07/blog-post_31.html); (2) Mathematical Analysis, Paul R. Smith (2003); and, (3) Hai Zhitao [海之涛] The correct prediction science of the 44th, 45th, 46th and 47th images of *Tui bei tu*" (https://blog.wenxuecity.com/myblog/79079/202110/22858.html).

[14] World Development Indicators, the World Bank. https://databank.worldbank.org/source/world-development-indicators

[15] Polynomial trend equation and forecast, https://www.officetooltips.com/excel_365/tips/polynomial_trend_equation_and_forecast#google_vignette.

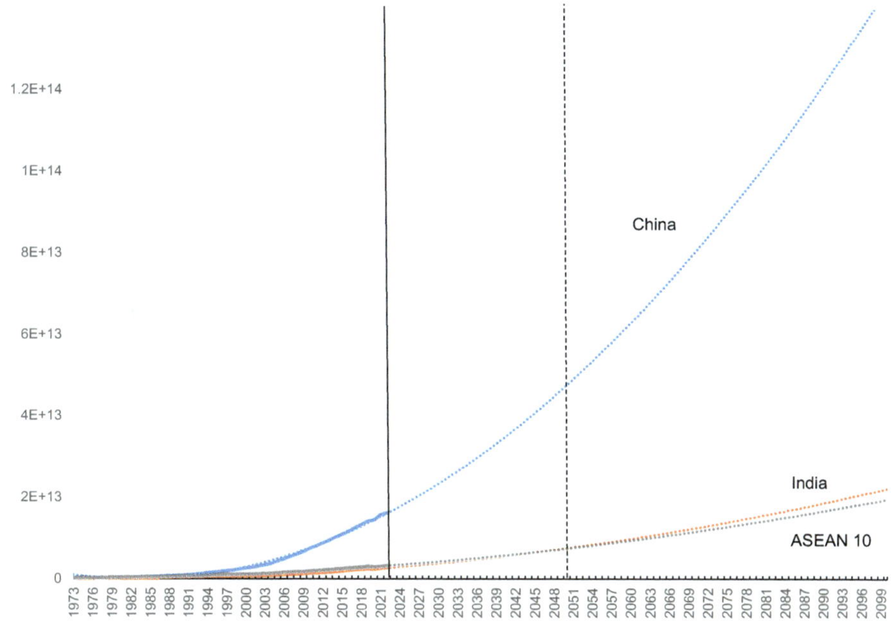

Fig. 1.2 Real GDP (at a constant 2015 prices, US$) for 1973–2022, and forested to 2050 and 2100

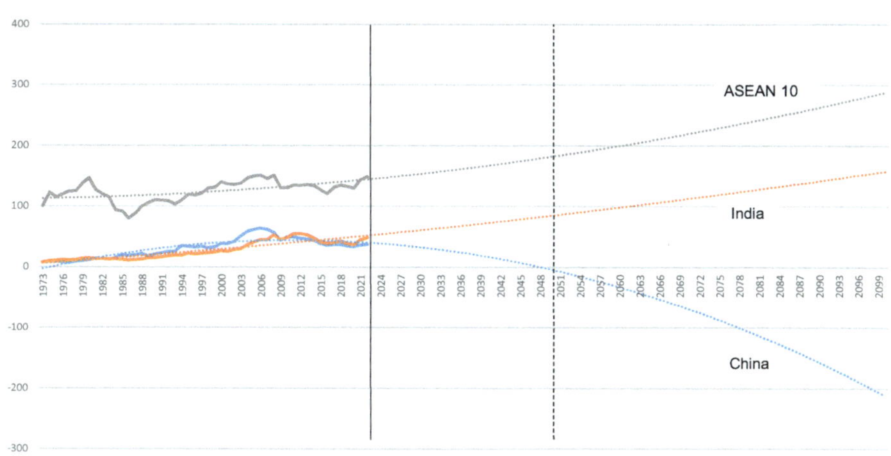

Fig. 1.3 Trade (% of GDP) for 1973–2022, and forested to 2050 and 2100

but the opposite is hold then, but their real GDP gap is narrow and both are moving together till 2100. Such real GDP gap is widening between China, and ASEAN and India approaching 2050 and 2100. A speedy increase in China's real GDP captures an intuition that China will become more advanced through technological progressions in which technology is a revolutionary factor production with no population growth is assumed. Perhaps, middle income trap that most of ASEAN, China, and India are now or next future failing to transition to high income economies due to rising costs including cost of living and declining competitiveness. Given changes to middle income classification thresholds, Fig. 1.2 signals that China can escape (moves out of) middle income trap by 2050 toward 2100 as a high income nation. However, it is not the case for ASEAN (except for, Singapore), and India. Asia is confronted with the challenge of maintaining rapid growth after transforming from low to middle income, and aims to reach high income status. The experience of newly industrialising economies highlight that innovation, human capital, and infrastructure are necessary to move beyond the middle income trap-from middle to high income (Asian Development Bank 2017).

Turning to the trade, Fig. 1.3 shows a gradual upward trend of ASEAN, China, and India within the observed period 1973–2022. It is interesting to note that the future data 2023–2050 and 2100) continuously exhibit an upward trend gradually by moving together between ASEAN and India. However, the future China's trade is divergence from others reaching a negative trade (as % of GDP) in 2049 and onward with is out of the conventional wisdom. Perhaps, it assumes a downward trend of China's trade which ultimately consistent with Fig. 1.2 in which China's real GDP increases extraordinary high after 2022 relative to a slower China's trade exposure. From now and approaching 2050, it may be a dramatically and evolutionally structural changes of the trade (exports and imports) which is resulted from production of technology (capital) goods, as well as trade and diplomatic policies in China ageist the West but favour to the East (i.e. Asia). As noted by Zhang and Shen (2012), the China-ASEAN FTA (multi-layered free trade agreement) has been completely fulfilled, and the ASEAN-India FTA has been concluded. Nevertheless, Sathyamoorthy and Tang (2019) found that the export-led growth does only hold for India and Thailand. No empirical evidence of the exports expansion to spur growth for other available ASEAN (Indonesia, Laos, Malaysia, Philippines, Singapore, and Thailand), and China. It probably explains the decreasing future trend of trade by shifting other engines of growth such as technology and innovation.

Net financial account of BoP is about claims on or liabilities to non-residents relating financial assets which includes direct investment, portfolio investment, and reserve assets. Figure 1.4 illustrates that they are tightly moving of each other between 1982 and 1997 before China 'taking-off' to the peak in 2008. The ASEAN recorded a positive NFA since 1997 till the end of observed period 2022, while India's NFA is negative most of the time. China's NFA is above ASEAN, and India. However, the future data (2023–2100) reveal that NFA of ASEAN and India moving together except for China with a substantially decrease. China, and ASEAN fail to negative NFA in 2047, and 2069, respectively, and India since 1982. A negative NFA indicates that the financial account experience a deficiency when a country's nationals purchase more foreign assets than foreign buyers purchase domestic assets. This will result in a negative financial account. China's negative NFA is predicted to move below ASEAN from 2036 and onward, while below India since 2066. As highlighted by Walsh (2014), Asian countries often have more financial sectors compared to other regions with same income level (p. 19), and it is expected that the Asian economies are expected to become more entwined, with investors more likely to invest assets

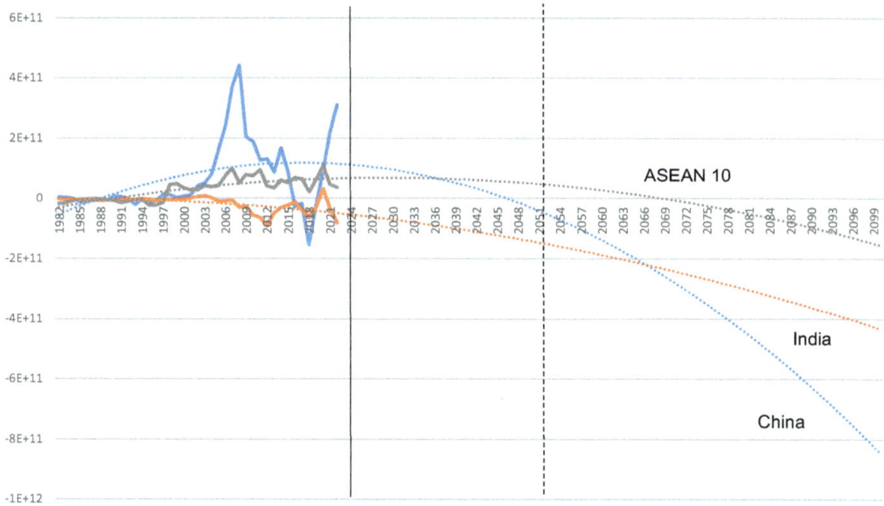

Fig. 1.4 Net financial account (BoP, current US$) for 1982–2022, and forested to 2050 and 2100

in other countries over the next 20 years. This observation is particularly notable in Southeast Asia, where ASEAN countries are working for further integrating their economies (p. 21). ASEAN remains a major host of foreign direct investment (FDI) (ASEAN Secretariat 2006, p. 1). Tang (2002) finds the financial liberalisation of five ASEAN founding economies (Malaysia, Singapore, Indonesia, Thailand, and the Philippines) are cointegrated based on annual data 1965–1998.

1.3.2 The Interconnectedness of Asian Economies–ASEAN, China and India

The 2013 IMF's (International Monetary Fund) recent 'spillover' analysis has reinforced the importance of the interconnectedness among countries i.e. how events in one country impact others. If the world's top five economies collaborated to carry a more rigorous, comprehensive and compatible policies, world GDP could increase by about 3% in the longer run.[16] This section captures such interconnectedness among ASEAN, China, and India within the observed sample period of 1973/1982–2022, and to 2050, and 2100 (the latter both are forecasted values illustrated as Figs. 1.2, 1.3 and 1.4). Two-time series econometric approaches are being conventionally and practically considered, namely cointegration that is, the underlying variables (real GDP, trade, and NFA) among ASEAN, China, and India are moving together in the long-run. The second is the Granger non-causality which to inform that one of the

[16] The Interconnected Global Economy: Challenges and Opportunities for the United States—and the World, by Christine Lagarde, Managing Director, International Monetary Fund. https://www.imf.org/en/News/Articles/2015/09/28/04/53/sp091913.

three countries/group) "*came first*". Interconnectedness is observed by the finding of bidirectional causality(ies).

Table 1.1 reports the number of [static] cointegrating relation(s) for both trace and maximum eigenvalue tests of the underlying eight deterministic case. The augmented Dickey-Fuller test (ADF) show all variables are non-stationary at levels, or $I(1)$. Their statistics are not reported here but available upon request. Given the observed period of 1973–2022, at least one cointegrating (long-run) relation does exist among ASEAN, China, and India for output (RGDP), and net financial account (NFA), except for trade. For RGDP, the maximum eigenvalue tests offer no cointegration for six deterministic cases. Such no long-run relation of trade (both trace and maximum eigenvalue tests of all eight deterministic cases) among ASEAN, China, and India signifies that the so-called ASEAN-China/India, and China-India Free Trade Area (FTA) which includes FTA in goods, services and investment, are insufficiently and not efficiently implemented to liberalise further for promoting the economic integration among them. However, when the 'future' data (2023–2050 and 2023–2100) are considered, a favourable finding of cointegration (at least one long-run relation) is fond for trade. Such finding is enhanced with at least two cointegrating relations with more future of forward data of 2023–2100. Similar observations are found for both RGDP and NFA variables. The interconnectedness among ASEAN, China, and India for RGDP, trade, and NFA reinforces real market (RGDP and trade), and financial market (NFA) is a 'necessary and sufficient' condition in realising Asian economic integration in the long-run. Hence, such a finding of cointegration and stronger evidence with future data implies more interactions between real market and financial market among ASEAN, China, and India in future, indeed 2100.

RGDP is real domestic product at constant 2015 US$, and the data is transformed into natural logarithm, *ln*. Trade is trade as percentage of GDP. NFA is net financial account balance in which the data is available from 1982. Rank selected at 0.1 level using critical values from MacKinnon-Haug-Michelis (1999). The lag length is chosen by Akaike information criterion. The reported figures are the number of cointegrating relations those are statistically significant at least at 10% level. Deterministic Case Descriptions: Case 1: No deterministic terms; Case 2: Cointegrating relationship includes a constant; Case 3 (Johansen-Hendry-Juselius): Cointegrating relationship includes a constant. Short-run dynamics include a constant; Case 4 (Johansen-Hendry-Juselius): Cointegrating relationship includes a constant and trend. Short-run dynamics include a constant; Case 4: Cointegrating relationship includes a trend. Short-run dynamics include a constant; and Case 5 (Johansen-Hendry-Juselius): Both the cointegrating relationship and Case 5: Short-run dynamics include a constant and trend.

Turning to the second approach, the [dynamic] Granger non-causality tests which have been graphically illustrated as in Fig. 1.5 for their findings of RGDP, trade, and NFA among ASEAN, China, and India. The figures presented are the end-of-sample i.e. 2022, 2050, and 2100. To note that their directions of causality are rather to be interpreted as in the short-run given that all of the macroeconomic variables are first-differenced (i.e. $Xt–X_{t-1}$), in order to ensure stationary in trivariate VAR (Vector Autoregressive) framework. For the period 1973–2022, no interconnectedness that only a unidirectional causality of RGDP is observed from India to China. While, the

Table 1.1 Cointegration among ASEAN, China, and India–RGDP, Trade, and NFA

		Deterministic case							
1973/ 1982–2022	Test type	1	2	3 JHJ	3	4 JHJ	4	5 JHJ	5
*ln*RGDP (*l* = 2)	Trace	3	3	3	3	2	2	3	3
	Max-eigen	3	3	0	0	0	0	0	0
Trade (*l* = 1)	Trace	0	0	0	0	0	0	0	0
	Max-eigen	0	0	0	0	0	0	0	0
NFA (*l* = 3)	Trace	2	1	1	1	1	1	3	3
	Max-eigen	1	1	1	1	1	1	1	1
1973/1982–2050									
*ln*RGDP (*l* = 2)	Trace	3	2	3	3	2	2	1	1
	Max-eigen	1	2	3	3	2	2	1	1
Trade (*l* = 1)	Trace	1	2	2	2	1	1	1	1
	Max-eigen	1	0	0	0	1	1	1	1
NFA (*l* = 3)	Trace	1	1	1	1	2	2	2	2
	Max-eigen	1	1	1	1	2	2	2	2
1973/1982–2100									
*ln*RGDP (*l* = 2)	Trace	3	3	3	3	3	3	3	3
	Max-eigen	3	3	3	3	3	3	3	3
Trade (*l* = 1),	Trace	2	3	2	2	3	3	2	2
	Max-eigen	2	3	2	2	3	3	2	2
NFA (*l* = 3)	Trace	2	3	2	2	3	3	2	2
	Max-eigen	2	3	2	2	3	3	2	2

direction of causality for NFA is from ASEAN, and China to India, respectively. There is no evidence of causality for trade over the observed sample period ends 2022. Hence, ASEAN, China, and India do not sufficiently integrate each other at their current stage. However, the trade interconnectedness is fully captured by the data ends with 2100, that is bidirectional causalities of trade among ASEAN, China, and India. In fact, with future data ends 2050, only unidirectional causalities are found from China to India, and from ASEAN to China. Looking at the possible economic integration (real GDP), the 1973–2050 data inform that ASEAN does Granger-cause China, but a bidirectional causality between China, and India. Interestingly, full economic integration is expected with bidirectional causalities of real GDP among ASEAN, China, and India when future data till 2100 are being considered. And, the ASEAN, China, and India are shown to be dynamically interconnected with NFA for the future data 1982–2050 and 2100.

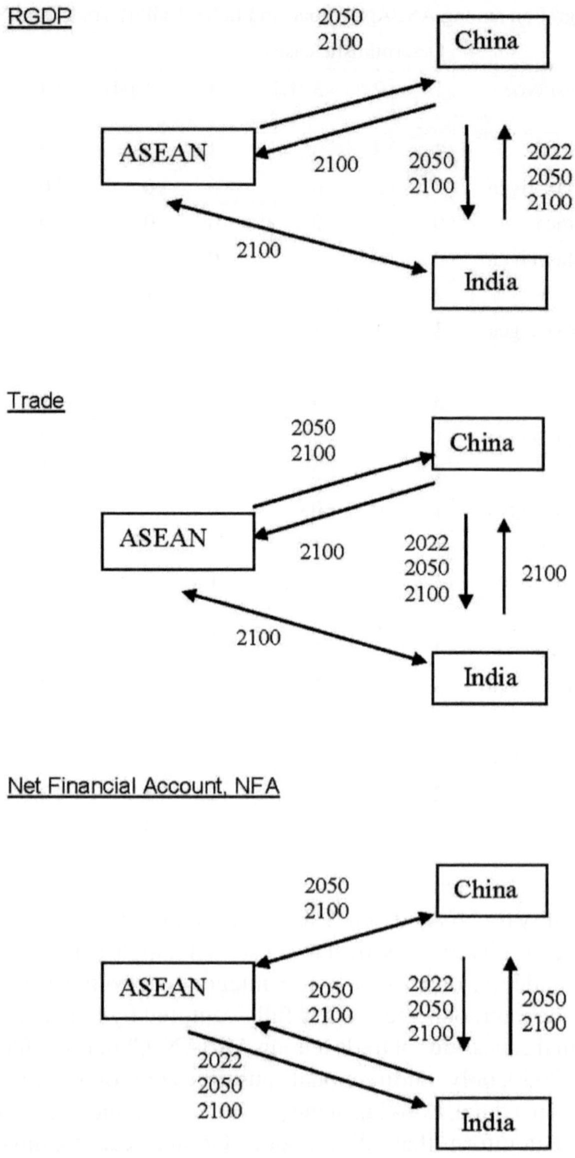

Fig. 1.5 Granger non-causality between ASEAN, China, and India[17]

[17] The variables are transformed in first-difference form to ensure stationary. The lag length is chosen by AIC, and the test statistics are statistically significant at least at 10% level. The reported years are the end of sample periods i.e. 1973 (1982, NFA) -2020 (observed data), 2050, and 2100 (both forecasted data).

1.4 Discussion of Implications

Foreseeing the Year 2050-as projected by the United Nations that by 2050, China's population will fall to around 1.32 billion, while India's population is projected to record 1.67 billion. Since the 1990s, China's fertility rate has decreased below the replacement level of 2.1.[18] By assuming various scenarios regarding the relative size of major economies by2050, Agarwal (2008) foresees a significant rise in the share of world income of China and partly India over the next four decades. This includes a major rise in China's influence over world economic management with coordination and cooperation among other BRICSAM countries (Brazil, Russia, India, China, South Africa and Mexico), which share similar interest in the functioning of the global economy.

"The World in 2050"[5] report foresees more than double in size the world economy by 2050 because of improvements in technology-driven productivity. The emerging economies in 2050 will be led by China (1st), India (2nd), and Indonesia (4th) in which the emerging economies need to enhance their institutions and their infrastructure substantially in order to achieve their long-term growth potential. Among ASEAN, Indonesia is only country projected to be dominate the world's top 10 economies in 2050 based on GDP at PPP (purchasing power parity). Accordingly, the soft power of China is the indirect and non-military influence of the People's Republic of China. Indeed, the BRI, an ambitious vision to establish two new trade routes connecting China with the global market by focusing on infrastructure development. Accordingly, China needs to develop an expanded and interdependent market, to spur its economic and political power, and lastly, to create the comfortable conditions for building a high technology economy.[19,20]

Similarly, India's soft power appeal continued to be influential even after the Cold War. In the 1990s, India was brought into Asian institutions by ASEAN, which saw the appeal of its growing economy and democratic values.[21] India and ASEAN are significant economic partners, with a promising future for enhancing e-commerce and

[18] For family and country: why China needs more babies. https://multimedia.scmp.com/infographics/news/china/article/3224346/china-population/index.html.

[19] What is China's Belt and Road Initiative (BRI)? https://www.chathamhouse.org/2021/09/what-chinas-belt-and-road-initiative-bri.

[20] The 43rd prophecy of '*Tui bei tu*' as be interpreted that "because the BRI will connect China with many countries, Chinese culture, politics and economy will surely influence the world, and Chinese culture will become the mainstream of the world after 2050". It visualises that China has already been a major economic power in Southeast Asia, with Chinese businesses investing in areas viz. manufacturing and information technology China snaps up ASEAN's green, digital investment opportunities.

https://asia.nikkei.com/Politics/International-relations/China-snaps-up-ASEAN-s-green-digital-investment-opportunities.

[21] India rising: Soft power and the world's largest democracy. https://www.brookings.edu/articles/india-rising-soft-power-and-the-worlds-largest-democracy/.

digital connectivity.[22] India's initiatives to strengthen relations with Southeast Asian countries should be collaborated on a long-term, sustained basis. By the same token, ASEAN may consider India as a potential counterbalance to China's influence in the region (Yahya 2003, p. 99). Generally speaking, it is about to further improve cooperation in various areas of mutual interest such as maritime activities, transnational crimes, cybersecurity, science and technology, transport and connectivity, digital economy, Fintech, tourism, agriculture, environment, sustainable development, and public health.[23] Rahul et al. (2004, pp. 3305–6) point out the future expansion of mutually beneficial economic relations between ASEAN and India which requires a well-planed exploration of emerging economic opportunities either by ASEAN as an organisation or individual ASEAN countries. India's vision of becoming a developed economy by 2020 has driven domestic reform initiatives and further integration with the world economies. ASEAN acknowledges the need to further diversify its growth engines beyond traditional approaches like the US, Japan and more recently, China and India.

Indeed, Gujrati and Uygun (2024) forward that India is continuously striving to strengthen its position as a "leading power" that India has met its political victories and advanced its national goals through its unique cultural diplomacy wealth i.e. diaspora, yoga, Buddhism, etc. The Ministry of External Affairs has adopted a soft power approach to successfully expand India's international presence. According to them, India has installed its influence and emerged as a prominent voice in the global stage. Also, Toyoda (2024) reports that India is outlooked to become the third-largest economy globally, and has a strong presence in international politics as she has made advancements in scientific achievements, but faces challenges in terms of regulations and infrastructure. Bloomberg Economics analysis, however foresees India could become the world's top contributor to GDP growth by 2028.[24] The World Economic Forum in January 15. 2024 adds that India is poised to have a defining role in imparting the future of the global economy in 2024 and beyond as India has been an important growth engine globally contributing 16% to the global growth in 2023. India's growth rate of 7.2% in fiscal year of 2022–2023, was the second-highest among the G20 countries and nearly double the emerging market economies.[25]

Zhang and Shen (2012, p. 93) consider ASEAN, China and India must address new challenges in particular. The "middle income trap" and approaching to a higher stage of economic development. Another old challenge returns as Wan and Tang (2021) reveal that COVID-19 data (31.12.2019–5.10.2020) is transitory (i.e. will be back) for

---

[22] ASEAN-India Relations: Potential for Further Growth. https://www.isas.nus.edu.sg/papers/asean-india-relations-potential-for-further-growth/.

[23] ASEAN, India strengthen comprehensive strategic partnership.
https://asean.org/asean-india-strengthen-comprehensive-strategic-partnership/.

[24] Can India eclipse China as world growth driver? By Scott, M., April 8, 2024. https://www.bloomberg.com/news/newsletters/2024-04-08/world-economy-latest-india-miracle.

[25] India could become the world's 3rd largest economy in the next 5 years. Here's how. January15, 2024. World Economic Forum. https://www.weforum.org/agenda/2024/01/how-india-can-seize-its-moment-to-become-the-world-s-third-largest-economy/#:~:text=On%20the%20economic%20front%2C%20India,emerging%20market%20economies%20that%20year.

China, probably India (inconclusive), and six of ASEAN member countries (Brunei, Laos, Malaysia, Myanmar, Thailand, and probably Vietnam) –another pandemics by 2050.[26,27,28] It is intuited that the future of ASEAN in 2050 (economic integration) will be remained as present given a short run of 26 years from now. Perhaps, ASEAN has to initiative the BRI dream of a complex network of mega infrastructure projects connecting China and other ASEAN member countries as well as India. The UN (United Nations) Global Biodiversity Framework which in equating human wellbeing with living in harmony with nature, setting goals to achieve it by 2050 (Doncaster and Bullock 2024) remains uncertain–indeed, just to take-off.

Foreseeing the Year 2100-For the next 77 years from now (2024), it is to visionalise what ASEAN, China, and India to be in 2100, rather than those what they have been achieved. For example, "The next 100 years: Vision 2100 for Malaysia" announced by the present Prime Minister of Malaysia that "*I think the country requires a new healthy culture (in order for it to develop)* via *exchanging of ideas throughout intellectual discourse.*"[29] The ongoing expansion of the global technological frontier and the convergence of labour productivity are positioning India to become the largest economy in the world, following by China, Nigeria, and the sixth for Indonesia -only ASEAN member country in 2100.[30] India's GDP is 51% larger than that of China,[31] and by the end of the century, India's GDP will be 90% exceeds China's and 30% surpassing the US, a new report has revealed.[32] India's 1.4 billion people making the country becomes the largest populated in the world surpassing China, its GDP is forecasted to grow significantly in the future. Goldman Sachs Research forecast that India will achieve the second-largest economy in the world by 2075.[33] To remember that Asia has emerged as an influential force in the world as well as politics due to the vast populations like China, India and Indonesia. These large populated economies have endowed their economic, market and financial influence, which has been fuelled

[26] Historical pandemics described since 541 (Plague of Justinian) to 2019 COVID-19. https://www.frontiersin.org/files/Articles/631736/fmicb-11-631736-HTML-r1/image_m/fmicb-11-631736-t001.jpg.

[27] 10 infectious diseases that could be the next pandemic. https://www.gavi.org/vaccineswork/10-infectious-diseases-could-be-next-pandemic.

[28] "*... Whether or not the coronavirus will return in 2030 as she claimed to predict, remains to be seen.*" From the Dream Diary of Ryo Tatsuki to the Eruption of Mount Fuji. https://medium.com/japan-blog/from-the-dream-diary-of-ryo-tatsuki-to-the-eruption-of-mount-fuji-a194022b9fd4.

[29] PM: The rot must stop here. https://www.nst.com.my/news/nation/2024/01/1004150/pm-rot-must-stop-here.

[30] As footnote 11.

[31] The World's Largest Economies in 2100–India No. 1. https://www.ubss.edu.au/articles/2022/september/the-world-s-largest-economies-in-2100-india-no-1/

[32] India to become world's top economic power by 2100, eclipsing US, China: report. https://www.indiaweekly.biz/india-to-become-worlds-top-economic-power-by-2100-report/#:~:text=Business-,India%20to%20become%20world%27s%20top%20economic%20power%20by%202100%2C%20eclipsing,infrastructure%20improvement%2C%20and%20environmental%20sustainability.

[33] How India could rise to the world's second-biggest economy, July 6, 2023. https://www.goldmansachs.com/intelligence/pages/how-india-could-rise-to-the-worlds-second-biggest-economy.html.

into economic, political and military influence. However, they are expected to face uncertainties and fragile superpowers (West, 2018, p. 323).

The population of Asia is 4.68 billion in 2100, while 3.92 billion for Africa, 0.59 billion for Europe, 0.45 billion for Northern America, and less for other continents.[34] West (2018) highlights that, while Asian Century hype has largely centered substantially on China's rise, Meanwhile, India's population is believed to surpass China's by 2022 and could be approximately 50% more r by 2100. If these observations persist, India's economic could surpass China's mostly in the second half of this century. This potential superpower shift might lead to political instability in Asia, since India has more allies in Asia than China (p.318), witnessing the rise of Elephant in 2100. The future of 2100 is a year for ASEAN, China, and India, perhaps Asia to re-think a vision of Utopia –an imagined place or state of things in which everything is perfect.[35] The *necessary and sufficient* condition of initiating the vision is, ASEAN as a group has to actively integrate their economic, culture, and mostly all aspects 'sufficiently' with China, and India.[36]

1.5 Future Directions

Last of this chapter concerns about a question that *Can Asia help Asia?* To note that Asia and Pacific are being considered as a relative bright spot amidst more gloomy context of the global economy's turbulent recovery. Asia is projected to contribute about 70% of global growth in 2023 in which 34.9% and 15.4% are from China and India. A 9.8% is from ASEAN (Thailand, Malaysia, the Philippines, Vietnam, and Indonesia) of which 4.4% by Indonesia.[37] Asian share of global GDP is likely around 57% by 2030, while 21% and 15% for Americas and Europe, respectively, and the rise of Asia is unlikely to halt anytime soon.[38] Acharya (2011) notes that limited recognition of about Asia's contribution to global governance. That is, the region's perspectives on international order are changing as Asia's economic and political influence grows. The leading Asian powers such as China, India and Japan, are increasingly focused on advancing and legitimizing their national power ambitions, rather than prioritizing contributions to global governance. According to West

[34] Statista-Forecast of the world population in 2022 and 2100, by continent (in billions). https://www.statista.com/statistics/272789/world-population-by-continent/.

[35] Oxford Reference. https://www.oxfordreference.com/display/10.1093/oi/authority.20110803115009560.

[36] As the 47th prophecy of '*Tui bei tu*' which is predicted to be onset 2100 -it gives up force and ruling the world with civilisation is to practice martial arts and cultivate culture-it is the key for "Eastern Galaxy Civilisation" to defeat "Western Galaxy Civilisation"-everyone feels that they should make some contribution to society, which is everyone's responsibility. See, footnote 12.

[37] Asia Poised to Drive Global Economic Growth, Boosted by China's Reopening. https://www.imf.org/en/Blogs/Articles/2023/05/01/asia-poised-to-drive-global-economic-growth-boosted-by-chinas-reopening

[38] The Future is Asian. https://www.worldeconomics.com/Thoughts/The-Future-is-Asian.aspx.

Table 1.2 The share of the global economy (rank), 2020–2100

	2020	2060	2100
Asia	47% (1st)	52% (1st)	44% (1st)
Africa	5% (4th)	16% (3rd)	39% (2nd)
America	25% (2nd)	18% (2nd)	11% (3rd)
Europe	22% (3rd)	13% (4th)	6% (4th)
Oceania	1% (5th)	1% (5th)	1% (5th)

(2018, p. 324), the prospects for Asian Century to achieve high income economies and middle-class societies will depend on how Asia responds to the following potential economic issues. They include maximizing the payoffs of global value chains, harnessing the potential urbanization, ensuring equal opportunities for all Asians, solving Asia's demographic dilemmas, correcting Asia's political flows, combating Asia's economic crime, and approaching peace and harmony. Indeed, many have anticipated significant benefits from such "Goldilocks scenario".

Table 1.2 informs that Asia's economy is the top (almost half) of the world in 2020, as well as the future of 2060 and 2100 conditioning the Asian Century. Interesting to note that Africa region is rising. Zhang and Shen (2012) consider the rise of ASEAN, China and India as economic powerhouses holds immense significance for both regional and global economic development. According to them, ASEAN, China and India are projected to promote an open, transparent and efficient regional including global environment. They should cooperate in shaping the architecture to ensure regional, to promote global free trade, investment and financial stability, and lastly to consider more active roles in future global governance and rule-making.

Duplicating as in What Will the World Economy Look Like in 2100? https://www.ubss.edu.au/articles/2022/july/what-will-the-world-economy-look-like-in-2100/, Table 5. The Global Economy, 2020–2100. The reported figures here are re-calculated as percentage of the world.

It is interesting to note that Indonesia positions itself as a leader within ASEAN serving as a significant mediator in disputes within the bloc such as rectifying intra-ASEAN disputes at the 2012 ASEAN Ministerial Meeting attests. Such consideration is based on its increasing economy powers throughout ASEAN countries. Indonesia's democracy is a role model for other ASEAN countries navigating authoritarianism or democratic transitions. Its dedication to peace and cooperation has contributed significantly in creating stability in Southeast Asia. Indonesia is emerging as a source of foreign policy ideas for the region, more generally as "dynamic equilibrium" with increasingly integrative, cooperative and peaceful.[39] Hence, the answer to the question is, yes–Asia can help Asia, an Asian Century indeed! It foresees that ASEAN will be led by Indonesia. China progressively strengthen the Asia's inter-countries infrastructures and so on. India also helps ASEAN in a later stage. Such intuition is to promote the so-called "Asia inclusion" that is, all the 47 countries of Asia

[39] Can Indonesia Lead ASEAN? https://thediplomat.com/2013/12/can-indonesia-lead-asean/.

(including ASEAN as a whole) not only have their basic [sound] economic fundamentals, but productive, innovation, openness, and spillover among them equally without constraining by culture, languages, and histories. It reflects the current policies and practices relating to diversity, equity, and inclusion (DEI). An implication considers the countries leaders' mindset of *oneness* in all aspects either through cooperation or collusion between countries to share for a greater, richer, and economic self-fulfilment Asia.

1.6 Conclusion

This future study aims to foresee ASEAN economy for 2050 and 2100 with the rise of China and India then considering three macroeconomic variables, namely real GDP, trade, and net financial account (BoP). Using quantitative methods of futures studies, the future data are forecasted from the available data of 1973 (1982)−2022 for 2023–2050, and 2051–2100 by polynomial trendline technique. The findings of cointegration and Granger non-causality tests support the intuition that ASEAN, China, and India become more connectedness in the future approaching 2050 and 2100. This chapter considers the future of ASEAN in 2050 remains as present, while 2010 is a take-off for ASEAN to integrate 'sufficiently' with China, and India toward a vision of Asian Century. Meanwhile, Indonesia is believed to play a crucial role to lead ASEAN in future. At last, this chapter foresees that Asia will help Asia in fostering their economic integration, in particular among ASEAN, China, and India.

The contribution of this chapter is unique by offering specific insights on ASEAN's future economic trajectory. First, it foresees a longer future of 2100 comparing to the available The World in 2050 report. Such informative futures studies as this for 2100 is relatively empty in the literature. Secondly, the futures data (to 2050 and 2100) of extrapolation are employed in this chapter for the countries' connectedness of real GDP, trade, and net financial account (BoP). Indeed, the existing empirical studies, most of the cases are based on the observed data. Lastly, this chapter foresees the economic related issues on, and visioning Asia in specific. Available future studies conventionally focus on either China, India, or Africa. For future research on this topic, more macroeconomic variables than of real GDP, trade, and net financial account are required for comprehensiveness. Randomly speaking, among them are inflation, interest rates, exchange rates, financial development, intuitional quality, and so on. In fact, the ASEAN member countries can be considered instead of aggregate ASEAN base data to be considered, and to include other Asia countries like Japan, South Korea, and so on.

Highlights

The future trends of the three selected core macroeconomic variables (real GDP, trade, and net financial account) are relatively different from the presently observed.

The connectedness of them among ASEAN, China, and India is supported by cointegration and Granger non-causality tests.

The ASEAN, China and India's connectedness of these variables are stronger toward the future of 2050 and 2100.

The study foresees that Asia can help Asia.

The year 2010 should remark a take-off for ASEAN to progressively integrate with China, and India.

References

Acharya A (2011) Can Asia lead? power ambitions and global governance in the twenty-first century. Int Aff 87(4):851–869. https://doi.org/10.1111/j.1468-2346.2011.01007.x

Agarwal M (2008) The BRICSAM countries and changing world economic power: scenarios to 2050. https://www.files.ethz.ch/isn/94033/WP%2039.pdf. Accessed 15 May 2024

Asian Development Bank (2011) Asia 2050: realizing the Asian century–executive summary. https://www.adb.org/publications/asia-2050-realizing-asian-century. Accessed 10 Oct 2023

Asian Development Bank (2013) Navigating a changing world economy: ASEAN, the People's Republic of China, and India. https://www.adb.org/publications/navigating-changing-world-economy-asean-peoples-republic-china-and-india. Accessed 10 Mar 2024

Asian Development Bank (2017) Asia's middle-income challenge: an overview. https://www.adb.org/sites/default/files/publication/381381/ewp-525.pdf. Accessed 10 Mar 2024

Chatterji, R (2021). China's relationship with ASEAN: an explainer. https://www.orfonline.org/research/china-relationship-asean-explainer. Accessed 16 Oct 2023

Doncaster CP, Bullock JM (2024) Living in harmony with nature is achievable only as a non-ideal vision. Environ Sci Policy 152:103658. https://doi.org/10.1016/j.envsci.2023.103658

Granger CWJ (1980) Testing for causality: a personal viewpoint. J Econ Dyn Control 2:329–352. https://doi.org/10.1016/0165-1889(80)90069-X

Gujrati, R, Uygun, H (2024) The rise of India as a global leader: an analysis of economic, cultural, and diplomatic strategies. NOLEGEIN-J Glob Mark 7(1):31–39. https://mbajournals.in/index.php/JoGM/article/view/1387

Hoontrakul P, Balding C, Marwah R (2014) The rise of China and India-its implications. In: Hoontrakul P, Balding C, Marwah R (eds) The global rise of Asian transformation. Palgrave Macmillan, New York, pp 133–161. https://doi.org/10.1057/9781137412362_6

Johansen S, Juselius K (1990) Maximum likelihood estimation and inferences on cointegraion-with applications to the demand for money. Oxford B Econ Stat 52(2):169–210. https://doi.org/10.1111/j.1468-0084.1990.mp52002003.x

Kuosa T (2011) What is futures studies? In: S Rajaratnam (ed) Practising strategic foresight in government: the cases of Finland. Singapore and the European Union, pp 14–17. http://www.jstor.org/stable/resrep05909.8

MacKinnon JG, Haug AA, Michelis L (1999) Numerical distribution functions of likelihood ratio tests for cointegration. J Appl Economet 14(5):563–577. https://doi.org/10.1002/(SICI)1099-1255(199909/10)14:5%3c563::AID-JAE530%3e3.0.CO;2-R

Rahul S, Asher MG, Rajan, RS (2004) ASEAN-India economic relations: current status and future prospects. Econ Polit Weekly 39(29):3297–3308. http://www.jstor.org/stable/4415291

Reddy, PKM (2020) India, ASEAN and China: some questions. World Aff 24(4): 120–135. https://www.jstor.org/stable/48609814

Sathyamoorthy V, Tang TC (2019) Does institutional quality matter for the success of export-led growth?. Int J Econ Manag 13(2):407–420. http://www.ijem.upm.edu.my/vol13no2/11)%20Does%20Institutional%20Quality%20Matter.pdf

ASEAN Secretariat (2006) ASEAN, China and India: comparative economic performance, issues and implications. https://asean.org/wp-content/uploads/images/2012/publications/19006.pdf. Accessed 15 May 2024

Shoemaker S (1969) Time without change. J Philos 66(12):363–381. https://doi.org/10.2307/2023892

Tang TC (2002) Long-run linkages of financial liberalisation among ASEAN economies: results from bounds testing approach. J Asia-Pa Aff 3(2):113–134

Toyoda, M (2024) India-the future of a powerful country. Econ, Cult & Hist Jpn Spotlight Bimon 43(2):11. https://www.rieti.go.jp/en/papers/contribution/urata/13.html. Accessed 1 Mar 2024

Walsh JP (2014) The future of Asian finance. https://www.imf.org/external/pubs/ft/fandd/2014/06/walsh.htm. Accessed 10 Jan 2024

Wan, CY, Tang, TC (2021) I'll be back–are shocks to COVID-19 total confirmed cases permanent or transitory? Labuan Bull Int Bus Financ 19(1):1–32. https://doi.org/10.51200/lbibf.v19i1.2726

West, J (2018) Can Asian countries live together in peace and harmony? In: Asian Century… on a Knife-edge. Palgrave Macmillan, Singapore. https://doi.org/10.1007/978-981-10-7182-9_10

Yahya F (2003) India and Southeast Asia: revisited. Contemp Se Asia 25(1):79–103. http://www.jstor.org/stable/25798629

Zhang Y, Shen M (2012) Emergence of ASEAN, China and India and the regional architecture. China World Eco 20(4):92–107. https://doi.org/10.1111/j.1749-124X.2012.01297.x

Tuck Cheong Tang joined Universiti Malaya on February 17, 2014, after serving at Monash University Malaysia since July 2001. He earned his Ph.D in Economics from Monash University, Australia, in 2012, under Dietrich Karl Fausten and received the prestigious Mollie Holman Doctoral Medal. Tang is an experienced educator, teaching a wide range of economics courses at both undergraduate and postgraduate levels. His research focuses on open economy macroeconomics, particularly topics related to balancing items in the balance of payments, and he has an extensive record of published articles and conference presentations worldwide. Tang is also an active contributor to the academic community, frequently serving as a reviewer for leading journals. Outside academia, he is dedicated to martial arts of Wing Chun.

Chapter 2
Vietnam: The Rise of a Future Global Economic Power in Asia

Clusters of Future Studies: Corporate Foresight; Past & Futures; Technological Trends

Kiet Le Hoang and Hiep Tran Xuan

Abstract In the 21st century, Vietnam has become one of the most dynamic emerging economies in the world. According to forecasts by many economic experts, Vietnam will become one of the twenty largest economies in the world by 2045. The rapid development of the Vietnamese economy stems from various causes. Among them, there are three main reasons: (1) Favorable geopolitical and geo-economic position, with abundant natural resources; (2) Deeply integrated economy, large market potential; (3) Independent, self-reliant and wise foreign policy of Vietnamese leaders. This chapter uses a qualitative analysis method through secondary data and primary data sources to focus on researching and evaluating the three aforementioned reasons that have been and are helping the Vietnamese economy grow rapidly after nearly 40 years of economic reform and international integration. From there, the article concludes that Vietnam has the full potential to become one of the top twenty economies in the world by 2045.

Keywords Vietnam · Economic power · Asia · Southeast Asia · Future

K. Le Hoang (✉)
Faculty of Basic Sciences, Yersin University, Lam Dong, Vietnam
e-mail: kietnckh1999@gmail.com; kietlh.tg@yersin.edu.vn

Faculty of International Studies, University of Social Sciences and Humanities, Vietnam National University, Hanoi, Vietnam

H. T. Xuan
Faculty of History – Geography – Politics, The University of Danang, University of Science and Education, Da Nang, Vietnam
e-mail: txhiep@ued.udn.vn

E. Lau et al. (eds.), *Future of Economy in Asia*,
https://doi.org/10.1007/978-3-031-95500-6_2

2.1 Introduction

In the 21st century, the world has witnessed the remarkable rise of several Asian economies, especially the "two Asian giants" of China, India, the East Asian countries, and the small developing countries located in Southeast Asia (Litsareva 2022). In that context, Vietnam has emerged as a notable economic phenomenon in Southeast Asia with an impressive average growth rate of 6.5–7% GDP/year and the potential to develop into an economic power in the future (Vietnam Ministry of Finance 2021). The leading Japanese scholar on economics and politics—Hamada Kazuyuki has predicted: *"Vietnam with its young population, growing middle class, rapidly developing domestic market, coupled with a spirit of overcoming adversity, and skillful diplomacy strategy will become one of the twenty largest economies in the world by 2045"* (Kazuyuki 2021: 72). In agreement with the above view, global economic organisations all forecast that Vietnam will become an economic powerhouse with strong influence in the region and the world (Caporale 2022). The assessment of the powerful rise of the Vietnamese economy in the future is based on solid evidence. This is not the result of random assessments, but rather an evaluation based on many key factors. Among them, there are three main factors to assess the future success of the Vietnamese economy.

First, Vietnam occupies a strategic geopolitical and geo-economic position in the East Sea and Southeast Asia region. With a long coastline, vast territory, and a position at the center of important maritime routes in the East Sea, Vietnam is considered an important gateway to access the Southeast Asian market and an attractive destination for international investors (Kiet and Tuyen 2023). In addition, Vietnam also has considerable economic development potential with a large population, a young labor force, and abundant natural resources such as oil and gas, minerals, fertile land, (Anh 2023). Second, Vietnam has successfully implemented socialist-oriented market economic policies and international economic integration. Since starting the Renewal process in 1986, Vietnam has transitioned from a centrally-planned economy to a market economy, opening up to foreign investment activities and participating in international trade agreements and multilateral economic organisations. These actions have created strong growth momentum for the Vietnamese economy, making *"this country one of the most attractive destinations for foreign investors in the Asian region"* (Tu 2024). Third, Vietnam has a stable political system, pursues an independent, self-reliant and wise foreign policy, diversifies its foreign relations, and actively participates in regional and international organisations. The "four no's" diplomatic strategy has helped Vietnam build a peaceful and stable environment for economic development (Communist Party of Vietnam 2022). At the same time, Vietnam has also actively participated in organisations such as ASEAN, APEC, WTO, to promote economic and trade cooperation with international partners. With strong economic growth, Vietnam is increasingly influential in both traditional and non-traditional security issues in the region and globally (Do 2022). Vietnam has played an active role in international forums such as the United Nations, ASEAN and has held many important positions in multilateral organisations. Therefore, Vietnam's growing international role has contributed to enhancing the position and influence of this country in the world. With these important factors, Vietnam is on track to become one of the twenty largest economies globally in the future.

However, Vietnam's economic development path has also not been without challenges and difficulties, especially in the current period when Vietnam is still assessed as a "processing economy" and has not been self-reliant in production technology, etc. (Trang 2023). Therefore, research on Vietnam's economic development process is very important, helping to better understand the potential and challenges that this country is facing. The chapter will focus on analysing the three main factors leading Vietnam's economic growth momentum: (1) Favorable geopolitical and geo-economic position, with abundant natural resources; (2) Deeply integrated economy with large market potential; (3) Vietnam's independent, self-reliant and wise foreign policy. The research results of this chapter are important in affirming Vietnam's potential and ongoing competitive advantages to rise as one of the leading global economies by 2045.

2.2 Literature Review

The existing literature presents compelling evidence for Vietnam's potential emergence as a global economic power, with scholars examining various factors contributing to this trajectory. Recent studies have highlighted three primary dimensions: geopolitical positioning, economic integration, and foreign policy strategy.

In terms of geopolitical significance, Shambaugh (2022) emphasises Vietnam's strategic location in the Indo-Pacific region, particularly highlighting its position along critical maritime routes and its role in addressing China's "Malacca Dilemma". This geopolitical advantage is further supported by Kiet and Tuyen (2023), who argue that Vietnam's extensive coastline and central position in Southeast Asia make it an essential gateway for accessing regional markets. Regarding economic integration, scholars have documented Vietnam's transformation since its economic reforms. Que (2019) places Vietnam among the notable economic phenomena in Southeast Asia, highlighting its impressive average growth rate of 6.5–7% GDP annually. This observation is reinforced by Caporale (2022), who predicts Vietnam's emergence as an economic powerhouse based on its successful integration into global markets and strategic trade partnerships. The literature also emphasises the role of foreign policy in Vietnam's economic rise. Ke (2014) identifies political stability and prudent foreign policies as crucial factors driving Vietnam's economic trajectory. This view is echoed by Do (2022), who examines Vietnam's emergence as a middle power in Asia, highlighting how its "four no's" diplomatic strategy has contributed to maintaining regional stability while fostering economic growth.

Recent empirical studies have documented Vietnam's economic progress. According to Tu (2024), Vietnam has become one of the most attractive destinations for foreign investors in Asia, while Trang (2023) points out remaining challenges, particularly in technological self-reliance and production capabilities. This balanced perspective is important for understanding both opportunities and obstacles in Vietnam's development path. The literature consistently emphasises the interconnection between political stability and economic development. As noted by Linh and Thao (2022), Vietnam's political stability under unified leadership has created a secure environment for foreign investment and sustainable economic growth. This stability, combined with what Kazuyuki (2021) describes as Vietnam's young population and growing middle class, positions the country favorably for continued economic expansion.

This chapter specifically addresses several key gaps in the existing literature. First, while previous research has examined individual aspects of Vietnam's development, such as foreign policy (Do 2022) or economic integration (Que 2019), this chapter provides a comprehensive analysis that integrates three critical factors: geopolitical positioning, economic integration, and foreign policy strategy. This integrated approach offers a more holistic understanding of Vietnam's potential trajectory toward becoming a global economic power. Second, whereas existing studies like Shambaugh (2022) and Kiet and Tuyen (2023) focus primarily on Vietnam's geopolitical significance, this chapter extends beyond geopolitical analysis to examine how Vietnam's geographic advantages translate into concrete economic benefits, particularly through its expanding seaport system and growing role in global supply chains. The study uniquely combines analysis of empirical data on trade volumes, FDI flows, and economic indicators with geopolitical considerations to provide a more comprehensive assessment of Vietnam's development potential. Third, this chapter addresses a significant gap in understanding how Vietnam's "four no's" diplomatic strategy contributes to its economic development. While scholars like Ke (2014) have noted the importance of political stability, this chapter specifically examines how Vietnam's balanced foreign policy approach enables it to maintain productive economic relationships with competing global powers, particularly in the context of United States—China rivalry.

These chapter contributions aim to provide a more nuanced and comprehensive understanding of Vietnam's path toward becoming a global economic power, addressing gaps in current scholarship while offering new insights into the interplay between geopolitical advantages, economic integration, and diplomatic strategy in shaping Vietnam's economic future.

2.3 Methodology

This study employs a qualitative research approach combining four complementary methodological strategies to examine Vietnam's potential emergence as a global economic power. First, policy document analysis was conducted through examination of Vietnam's economic development frameworks, foreign policy statements, and national strategic plans during the 2010–2023 period. This process involved identifying recurring themes, policy continuities, and strategic pivots that shape Vietnam's economic trajectory.

Second, secondary data analysis utilised existing statistical compilations from Vietnamese government sources (General Statistics Office, Ministry of Planning and Investment) and international organisations (IMF, World Bank) to identify broad economic trends without requiring original quantitative modeling. This approach provided a contextual understanding of Vietnam's economic development patterns, growth sustainability, and sectoral transitions.

Third, expert perspective synthesis integrated insights from published works by Vietnamese and international scholars specialising in Southeast Asian economics, geopolitics, and development studies. This method facilitated the triangulation of diverse viewpoints on Vietnam's comparative advantages, strategic challenges, and future economic potential. We analysed scholarly assessments of Vietnam's geopolitical positioning, economic integration processes, and diplomatic strategy to derive consensus views on the country's development trajectory.

Fourth, contextual forecasting methods were employed to evaluate Vietnam's future economic positioning. Unlike quantitative forecasting requiring extensive modeling, our approach utilised qualitative scenario development based on identified developmental patterns, regional economic trends, and policy continuities. This interpretive forecasting examined how Vietnam's strategic advantages might translate into economic outcomes through 2045.

Our methodological framework balances documentary evidence with expert assessments while acknowledging the inherent limitations of pure prediction. By focusing on strategic factors influencing economic development rather than precise numerical projections, this approach provides a conceptually robust evaluation of Vietnam's potential to emerge as a significant economic power. The methodology emphasises identifying the foundational elements supporting Vietnam's economic rise rather than quantifying specific growth trajectories, thereby offering insights into the qualitative factors shaping the country's development potential.

2.4 Findings

2.4.1 Vietnam's Geopolitical, Geo-Economic Position and Natural Resources

Vietnam is one of the countries with a very important geopolitical position located in Southeast Asia and the Indo-Pacific region, a land bridge connecting the largest part of the Eurasia continent with Southeast Asia. Vietnam has an area of 331,212 km^2, located in the center of Southeast Asia; Vietnam's land borders adjoin China to the north, and Laos and Cambodia to the southwest; along with abundant continental natural resources (see Table 2.1). Vietnam has a coastline stretching from north to south up to 3,260 km (not including islands), with a sea area under Vietnam's sovereignty and jurisdiction covering an area of about 1,000,000 km^2. Vietnam's coastline length to land area ratio is approximately 0.01 (meaning for every 100 km^2 of land there is 1 km of coastline), ranking first among the countries of the Indochina Peninsula, above Thailand and nearly equivalent to Malaysia. In Vietnam's sea and continental shelf areas, there are up to 3.6 million tons of marine fish reserves, many sedimentary basins with oil and gas potential have been discovered, of which the Cuu Long and Nam Con Son basins are assessed to have the greatest oil and gas potential and relatively favorable exploitation conditions. The total geological predicted reserves of oil and gas on Vietnam's entire continental shelf is nearly 10 billion tons of oil equivalent, with recoverable reserves of around 2 billion tons and predicted gas reserves of around 1,000 (m^3). In addition, the Paracel and Spratly Islands areas also contain frozen gas hydrate resources, considered a future alternative energy source to oil and gas (Cam 2022). Vietnam has 2,773 coastal islands, with a total area of about 1,630 km^2. The islands are scattered unevenly, with about 2,300 islands concentrated in the Quang Ninh—Hai Phong and Southwest sea areas. Vietnam is also a transit country for landlocked nations like Laos, and territories located inland such as Cambodia, Thailand, and Yunnan (China).

Vietnam has an archipelagic position astride the vital sea lanes between the Indian and Pacific Oceans. Out of the 10 largest shipping routes in the world today, 5 pass through or are related to the East Sea. Studies show that nearly 1/3 of world merchant ships transit the East Sea, and if the East Sea shipping lanes were blocked, merchant ships would have to go around the eastern islands of Southeast Asia, lengthening voyages by many days compared to passing through the Strait of Malacca. In addition, the East Sea region also has important straits for many countries, of which the Strait of Malacca is of foremost strategic importance worldwide because all maritime trade of Southeast Asia and Northeast Asia must pass through it. This is also the location that China views as the "Malacca Dilemma", referring to the possibility that in the event of war, countries could control the strait, thereby blocking China's energy imports and goods exports (Shambaugh 2022: 10). This is also the strait with the highest volume of ships passing through and ranks second in the world for annual oil shipments after the Strait of Hormuz. Therefore, with its long coastline, Vietnam has strengths in developing seaports and maritime services. The exploitation of seaports

Table 2.1 Some natural resources that Vietnam regularly exploits and uses for economic growth

No	Natural resources	Reserves
1	Land	39 million hectares of natural land
2	Groundwater	About 130 million m^3/day
3	Coal	About 3.5 billion tons
4	Bauxite	6–8 billion tons
5	Tin	129,000 tons
6	Iron	Nearly 1 billion tons
7	Apatite	Over 1 billion tons
8	Copper	About 600 thousand tons
9	Chromium	About 10 million tons
10	Gold	About 100 tons

Source Ministry of Natural Resources and Environment (Vietnam)

is considered one of the spearhead priorities for Vietnam to become a maritime power, and cargo volume passing through Vietnam's seaports has grown rapidly, contributing to Vietnam becoming an economic powerhouse in Asia in the future (see Table 2.2). Currently, Vietnam has 34 seaports, including 3 seaports ranked among the busiest in the world: Saigon Port ranked 22nd with a throughput of 7.9 million TEUs in 2021 and a growth rate of about 1.3% compared to 2020; Hai Phong Port ranked 28th with an annual throughput of 5.69 million TEUs in 2021, growing about 10.8% compared to 2020; Cai Mep Port ranked 32nd with throughput reaching 5.38 million TEUs in 2021. Cai Mep Port is considered a bright spot in the Southeast Asian region with a post-COVID-19 growth rate of around 22% (Le 2023).

Vietnam has a geo-economic position located in the center of global growth, with an increasing role due to the development of the middle class in Asia. This region is becoming the engine of global economic growth. In the "pivot to Asia" development trend, Vietnam enjoys a great geo-economic advantage as current economic powers like South Korea, Japan, China, and Australia are geographically close and dependent on the sea lanes passing through the East Sea. Along with that, the recent "China+1" trend provides a premise for enterprises to invest and set up production bases in Vietnam, thereby attracting foreign investment and modern production technology. Vietnam is also a connecting country and logistics hub serving nations inside and outside the region. In China's BRI initiative, Vietnam plays an important role in this grand strategy, especially in the "Maritime Silk Road" (MSR). Moreover,

Table 2.2 Cargo volume passing through Vietnam's seaports in the period 2015–2022. *Unit* thousand tons

Year	2015	2016	2017	2018	2019	2020	2021	2022
Cargo volume	427.8	459.8	519.3	606.6	664.6	692.3	706.1	733.2

Source General Statistics Office (Vietnam)

Vietnam also lies on the main axis of the East-West Economic Corridor connecting the two great oceans of the Indian and Pacific, creating great potential and advantages for economic development. However, Vietnam's current geo-economic position is also facing great challenges from the impact of many contemporary factors. In the context of intensifying global competition, the outbreak of the 4.0 industrial revolution; globalisation trends; traditional and non-traditional security issues have posed challenges for Vietnam to make overall, long-term, flexible strategic adjustments to adapt to regional and global trends. Challenges that can affect the geo-economic factor often have a multinational nature, so geo-economic risks are very high and have a strong impact on Vietnam's overall development strategy and policies.

In general, the favorable geopolitical and geo-economic position coupled with abundant natural resources is one of the key advantages enabling Vietnam's potential to become an economic power in the future. Located at the center of Southeast Asia with a long coastline, diverse terrain and abundant resources, Vietnam plays an important role on the strategic maritime routes linking the Indian and Pacific Oceans. Additionally, developing a modern seaport system is also an advantage for Vietnam to become a logistics hub, providing services for countries in the region. The geo-economic position at the center of global growth also opens up opportunities for Vietnam to attract foreign investment and modern technology. However, challenges related to traditional and non-traditional security are also posing issues that need to be addressed for Vietnam to continue leveraging its geopolitical and geo-economic advantages. If these difficulties can be overcome, Vietnam will have sufficient potential to develop into an economic powerhouse in the future.

2.4.2 *Vietnam's Economy in the Process of International Integration*

Vietnam's economy has undergone a process of deep international integration, laying a solid foundation for the country to become an economic powerhouse in the future. Since the start of economic reforms in 1986, Vietnam has transitioned its economy from a centrally planned model to a socialist-oriented market economy. These economic reforms opened up Vietnam's economy, attracted foreign investment, and enabled deep integration into the global economy. Especially since Vietnam joined the WTO in 2007, this has allowed Vietnam to access major markets and participate in bilateral and multilateral free trade agreements (FTAs) with other partners around the world. As a result, Vietnam has witnessed a rapid increase in FDI inflows over time, contributing to driving Vietnam's economic growth at a rate of 6.5–7% per year.

Figure 2.1 shows that Vietnam has experienced significant fluctuations in FDI over time. Since joining the WTO in 2007, Vietnam attracted a large amount of FDI in 2008, with a registered figure of up to $64 billion, three times higher than the previous year. However, due to factors such as the 2008 global financial crisis and the

European sovereign debt crisis in 2010, FDI inflows into Vietnam during the 2009–2012 period declined significantly before recovering and remaining relatively stable from 2013–2019. In 2020, due to the severe impact of the COVID-19 pandemic on the global economy, outflows of investment capital, especially FDI, plummeted, and Vietnam was no exception to this trend. Specifically, total FDI inflows into Vietnam in 2020 decreased by 6.7% compared to 2019, with a value of around $21 billion, of which newly registered capital was $14.6 billion and adjusted registered capital was $6.4 billion. During this period, the value of newly registered capital was always higher (about 2–3 times) than adjusted registered capital, showing that Vietnam continuously attracted new investors to the market. By 2021, despite the COVID-19 situation in Vietnam becoming more complicated than in 2020, FDI attraction results in Vietnam grew strongly, continuing into 2023 with an average registered capital of around $30 billion/year. The biggest reason for this growth stemmed from the shift of production chains from China to Southeast Asian countries and India, with Vietnam being the largest recipient of FDI inflows in the region.

Currently, Vietnam is the most successful country in trade exchange and FDI attraction, improving the growth rankings of its economy while maintaining stability and long-term development in its economic-social structure. Vietnam is also actively participating in bilateral and regional FTAs. Vietnam has an advantage in being part of a high-quality economic integration network, having signed 17 FTAs with most major economies in the world. In particular, the Comprehensive and Progressive Agreement for Trans-Pacific Partnership (CPTPP) and the EU-Vietnam Free Trade Agreement (EVFTA) are expected to bring new opportunities, helping Vietnam attract more foreign investment, boost exports and facilitate technology transfer. These FTAs not only expand markets for Vietnam's goods and services, but also

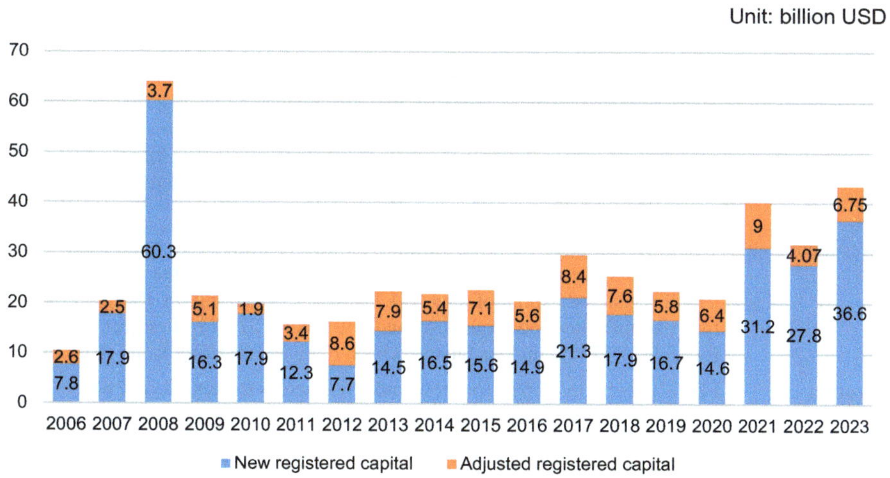

Fig. 2.1 FDI Investment into Vietnam in the period 2006–2023. *Source* General Statistics Office (Vietnam)

contribute to improving the business environment and enhancing the competitiveness of the economy. The process of international economic integration has helped make Vietnam an attractive destination for foreign investors.

As shown in Table 2.3, as of November 2023, Vietnam has attracted $462.4 billion in FDI capital with 38,844 valid projects. The growth of FDI capital and international investment inflows has contributed to promoting economic development, technology transfer, and job creation for Vietnam. Thanks to the process of deep economic integration, Vietnam has also expanded its export markets and diversified its key export items. Previously, Vietnam mainly exported agricultural products, but now industrial and electronic products are gradually accounting for an increasing proportion. In 2023, Vietnam's total merchandise export turnover reached $355.5 billion, of which exports of electronics, computers and components accounted for $57.3 billion (General Statistics Office of Vietnam 2023). The process of economic integration has also brought great benefits to the domestic market, enabling Vietnam to access modern technology, high-quality human resources, and expand consumer markets. With a population of around 100 million, of which 62.2% is young, an abundant workforce, and a growing middle class, Vietnam is home to an attractive domestic market with great growth potential.

Figure 2.2 shows that Vietnam's GDP growth rate has been stable and sustainable, with an average GDP growth of around 6.5% per year, despite the peak of the COVID-19 pandemic in 2020 and 2021, growth remained positive at 2.87% and 2.55% respectively while most other economies experienced negative growth. By

Table 2.3 Effective accumulated foreign investment capital as of November 2023

Total valid FDI projects	There are 38,844 valid FDI projects with a total registered capital of $462.4 billion. The accumulated implemented capital of FDI projects is estimated to reach $294.2 billion, equivalent to 63.6% of the total valid registered investment capital
Industry	Foreign investors have invested in 19 out of 21 sectors in the national economic sector system, in which the manufacturing industry accounts for the highest proportion with over $280.5 billion, accounting for 60.7% of the total investment capital. Following is the real estate sector with over $67.6 billion (accounting for 14.6%); electricity production and distribution with nearly $38.6 billion (equivalent to 8.3%)
Investment	There are 143 countries with valid investment projects in Vietnam since Grenada dissolved its only project. Among them, South Korea ranks first with a total registered capital of over $84.1 billion (accounting for 18.2% of the total capital). Singapore ranks second with nearly $73.7 billion (accounting for 15.9%). Following are Japan, Taiwan (China) and Hong Kong (China)
Location	FDI is present in all 63 provinces and cities across the country, in which Ho Chi Minh City remains the locality leading in attracting foreign investment with over $57.25 billion (accounting for 12.4% of total investment capital), followed by Binh Duong with over $40.3 billion (accounting for 8.7% of total investment capital), Hanoi with over $39.58 billion (accounting for 8.6% of total investment capital)

Sources Ministry of Planning and Investment (Vietnam)

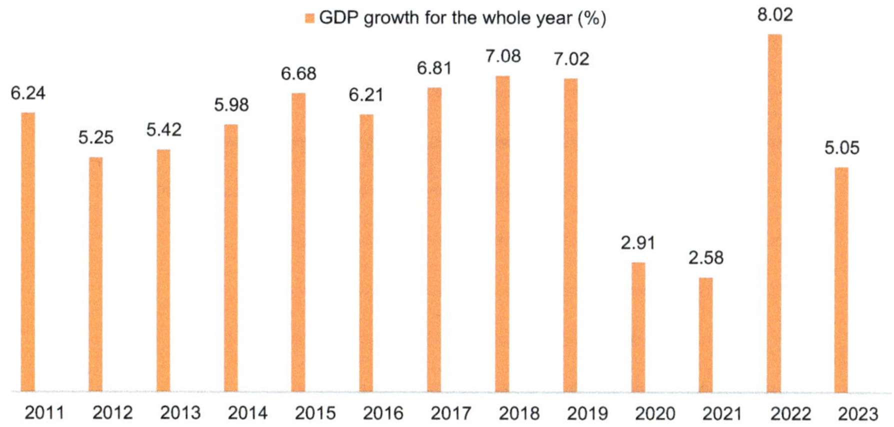

Fig. 2.2 Vietnam's GDP growth in the period 2011–2023. *Source* General Statistics Office (Vietnam)

2022, this figure had increased to 8.02% thanks to the recovery of consumption and production as the COVID-19 situation was brought under control. By 2023, this figure had decreased to 5.05% due to the impact of global conflicts; supply chain disruptions; and weaker domestic demand compared to 2022. However, according to IMF forecasts, Vietnam's GDP growth rate will quickly recover to previous levels, and by 2025 it will become the third largest economy in Southeast Asia.

However, besides the achievements, Vietnam still faces some challenges such as the quality of human resources, outdated technology, low labor productivity, and uneven infrastructure development. In addition, Vietnam also needs to make more efforts in institutional reform, improving the legal framework, and enhancing national governance capacity to better adapt to the process of deep integration. Particularly, the adaptation and transition of the economic growth model over time has had positive impacts, but the transition remains slow, and Vietnam's economic growth is still focused on money supply and credit growth. Therefore, Vietnam's GDP debt ratio has been increasing over the years, from only 89.7% in 2015, but this figure reached 125% in 2022 (see Fig. 2.3). The rising GDP debt ratio poses major challenges for Vietnam in terms of human resources, qualifications, and mindset in new fields such as digitalisation, semiconductor technology, green economy, etc. At the same time, the transition to a new growth model faces challenges that will create social instability, so flexible policies are needed to adapt to each cycle in the implementation of macroeconomic policies. On the other hand, Vietnam is also a country with low competitiveness, low average income, at risk of falling into the middle-income trap, with weak business linkages, increasing wealth gaps, environmental pollution issues. These factors have had and are having a huge impact on Vietnam's economic development process.

Overall, the process of international economic integration has brought great opportunities and benefits to Vietnam. Participating in FTAs, boosting imports and exports,

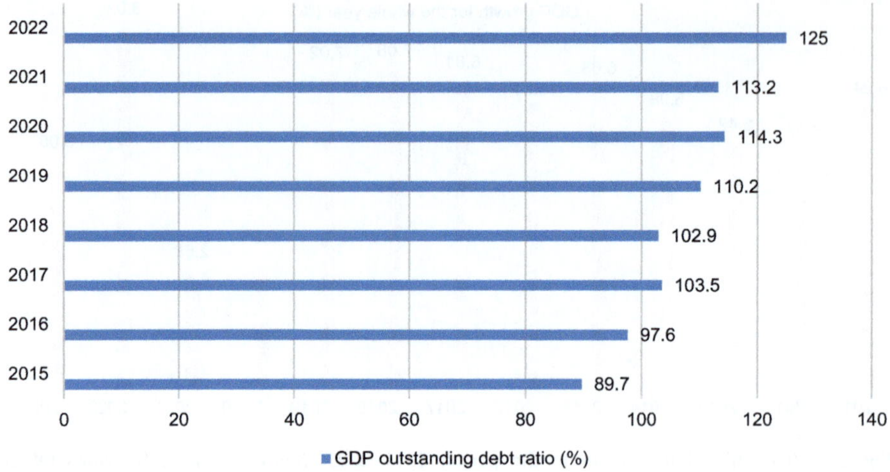

Fig. 2.3 Vietnam's GDP debt ratio in the period 2015–2022. *Source* General Statistics Office (Vietnam)

and attracting foreign investment have driven economic growth, enhanced competitiveness, and expanded development space for the economy. At the same time, this process has created momentum for Vietnam to shift its economic structure towards modernisation and increase the proportion of high-tech industries. With continuous efforts in economic reform, an open-door policy, and deep integration, Vietnam is gradually asserting its position as a potential market, ready to become an important trading gateway in the Asia-Pacific region. A deeply integrated international economy, combined with a populous domestic market, a young workforce, and an increasingly strong middle class, are the foundational factors helping Vietnam rise as an economic powerhouse in the future.

2.4.3 Vietnam's Independent, Self-reliant and Prudent Foreign Policy Strategy

Vietnam's independent, self-reliant and prudent diplomatic strategy has played and is playing an important role in affirming Vietnam's position, consolidating a peaceful and stable environment for national economic development. Scholar Luong Van Ke has argued that: *"The stable politics with consistent and prudent foreign policies is one of the brightest points helping Vietnam to be on track to become an economic power in the future"* (Ke 2014: 364). Sharing the same view, scholar Bui Van Hieu stated that: *"Vietnam's remarkable development achievements first come from political stability. In the context of conflicts and turmoil occurring in many places around the world, Vietnam is considered a country with stable politics under the united leadership of the Communist Party of Vietnam, which is the foundation for foreign partners to feel secure when investing in Vietnam"* (Linh and Thao 2022). Thereby, Vietnam's political stability and independent and self-reliant foreign policy strategy is an important foundation for Vietnam to converge sufficient capabilities to become an economic power in the future. Since Vietnam's economic reforms and renovation in 1986, the 6th National Congress of the Communist Party of Vietnam affirmed: *"Vietnam is determined to take the path to socialism with a focus on peace, national independence, democracy and progress"* (Communist Party of Vietnam 2020). Therefore, Vietnam's foreign policy strategy is built on the foundation of an independent, self-reliant, peaceful, friendly, cooperative and development-oriented foreign policy. At the 32nd Diplomatic Conference, General Secretary Nguyen Phu Trong reaffirmed Vietnam's foreign policy principle: *"Vietnam is always ready to be a trusted and responsible friend and partner with all countries in the international community, regardless of political regime, on the basis of mutual respect for independence, sovereignty and territorial integrity"* (Tuoi Tre Newspaper 2023). This independent and self-reliant foreign policy has helped Vietnam both protect national interests and create a peaceful and stable environment for development. As a result, Vietnam currently has a Comprehensive Strategic Partnership (Vietnam's highest diplomatic level) with both superpowers the United States and China, a rare occurrence amidst the increasingly unpredictable and complex United States-China competition.

Contributing significantly to Vietnam's prudent and balanced foreign policy strategy is Hanoi's persistent implementation of the "four nos" policy in the field of defense and security. The "four nos" policy principles are *"not joining any military alliances; not aligning with one country to counterbalance another; not allowing any foreign countries to set up military bases or use its territory to carry out military activities against other countries; not using or threatening to use force in international relations"* (Que 2019: 322). This policy is considered a crucial pillar for Vietnam in balancing relations with major powers, especially the current United States-China rivalry in Southeast Asia. According to data from Vietnam Customs (2023), China is Vietnam's largest trading partner with a two-way trade value of $171.85 billion, followed by the United States ($110.84 billion), South Korea ($76.01 billion), Japan ($44.95 billion). The specific figures above show that this policy has helped Vietnam maintain "strategic autonomy" as a small country with major powers, avoid getting entangled in great power rivalries, preserve peace and stability, build trust in the international community, thereby promoting development to become an important economic partner for major countries.

Vietnam's prudent, flexible and self-reliant foreign policy is reflected in the diversification and multilateralisation of foreign relations on the basis of equality and mutual benefit. Vietnam has established relations with 7 Comprehensive Strategic Partners, including: China (2008), Russia (2012), India (2016), South Korea (2022), the United States (2023), Japan (2023) and Australia (2024). At the same time, it has 30 Strategic Partners (including the 7 Comprehensive Strategic Partners), and has established bilateral economic and diplomatic relations with more than 192 countries and territories around the world. In addition, Vietnam is an active and responsible member of over 70 major international and regional organisations such as the

United Nations, ASEAN, APEC, WTO. Notably, the event where the United Nations General Assembly voted 192/193 for Vietnam as a non-permanent member of the United Nations Security Council for the 2020–2021 term affirmed Vietnam's rising position and prestige in the international arena. Therefore, amidst intensifying great power rivalry in the region, Vietnam has become an important country for the relocation of production chains of multinational corporations from countries in the region such as South Korea and Japan. As a result, Vietnam's robust rise has made all major powers increasingly desirous of promoting comprehensive cooperation with Hanoi. In general, Vietnam's independent, self-reliant and prudent foreign policy has significantly contributed to creating a peaceful and stable environment for Vietnam to focus on economic development. This will continue to be one of the great advantages helping Vietnam affirm its new role and position in the region and internationally, consolidating the foundation to become an economic power in the future.

2.5 Discussion of Implications

The empirical evidence presented in this study suggests several significant implications for Vietnam's ascent as a future economic power. First, Vietnam's geopolitical positioning creates a strategic advantage in the reconfiguration of Indo-Pacific supply chains. As Caporale (2022) notes, Vietnam's position along critical maritime routes addresses China's "Malacca Dilemma", allowing Vietnam to function as a pivotal node in regional trade networks. This geospatial advantage, coupled with Vietnam's expanding seaport infrastructure, positions the country to capitalise on the post-pandemic restructuring of global supply chains.

Second, Vietnam's economic integration trajectory demonstrates the efficacy of a calibrated approach to market liberalisation while maintaining political stability. The data on FDI inflows (Fig. 2.1) illustrates Vietnam's resilience to global economic shocks, evidenced by its rapid recovery following the 2008 financial crisis and the COVID-19 pandemic. This resilience aligns with Linh and Thao's (2022) observation that political stability under unified leadership has created a secure environment for foreign investment, suggesting that Vietnam's governance model offers a viable alternative development pathway for emerging economies.

Third, the GDP debt ratio analysis (Fig. 2.3) reveals a critical challenge in Vietnam's transition to a higher value-added economic model. The increasing debt-to-GDP ratio (from 89.7% in 2015 to 125% in 2022) signals potential vulnerabilities in sustaining growth momentum through traditional channels of money supply and credit expansion. As Trang (2023) argues, Vietnam remains predominantly a "processing economy" with limited technological self-reliance, indicating the necessity for strategic investments in indigenous innovation capabilities and human capital development.

Fourth, Vietnam's "four nos" diplomatic strategy offers important lessons for small and medium powers navigating great power competition. The success of this approach is evidenced by Vietnam's ability to maintain productive economic relationships with competing powers, as demonstrated by trade data showing China ($171.85 billion) and the United States ($110.84 billion) as Vietnam's top trading partners. As Do (2022) notes, this balanced approach has enabled Vietnam's emergence as a middle power in Asia, suggesting that strategic autonomy rather than alignment may prove more beneficial for developing economies in a multipolar world order.

These implications collectively suggest that Vietnam's economic trajectory represents more than just national development-it potentially offers an influential model for navigating the complex interplay between geopolitical positioning, economic integration, and diplomatic strategy in the evolving global economic landscape. As Kazuyuki (2021: 72–74) predicted, Vietnam's young population, growing middle class, and skillful diplomatic strategy position it favorably to become one of the twenty largest economies by 2045, though significant challenges in technological capacity and human resource development remain to be addressed.

2.6 Future Directions

The analysis suggests several key directions for Asia's economic landscape, with Vietnam positioned as an emerging economic force within this evolving framework. First, there is a clear trend toward the reconfiguration of maritime trade networks in Asia, with ASEAN countries, particularly Vietnam, emerging as crucial connectors in global supply chains. This shift is evidenced by Vietnam's expanding seaport system, with major ports like Saigon, Hai Phong, and Cai Mep showing significant growth in cargo throughput (see Table 2.2).

The rise of Asia's middle class represents another pivotal trend, marking a gradual shift in economic gravity from China toward Southeast Asian manufacturing hubs. This demographic transition creates substantial opportunities for countries like Vietnam to develop their domestic markets while serving as production bases for regional and global consumption (Source of Asia 2022). The "China+1" strategy adopted by many multinational corporations further accelerates this trend, suggesting a more distributed manufacturing landscape across Asia.

A new Asian economic model appears to be emerging, characterised by a balance between FDI-led growth and domestic technological capacity building. This model reflects the need for Asian economies to move beyond their traditional role as manufacturing centers and develop indigenous technological capabilities. Vietnam's experience, particularly its challenges in technological self-reliance and the need to upgrade its industrial base, exemplifies this transition. Therefore, the future geopolitical landscape suggests a more multi-polar Asia, where economic powers must maintain strategic autonomy amid US-China competition. Vietnam's "four nos" policy and its success in maintaining productive relationships with both superpowers may serve as a model for other emerging Asian economies (Do 2022). This balanced approach becomes increasingly relevant as regional economic integration deepens through production networks centered around strategic locations like Vietnam.

Looking ahead, Indo-Pacific economic integration is likely to accelerate, driven by new trade agreements and the reorganisation of supply chains. Vietnam's network of FTAs, including high-standard agreements like the CPTPP and EVFTA, indicates a broader regional trend toward deeper economic integration. However, this integration must address challenges such as technological gaps, workforce development, and environmental sustainability to ensure long-term economic resilience. These future directions suggest that Asia's economic landscape will be characterised by more distributed growth centers, enhanced regional integration, and a delicate balance between cooperation and strategic autonomy. Success in this environment will require careful management of both economic development and geopolitical relationships, as exemplified by Vietnam's emergence as a significant regional economic player.

2.7 Conclusion

The comprehensive analysis of Vietnam's potential trajectory toward becoming a global economic power reveals several compelling factors that substantiate this possibility. First, Vietnam's strategic geopolitical and geo-economic position, coupled with abundant natural resources, provides a solid foundation for sustained economic growth. The country's extensive coastline, strategic location along vital maritime routes, and rich natural endowments position it advantageously within global supply chains and trade networks. Second, Vietnam's deeply integrated economy, characterised by extensive participation in international trade agreements and successful attraction of foreign direct investment, demonstrates its increasing economic sophistication and global market integration. Third, the country's independent, self-reliant, and prudent foreign policy has created a stable environment conducive to economic development while maintaining balanced relationships with major global powers.

Looking ahead to 2045, Vietnam's ascension to become one of the world's twenty largest economies appears achievable, though not without challenges. The country must address several critical areas, including upgrading workforce skills, modernising production technologies, improving labor productivity, and strengthening infrastructure. Additionally, Vietnam needs to navigate the complex geopolitical landscape while maintaining its independent foreign policy stance and fostering beneficial relationships with major global powers. The success of Vietnam's economic transformation will depend largely on its ability to execute these strategic imperatives while maintaining political stability and social cohesion. The evidence suggests that Vietnam possesses the fundamental elements necessary for sustained economic growth and development. However, the realisation of its economic potential will require continued commitment to reform, strategic policy implementation, and adaptive responses to emerging global challenges. As Vietnam continues its development journey, its experience may offer valuable lessons for other emerging economies seeking to navigate the complex path of economic transformation in an increasingly multipolar world.

Highlights

- Asia's future maritime trade networks shifting toward ASEAN, with Vietnam emerging as key connector
- Rising Asian middle class driving economic gravity shift from China to Southeast Asian manufacturing
- New Asian economic model emerging: balancing FDI-led growth with domestic technological capacity
- Multi-polar Asia requiring economic powers to maintain strategic autonomy amid US-China rivalry
- Indo-Pacific economic integration accelerating through Vietnam-centered production networks.

Acknowledgements Thanks to my family for always supporting me in pursuing this career.

References

Anh LT (2023) Potential for digital economy development in Vietnam and some policy implications. Tap Chi Cong Thuong. Available https://tapchicongthuong.vn/bai-viet/tiem-nang-phat-trien-nen-kinh-te-so-o-viet-nam-va-mot-so-ham-y-chinh-sach-102900.htm. Accessed 21 March 2024

Cam V (2022) Economic weight of the South China Sea, Sai Gon Giai Phong Newspaper. Available https://dttc.sggp.org.vn/suc-nang-kinh-te-bien-dong-post91025.html. Accessed 21 March 2024

Caporale V (2022) Why Vietnam's economic future is bright—and growing brighter. The Diplomat. https://thediplomat.com/2022/09/why-vietnams-economic-future-is-bright-and-growing-brighter/. Accessed 21 March 2024

Communist Party of Vietnam (2020) The 6th National party congress. https://daihoi13.dangcongsan.vn/cac-ky-dai-hoi/tu-dai-hoi-den-dai-hoi/dai-hoi-dai-bieu-toan-quoc-lan-thu-vi-250. Accessed 21 March 2024

Communist Party of Vietnam (2022) Combating distorted arguments about the Four Nos defense policy. https://dangcongsan.vn/bao-ve-nen-tang-tu-tuong-cua-dang/dau-tranh-phan-bac-cac-luan-dieu-sai-trai-ve-chinh-sach-quoc-phong-bon-khong-609792.html. Accessed 21 March 2024

Do TT (2022) Vietnam's emergence as a middle power in Asia: unfolding the power-knowledge nexus. J Curr Southeast Asian Aff 41(2):279–302

Kazuyuki H (2021) Future powers—remapping the world toward 2030, translated. The World, Hanoi

Ke LV (2014) Vietnam's geopolitics in the 21st century. Vietnam National University, Hanoi

Kiet LH, Tuyen NV (2023) Vietnam's geopolitical position towards the United States in deterring China's hegemonic ambition. Da Nang J Sci Technol 21(8):63–69

Le P (2023) Vietnam's seaports have the highest cargo throughput and allow access to the World's largest container ships. Cafef. https://cafef.vn/cang-bien-viet-nam-co-luong-luu-thong-hang-hoa-va-cho-phep-tiep-can-tau-container-lon-nhat-the-gioi-188230325202315732.chn. Accessed 21 March 2024

Linh P, Thao P (2022) Political stability brings economic development in Vietnam. Quan Nhan Dan Newspaper. https://www.qdnd.vn/kinh-te/cac-van-de/su-on-dinh-chinh-tri-mang-lai-phat-trien-kinh-te-o-viet-nam-708960. Accessed 21 March 2024
Litsareva E (2022) Success factors of Asia-Pacific fast-developing regions, technological innovation development and economic growth. Int J Innov Stud 1(1):72–88
Ministry of Finance of Vietnam (2021) Vietnam's economic internal strength has strong development potential. https://mof.gov.vn/webcenter/portal/btcvn/pages_r/l/tin-bo-tai-chinh?dDocName=MOFUCM196381. Accessed 21 March 2024
Que NT (2019) Vietnam's relations with some major Countries in the early 21st century. Truth National Politics, Hanoi
Shambaugh D (2022) Southeast Asia: the US-China Pivot Point. Truth National Politics, Hanoi
Source of Asia (2022) The rise of the middle class in Southeast Asia. https://www.sourceofasia.com/rising-middle-class-in-vietnam-driving-new-trends-of-consumption/. Accessed 21 March 2024
Trang VTH (2023) Risks, challenges of Vietnam's economy and some recommendations, solutions. Banking Review. https://tapchinganhang.gov.vn/ru-i-ro-tha-ch-thu-c-cu-a-kinh-te-vie-t-nam-va-mo-t-so-kie-n-nghi-gia-i-pha-p.htm. Accessed 21 March 2024
Tu M (2024) Vietnam remains an attractive destination for FDI inflows. VnEconomy Newspaper. https://vneconomy.vn/viet-nam-tiep-tuc-la-diem-den-hap-dan-voi-dong-von-fdi.htm. Accessed 21 March 2024
Tuoi Tre Newspaper (2023) General Secretary Nguyen Phu Trong: Vietnam is always ready to be a trusted friend and partner to all Countries. https://tuoitre.vn/tong-bi-thu-nguyen-phu-trong-viet-nam-luon-san-sang-la-ban-la-doi-tac-tin-cay-voi-tat-ca-cac-nuoc-20231219162922994.htm. Accessed 21 March 2024
Vietnam Customs (2023) Vietnam's two $100 Billion trade partners. Hai Quan Newspaper. https://haiquanonline.com.vn/2-doi-tac-thuong-mai-tram-ty-do-cua-viet-nam-183168.html. Accessed 21 March 2024

Kiet Le Hoang has a Master's degree in Public Policy and is a PhD candidate in International Relations. He works at Yersin University, Vietnam. He has research experience in international relations and public policy with published research in Vietnam and some internationally. Additionally, he often participates in and collaborate on international projects by governmental organizations in Southeast Asia. His research interests focus on Europe and the Indo-Pacific. It focuses on issues of power competition between great powers.

Hiep Tran Xuan is a scholar and researcher currently working at The University of Danang, University of Science and Education, Vietnam. He is a leading International Relations researcher in Vietnam. With a prolific academic record, he has published over 160 scholarly works including books, book chapters, and articles, with nearly 40 publications indexed in SCOPUS/ISI databases.

Chapter 3
A Review of the Sri Lankan Economic Crisis: Lessons for Future Economic Policy-Making to Other Asian Nations

Clusters of Future Studies: Past & Futures

Robert Dygas, Kelaniyage Shihan Dilruk Fernando, Tharindu Udayanga Kamburawala, and Walisundara Mudiyansela Dhananjaya Madusanka Dissanayaka

Abstract The main goal of this research is to explain how the structural issues, poor governance, an ineffective public administration system, and civil society together create a socioeconomic hellfire and gradually burn the pearl of the Indian Ocean (Sri Lanka). Besides that, the main conclusion of the research is the ongoing Sri Lankan economic crisis and socioeconomic downfall as a long-term domino effect implied the economic factors, multiple interrelated and interdependent non-economic factors. The article uses qualitative research methods: studying literature and legal acts, analysis and synthesis, the descriptive method, and deductive inference. The Mendeley.com platform was used to identify the most relevant positions for the literature review. The important aspect is that some of the authors are Sri Lanka's citizens and have the current problem insight overview. The value of this research

R. Dygas
The East Asian Economic Studies Department of The World Economy Research Institute, SGH Warsaw School of Economics, Warsaw, Poland
e-mail: rdygas1@sgh.waw.pl

K. S. D. Fernando (✉)
The Institute of Management, Faculty of Economics, Finance and Management, University of Szczecin, Szczecin, Poland
e-mail: kelaniyage.fernando@phd.usz.edu.pl

T. U. Kamburawala
Department of Economics, Faculty of Humanities and Social Sciences, University of Sri Jayewardenepura, Nugegoda, Sri Lanka
e-mail: kktudayanga@sjp.ac.lk

W. M. D. M. Dissanayaka
Department of Public Administration, Faculty of Management Studies and Commerce, University of Sri Jayewardenepura, Nugegoda, Sri Lanka
e-mail: madusanka@sjp.ac.lk

E. Lau et al. (eds.), *Future of Economy in Asia*,
https://doi.org/10.1007/978-3-031-95500-6_3

is to show both economic and non-economic roots of Sri Lanka's worst macroeconomic turmoil since gaining independence in 2020. This article is the base for further quantitative research on both economic and non-economic factors of this crisis. Nevertheless, the entire chapter and its insights can be considered a great socioeconomic learning experience for future economic policy-making for other South and East Asian countries.

Keywords Sri Lanka · Economic crises · Structural issues · Governance · Public administration · Socioeconomic framework · Economic policy-making

3.1 Introduction

The worst macroeconomic turmoil that Sri Lanka has experienced since gaining independence began in 2020. The COVID-19 pandemic is not the sole factor causing this instability, as several structural issues have long plagued the economy. During the pandemic, a significant public debate ensued regarding whether Sri Lanka should receive financial aid from the International Monetary Fund. President Gotabhaya Rajapaksa's administration initially decided against seeking a bailout from the IMF due to the unpopular policies that would have been adopted. However, facing continuous pressure from the opposition as the economic crisis unfolded, the Sri Lankan government took the initial steps to seek assistance from the IMF before Rajapakse's resignation. After months of negotiations, Sri Lanka gained approval from the IMF board for an extended fund arrangement in March 2023. The most recent IMF bailout program now influences the ongoing strategies to address the Sri Lankan economic crisis. By closely monitoring the ongoing policy adjustments associated with the bailout program, this chapter examines the history of Sri Lanka's involvement with the IMF. Furthermore, it emphasises the importance of restructuring the nation's loss-making state-owned businesses. Nevertheless, this paper illustrates how non-economic factors, such as political culture, the public administration system, and civil society in Sri Lanka, have played a role in contributing to the economic downturn of the country.

3.2 Review Methodology

This chapter employs a qualitative research approach to explore the underlying causes and implications of Sri Lanka's ongoing economic crisis. Particularly, literature review, document analysis, and deductive reasoning. To ensure a structured review process, the Mendeley.com platform facilitated the identification and organisation of relevant scholarly literature.

Key themes such as governance challenges, ineffective public administration, and external economic dependencies were identified using thematic coding. The study integrates data from academic literature, policy documents, and legal acts to analyse the crisis's structural roots and implications. Descriptive methods outline the current economic conditions, while deductive reasoning infers broader lessons for policy-making in South and East Asia. This qualitative approach provides a comprehensive understanding of the crisis's multidimensional nature and offers insights for regional economic governance.

3.3 Findings

3.3.1 The Historical Background of the Sri Lankan Economic Crisis

This section reviews the literature about Sri Lanka's economy from 2010 to 2022, utilising the platform Mendeley.com for filtering the most relevant positions. The primary objective is to identify positions that elucidate Sri Lanka's economic collapse in 2022. This is regarded as the knowledge gap that this monograph seeks to address by delineating the roots and causes of the bankruptcy. Special emphasis is placed on the economic condition of private companies and their adherence to corporate governance practices, which may exhibit a significant positive relationship with the corporate financial performance, specifically the profitability of publicly listed companies in Sri Lanka for the period 2010–2022. The authors will draw on previously written documents discussing Sri Lanka's economy to propose the most probable causes of the collapse.

The literature on Sri Lanka's financial distress problems is limited, but it provides a starting point for economic analysis. It is important to note that the issue of bankruptcy is attributed to the government rather than the country as a whole. Recently, more authors have attempted to elucidate the causes and propose solutions for Sri Lanka. The primary factors identified include internal conflict, high inflation, the depreciating value of the Sri Lankan rupee, and a shortage of foreign reserves, rendering the Sri Lankan government unable to meet its debt obligations (Mehta 2023; Padda 2011; Zhang et al. 2022). One proposed solution to alleviate Sri Lanka's economic challenges was seeking short-term economic assistance from the International Monetary Fund (IMF) (Jiang 2022). Given the IMF's collaboration with China in establishing Financial Help Centres, this approach may pose challenges. Many economies currently grapple with debt issues and corruption. However, in Sri Lanka's case, it has been observed that even financially sound companies can face bankruptcy, eroding stakeholder confidence and causing substantial damage to the country's economy (Madhushani 2019; Senevirathne and Kuruppu 2018).

A total of 182 economists have asserted that Sri Lanka's debt is considered "odious debt" and, as such, should be deemed illegal, advocating for its cancellation. This perspective suggests that cancelling the debt could provide Sri Lanka with a new path to recovery (Colombo Telegraph 2023). According to Alexander Sack's doctrine,[1] *the odious debt* meets two criteria: (1) it is contracted against the interests of the Nation, or the interests of the People, or the interests of the State, and (2) the creditors cannot prove they were unaware of how the borrowed money would be used. The findings reveal that 71% of the companies in the Trading Sector were in financial distress in 2010–2014 and the rest of the 29% were in the grey zone. The fact that none of the companies lies under the safe zone highlights that as a whole the sector is in a menace (Niresh and Pratheepan 2015).

3.3.1.1 Twin Deficit Scenario

A detailed examination of Sri Lanka's debt history was conducted by Soumya Bhowmick, an Associate Fellow at the Centre for New Economic Diplomacy, Observer Research Foundation. Bhowmick demonstrated that Sri Lanka experienced a prolonged "twin deficit" scenario where the economy's current account (CA) and fiscal account (FA) moved in the same direction. In Sri Lanka's case, excess demand led to increased imports and inflation, rendering domestic goods less competitive. Furthermore, Bhowmick argued that the crisis in Sri Lanka resulted from the government's mismanagement during the implementation of ecological agriculture, aligned with the sustainable development framework of Agenda 2030. This mismanagement contributed to a decline in the exports of agricultural products, such as tea. Notably, the crisis compelled Sri Lanka's government to even import rice, despite it being one of the primary national products cultivated in the country (Bhowmick 2022). The escalating debt of Sri Lanka was also influenced by its participation in Chinese investment initiatives. It is crucial to note that Sri Lanka actively engaged in The Belt and Road Initiative spearheaded by the Chinese government, which played a significant role in the economic downturn of Sri Lanka.

Literature on this subject is somewhat limited, possibly influenced by political considerations. As a result, there is no direct evidence suggesting that China exploited the weaknesses in Sri Lanka's corporate governance to gain economic control over the country's investment decisions. Certainly, if authors have indeed demonstrated that compliance with corporate governance practices has a significant positive relationship with corporate financial performance, particularly the profitability of publicly listed companies in Sri Lanka for the period 2009–2013, it underscores the importance of effective corporate governance in contributing to financial success. This finding suggests that companies adhering to robust corporate governance standards during that time frame experienced enhanced financial performance, emphasising the potential impact of governance practices on business outcomes. (Mapitiya 2015). According to a report from KPMG (2007), it was highlighted that independent directors were not effectively challenging executive directors and management in the fulfilment of their governance responsibilities.

---

[1] See effects of the transformations of the states in their public debts and other financial obligations, published in Paris in 1927.

Additionally, the report noted that existing regulations on corporate governance permitted elected directors to make new appointments to the board. The study concluded that augmenting the representation of independent non-executive directors on the board played a role in a substantial reduction in corporate risk. This underscores the importance of effective checks and balances within corporate governance structures to mitigate risks and ensure responsible decision-making (Sameera 2020). Hence, there exists a provision allowing new appointments to be influenced by the close relationships of majority owners and existing board members. The evidence presented in this study aligns more closely with the stewardship theory, positing that inside directors may possess insider information about the firm, as opposed to external directors. Stewardship theory suggests that directors, especially those with close ties to the company, act as stewards and work in the best interest of shareholders, leveraging their inside knowledge for effective governance (Kiel and Nicholson 2006).

3.3.1.2 Good Corporate Governance

The prevalence of multiple directorships in Sri Lankan listed companies is notably high, primarily due to related entities that share common directors and substantial institutional ownership. An important characteristic highlighted in the literature provides clear evidence that corporate control is frequently consolidated in the hands of a select few individuals, families, or corporate groups who collectively possess the majority of ownership. This concentration of ownership underscores the influence wielded by a limited number of entities in shaping corporate decisions and policies (Koperunthevy and Vijayarani 2014). This situation fosters an environment conducive to nepotistic management decisions, which may primarily serve the interests of the directors or owners. Some authors contend that the diminished corporate governance control results from insufficient support by insiders, and a higher shareholding by outside directors is correlated with a reduced occurrence of financial statement fraud, thereby lowering agency costs. This implies that a direct and substantial involvement of directors in the corporate governance decision-making process is crucial for effective oversight and minimising potential conflicts of interest (Beasley 1996; Ahmed and Duellman 2007). Another noteworthy piece of evidence from the literature study indicates that a significant percentage of shares in companies listed on the Sri Lankan stock market are owned by institutional investors (Suren 2020). The evidence suggests that if Chinese investment interacts with the corporate governance of the Sri Lankan government, it could potentially influence decisions at the highest level, particularly regarding the financing of strategic investments.

The financial collapse of Sri Lanka is attributed, in part, to the mismanagement of state affairs and internal conflicts, fuelled by corruption and a decline in the standard of living for citizens. Some authors propose that addressing internal conflicts could be achieved by reallocating expenditures from the defence sector towards improving the standard of living for Sri Lankan citizens. This approach aims to address social and economic issues, potentially fostering stability and mitigating conflicts within the country (Padda 2011). It is indeed noteworthy that certain experts had provided recommendations to the Sri Lankan government policy many years ago, and these suggestions were not implemented. This lack of implementation has contributed to Sri Lanka facing economic challenges, particularly in areas such as the trade deficit, lower return on investment, high tariffs for trade protection, an inflexible labour market, inefficient education, high inflation, and communication problems between the government and the private sector.

The delayed or non-implementation of expert recommendations underscores the importance of policy responsiveness and adaptability to address ongoing economic issues and foster sustainable development (Sadiq and Priya 1995). Besides that, the financial conditions of the stock-listed companies were analysed by the experts and they stated clearly in their report that to save the companies from total bankruptcy, their financial position should be improved without any further delay (Balasundaram 2009) and that accounting has failed to effectively facilitate good governance of those organisations (Ekenayake et al. 2010). For a better understanding of the current economic crisis in Sri Lanka, it is worthwhile to inquire about the nature of the business-government-society relationship in the country. It was revealed that economic performance is highly influenced by the nature of the relationship between government in an economy (Schneider et al. 1998).

In their book, Steiner and Steiner (2016) presented four models to explain the nature of the business-government-society relationship (1) the market capitalism model, (2) the dominance model, (3) the countervailing forces model, and (4) the stakeholder model. Out of these four models, authors believe that the dominance model is well suited to explain the nature of the business-government-society relationship in Sri Lanka.

3.3.2 The Role of Non-economic Factors in the Economic Crisis of Sri Lanka

The study has revealed that various non-economic factors have contributed to the current economic crisis in Sri Lanka (see Fig. 3.1). Non-economic factors refer to various socio-political and cultural-related aspects that are often non-quantifiable in nature (Khan 2000). Political culture (Friedman 2009), loopholes in the public administration system (Cruz et al. 2023), and the role of civil society and other institutions (Marmot et al. 2013) play a significant role in an economy. It is crucial to analyse these factors in the Sri Lankan economy to gain a more realistic understanding of the current economic crisis.

Four main non-economic factors impacted Sri Lanka's bankruptcy: (1) the existing political culture, (2) the public administration system, (3) Civil Society and institutions, and (4) the nature of the Business-Government-Society relationship.

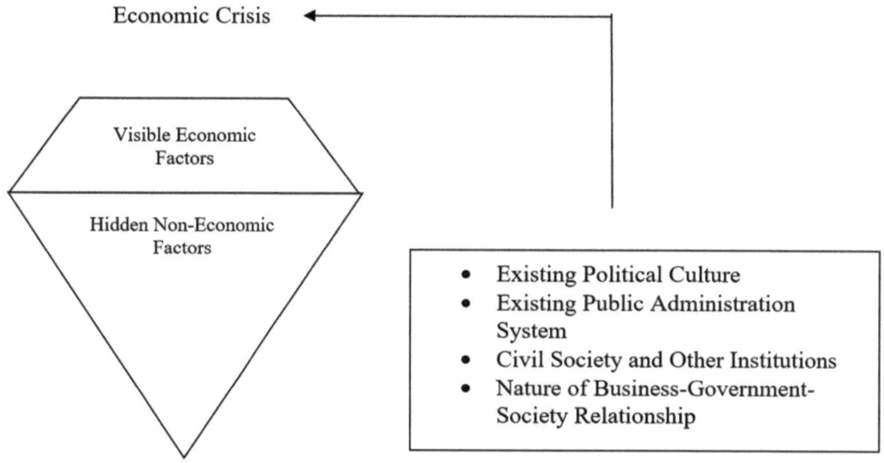

Fig. 3.1 Hidden non-economic factors of the economic crisis of Sri Lanka (Reproduced from (Mills 1983))

3.3.2.1 The Existing Political Culture

Political culture refers to the values and political conduct of individual or collective agents in an economy (Da Silva et al. 2016). As seen in many countries, the economic crisis in Sri Lanka also has been shaped by its political culture. After its independence in 1948, the economic policies of the country often fluctuated between free-market reforms and state control (Stokke and Uyangoda 2011). Seligson and Muller (1987) emphasised the need for stable economic policies for economic development. Thus, the frequent shifting of economic policies has affected investor confidence and long-term economic planning in the country (Stokke and Uyangoda 2011).

Although Sri Lanka is a democratic country with modern governing principles, traditional values, and practices mostly shape the framework of the modern political culture in the country. It was found that certain superstitious beliefs still exist and play a vital role in the Sri Lankan political culture. According to Thalpawila (2021), it was revealed that astrology plays a significant role in the electoral process as well as in shaping the opinion of voters during elections. It is because of a common practice of many astrologers who forecast the election results of any approaching elections (Thalpawila 2021). This indicates the insignificance of sound public policy discussions to win the elections in the country. However, long-term oriented sound public policies are essential to reach the development targets of an economy (Al-Thaqeb and Algharabali 2019). Therefore, the existing political and cultural beliefs and practices regarding the election process have contributed to hindering the demand for sound public policies in the economy.

In a representative democratic system, elected politicians are granted power to serve the people to maximise the general welfare of the citizens (Seligson and Muller 1987). However, this power is seen often as abused by the politicians in Sri Lanka.

This is happened in various forms in practice. In 2015, it was reported a huge bond scam in the Central Bank of Sri Lanka the involvement of politicians engaging in insider dealing (Abeyagoonasekera 2023). Corruption is considered a significant barrier to economic development. Corruption practices of less educated politicians have been identified as a common feature in Sri Lankan Politicians (Kumar and Sharma 2023). Furthermore, public service is extremely politicised in the country. Some key government positions are staffed by political appointees. Consequently, technical competence in public service has been compromised by eras by politicians (Stiftung 2022). Therefore, the existing political culture of the country has contributed to various aspects of the economic crisis in Sri Lanka.

3.3.2.2 The Public Administration System

The quality of the governance process has been identified as an essential requirement for the development of any country (Ramasamy 2020). Holmberg and Rothstein (2012), further argue that various development outcomes of economies are strongly correlated with the quality of public services. However, the public administration system in Sri Lanka is often blamed for its inherited inefficiencies in service delivery. The administrative system in Sri Lanka has five levels of governance structures including national, provincial, district, divisional, and village (De Alwis 2013). The existing multi-level governance structures have resulted in complications and crises in governance and administration in the country (Ramasamy 2020).

The reforms introduced to the public administration system with the view of upgrading public sector performance and efficiency were never achieved in practice, and ultimately, it became another source of corruption (Samaratunge and Pillay 2011). Bribery and corruption of public officials are also vastly identified as a cause of poor economic performance in Sri Lanka. Madhushani (2019) explains that corruption takes place in the sports sector of Sri Lanka through various governmental institutional frameworks. The level of corruption is very high in the public procurement sector (Ramasamy 2020). According to the Corruption Perceptions Index (CPI) 2022, Sri Lanka scored 36 on a scale from 0 ("highly corrupt") to 100 ("very clean") (Transparency International, 2022). Furthermore, there is a declining trend of this score from 2012 (40) to 2022 (36) indicating an increase in corruption in the country. Abeyagoonasekera (2023), also indicated corruption in the public sector as a reason for the current economic crisis in Sri Lanka.

Another distinctive feature of the existing public administrative system of Sri Lanka is its old administrative guidelines which often do not cater to the modern needs of the organisational environment. For instance, the public health sector still operates on mental health laws that were first formulated in 1873 and were enforced when the country was a colony of the United Kingdom (Hapangama et al. 2023). Currently, public financial management in the public sector is managed using public administration system the Financial Regulations issued in 1992 (Jayasinghe 2015). Hence, poor performance in the is inevitable due to these older guidelines. Therefore, certain characteristics of the existing public administration system such as inefficiencies, corrupt practices of officials, and outdated guidelines have also steered poor economic performance in the country. Most democratic countries are faced with a "democratic deficit" in the contemporary political systems (Fischer 2012). This is a situation where an insufficient level of democracy is seen in practice when compared to the theoretical model of a democratic government (Moravcsik 2004). Against this backdrop, the active participation and attention to political matters of citizens and other parties have deteriorated in practice. Nonetheless, the role of civil society and other institutions has been identified as critical for proper governance in a country (Ansell and Gash 2008). Therefore, the concept of "Participatory Governance" has been developed as a solution to this challenge (Fischer 2012). Beyond the citizen's role as a voter, participatory governance includes activities of direct deliberative involvement with other institutions in society.

3.3.2.3 Civil Society and Institutions

It is worthwhile to inquire about the behaviour of civil society and other institutions toward the democratic process in Sri Lanka. As observed in many democratic systems, the active participation of citizens has declined in the country (Nadaradjane 2022). Some voters use their vote for very short-term benefits. This is highly visible in the rural sector of the economy. Distributing clothes, dry foods, money, chairs, roofing sheets, etc., are very commonly observable incidents when approaching an election (Rajakarunanayake 2014). Another group of people vote to get government jobs for their family members (Alwis 2020). Similarly, most people consider narrow benefits when using their vote. This appears as an obstacle to appointing wise members as their representatives.

Some other institutions such as media and religious institutions are also playing a crucial role in the election process. There is evidence of biased behaviours of these groups towards certain political groups during elections. For instance, promoting a story of a cobra that came out of Kelani River was very viral during the last presidential election (Wijewardene 2020). However, the key parties who were involved in these incidents made controversial statements later after the election process (Gunasekara 2019). However, independent media is considered an essential requirement for the proper functioning of economies (Steiner and Steiner 2016). Since the misinformation should be avoided at their best by the media in the country.

Additionally, the irresponsible behaviours and conduct of civil society and other institutions also cannot be ignored in this analysis. As this analysis describes, their decisions also have contributed to the current economic crisis in the country. As explained the economic crisis is a result of various interconnected incidents. As per the above analysis, the existing political culture, existing public administrative system, and the role of the civil society and other institutions also have contributed to the current economic crisis in Sri Lanka. Unless we attempt a favourable change in these factors, the economic recovery journey will be less productive in the country. In such a manner, these factors will act as challenges for the economic recovery process of Sri Lanka.

3.3.2.4 The Dominance Model of the Business-Government-Society Relationship

This model assumes that market forces are inadequate to ensure ethical management in businesses (Steiner and Steiner 2016). Hence, the government works very closely with the businesses. The government tends to play a major role as a controller, price setter, and regulator in various aspects of the businesses. Those who subscribe to the model believe that corporations and a powerful elite control a system that enriches a few at the expense of the many (Steiner and Steiner, 2016). In that way, business and government dominate the great mass of people (society). See the model presented in Fig. 3.2.

In Sri Lanka, government intervention in markets can be considered a common practice across many sectors of the economy. The government plays its role as a player in some sectors such as banking, transportation, utility, etc. (Athukorala and Rajapatirana 2000). Meanwhile, subsidies and price controls have often been employed to ensure the affordability of essential goods like rice and fuel (Wijesooriya 2017). Previous studies show that election campaign finance laws in Sri Lanka are not adequately regulated (Parathalingam 2022). This has resulted in big companies and the wealthy spending millions to grease the hands of politicians to push policies most beneficial to them (Centre for Progressive Network 2022). This has resulted

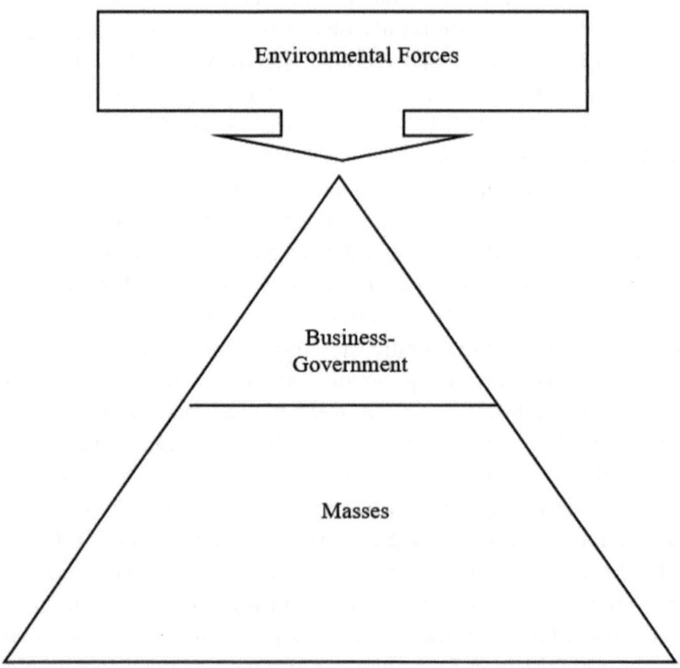

Fig. 3.2 The dominance model of the business-government-society relationship (Reproduced from Steiner and Steiner 2016)

in drowning out the voices of ordinary Sri Lankans. Accordingly, it can be argued that the wealthy have much more influence than ordinary civil society and small corporations in Sri Lanka. However, this kind of business-government-society relationship is regarded as undemocratic (Steiner and Steiner 2016). Moreover, such a system will influence badly the overall welfare of the economy. Therefore, the understanding of the current economic crisis will not be completed unless we realise this business-government-society relationship exists in the country. This might have a direct influence on the dwindling foreign reserves of the country within a very short period. On the way to recovering from the present economic crisis, Sri Lanka would be concerned about altering the existing business-government-society relationship in the country. Otherwise, the journey may either be delayed or ineffective in the long run.

3.3.3 The Ongoing Policy Reforms Overview

The Sri Lankan government approached the International Monetary Fund for assistance with restoring economic stability. 16 IMF finance programs have already been granted to Sri Lanka, and the government is now managing the 17 Extended Fund Facilities. The present policy initiatives are all in line with the most recent IMF policy package. The government prioritises six policy areas: achieving fiscal consolidation and institutional strengthening; ensuring financial stability; eliminating corruption risks; and addressing the sustainability of public debt. Among these policy areas, debt restructuring and state-owned enterprise reforms have drawn more domestic and foreign attention. These policy changes have a great deal of potential to speed up the process of economic recovery, but they must factor in the distributive impacts of such policies and ensure that safety nets are in place to protect the most vulnerable groups in the economy. Without this, the government stands to further exacerbate inequality and cause other structural problems in the long run.

The IMF's Executive Board granted the Sri Lankan government's request for a 36-month loan agreement under the EFF financing facility on June 3, 2016. This was less than the credit agreement reached in 2009 and was worth SDR 1.1 billion, comparable to approximately USD 1.5 billion (IMF 2016a). The economic developments in 2014 and 2015 pushed the Sri Lankan government to apply for this IMF EFF program in 2016.

In 2015, the government's budget deficit increased to 6.9% of GDP. Although revenue climbed by 1.5%age points to 13.1% of GDP, this was driven by past actions and tax receipts from a brief spike in automobile imports. The post-election salary increases, a larger interest bill, more expenditure on goods and services, and an increase in income transfer programs all contributed to an increase in spending of 2.1% age points, to 19.9% of GDP. 2015 saw a considerable decline in the overall balance of payments despite an improvement in the terms of trade. In 2015, the current account deficit was controlled at 2.5% of GDP, the same as in 2014. Capital movements significantly fuelled the deterioration in the balance of payments. Due to foreign withdrawals from government securities, decreased FDI inflows, and the sluggish execution of externally financed public and private projects, the capital and financial account position weakened. Investor confidence worsened due to instability in the global markets and concerns about local initiatives. The Sri Lankan government requested financing to address three structural challenges in the economy. The first was to address the pressure on the balance of payments, the second was to access additional finance for investments, and the third was to reassure private investors and improve confidence in the markets (LKI 2018). The 2016 EFF program contained six pillars: lower budget deficits, monetary policy reform, higher government revenues, state enterprise reforms, stronger public financial management, and supporting higher trade and investment (IMF 2016b).

Six goals are outlined in the 17th IMF program. The primary goal is to advance revenue-based fiscal consolidation and restructure social safety nets, financial institutions, and state-owned enterprises. Second, it aims to re-establish the viability of the public debt. The program also aims to restore price stability by rebuilding external buffers. Additionally, the initiative aims to protect against corruption and provide public financial stability. Another broader goal of the program is to raise sustainable economic growth. Debt reduction is a top priority in Sri Lanka's effort to restore its economy. The government established an interim policy in April 2022 to halt the service of external public debt for a temporary period while awaiting an orderly and mutual restructuring of the debts following the IMF program. The government agreed to hire renowned international financial and legal consultants to help restructure the country's external public debt, which would require negotiating an agreement with creditors to adjust the debt repayment schedule to a sustainable level.

The Sri Lankan government appointed M/s. Clifford Chance LLP as its legal advisory firm and M/s. Lazard, France, as its financial advisory firm for the debt-restructuring process (MOF 2022). The government is also implementing extensive institutional reforms to enhance efficiency, coverage, and targeting of social safety net programs. To shield the poorest households from the financial constraints imposed by IMF conditions, the government must make sure that robust social safety nets are implemented. Food security, affordable medicine, fuel, and effective public services are essential to minimise the negative impacts of austerity policies (Iveson 2023). In Sri Lanka, restructuring State-Owned Enterprises has been one of the main policy priorities. There are an astounding 527 state-owned businesses in Sri Lanka. 10% of the public sector employment is employed by the 55 SOEs designated as "strategically important." Many SOEs, especially those in Sri Lanka, have a history of being inefficient, loss-making, and a burden on the taxpayer. Only 10.4% of the 527 State-Owned Enterprises (SOEs) are SOEs for whom financial information is available. Most critically, between 2006 and 2017, the strategically significant SOEs lost 795 billion Sri Lankan rupees (Advocata 2019).

The Welfare Benefit Board, a legal institution under the Ministry of Finance, has been operationalised as the first step by the government to ensure that all SSN initiatives and reforms are coordinated. The Welfare Benefit Board is creating a new Social Registry, an electronic database of applicants, with assistance from the World Bank to increase coverage and targeting.

Several policies have been implemented to rebuild external buffers and restore price stability. By the end of 2024, the government hopes to lower headline inflation and stabilise it at a target range of 4–6%. As a result, the Central Bank implemented a contractionary monetary policy and increased the policy rates. It was intended to restrict government interference in exchange rate policy and to allow the Sri Lankan rupee to adapt to shifting fundamentals. For a limited time, the Central Bank of Sri Lanka (CBSL) implemented market advice to keep daily currency rate fluctuations within a specified range. However, given the intricate nature of the debt restructuring process, the harsh external environment, and the difficult political and social climate, there is a very high policy implementation risk. The program faces high risks of slippages in fiscal consolidation, revenue mobilisation, and reserve building due to Sri Lanka's poor track record of reform implementation. Implementing policy could be made more difficult by a deeper crisis brought on by more economic repercussions, a weakened banking sector, pressure from currency rates, and a decline in market confidence.

3.4 Future Directions

Sri Lanka due to its strategic position in the Indo-Pacific region has an important role in the trade transit and also military aspects. The problem is that a major part of the Sri Lanka economy due to the high level of debt is controlled by China. As a consequence, the impact of Sri Lanka on the development of the South Asia Economy is related to the strategy of China in this region. Sri Lanka's central bank is projecting higher economic growth for the next five years with output recovering to 5.0% by 2027 compared to 3.1% by the International Monetary Fund. If the projection of the Sri Lankan government is right the overall government debt can be stabilised in 2027 reaching 97.6% of the GDP level (Fig. 3.3). Significant debt restructuring is essential to restore a sustainable debt level.

It can be seen the process for debt reconstruction can be long in the case of Sri Lanka and the economic dependency on China's loans may create a "jo-jo" effect that economic problems will come back to bring the debt level to a position above 100% of GDP again. As the project Belt and Road (BRI) continues the significance of the international sea transport route through Hambantota, especially considering the war in the Middle East will increase. The future state of Sri Lanka's economy is driven not only by the speed of GDP growth but also by the living standard improvement of the citizens after the economic crisis ends. Considering the current corruption level in Sri Lanka and the strong economic dependency on China the chances that the economic benefits of the GDP growth will be distributed to Sri Lanka's citizens are very limited.

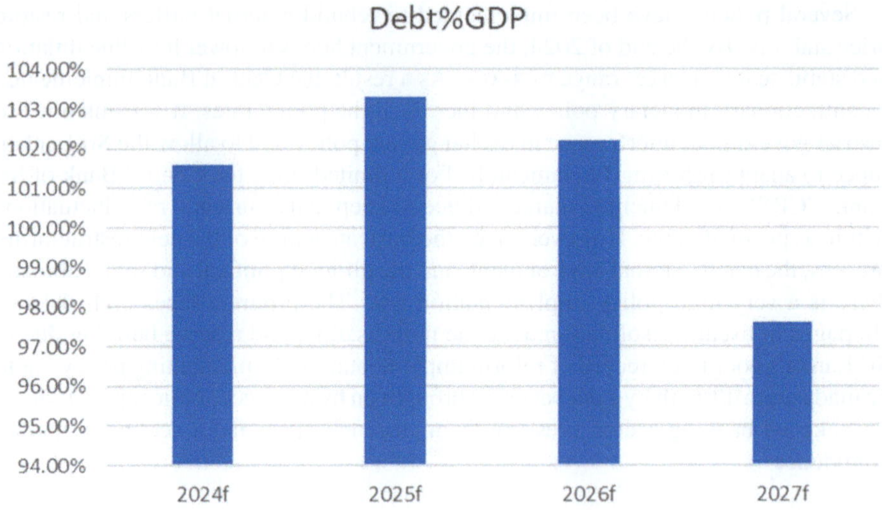

Fig. 3.3 Sri Lankan government debt as % of GDP–the forecast for 2024–2027 (Based on the projections of The Government of Sri Lanka, Ministry of Finance 2023)

3.5 Conclusion

3.5.1 Ambiguity of Future Economic Direction

Throughout history, evidence shows that great civilisations have vanished from the Earth. The ultimate truth is that nothing is constant. The authors have synthesised the phenomena of the recent Sri Lankan economic crisis as a long-term domino effect. In addition to economic factors, multiple interrelated and interdependent non-economic factors are responsible for the existing Sri Lankan socioeconomic downfall. The authors have identified these factors as the main challenges to the economy's recovery process. After analysing the broader non-economic factors, the authors prefer to categorise them under two main themes: challenges relating to political culture and the public administration system that is directly responsible for economic policy-making and the implementing process of the country. Thus, the authors have concluded that the root cause of the ongoing Sri Lankan socioeconomic crisis lies with its people and their irrational decisions and actions under the ineffective, unaccountable, irresponsible socio-economic framework.

3.5.2 IMF Reforms and Collaboration with FTA

Moreover, several key factors influence the future trajectory of Sri Lanka's economy. Currently, Sri Lanka is working to stabilise its economy under the 17th financial assistance program provided by the International Monetary Fund (IMF). The policy package under the IMF's Extended Fund Facility program identifies the restructuring of loss-making state-owned enterprises (SOEs) as a priority. These SOEs include sectors of strategic importance to Sri Lanka's economy, such as the Ceylon Petroleum Corporation, Ceylon Electricity Board, and Sri Lankan Airlines. Sri Lanka should exercise caution in restructuring these entities, as transferring ownership or management to foreign-assisted companies could pose risks to the nation's economic sovereignty. While the full privatisation or transfer of management authority of these entities might yield short-term financial relief, it could also lead to detrimental outcomes in the medium to long term if the nation loses control over critical sectors such as power and energy.

Therefore, the Sri Lankan government should carefully balance its adherence to the policy priorities within the Extended Fund Facility program with the need to preserve the country's long-term sovereignty over strategically significant sectors, safeguarding them against foreign influence. Complementing the IMF program, Sri Lanka's economic policy direction has been further defined by recent legislative efforts. Even if Sri Lanka maintains the existing framework, the economy remains vulnerable due to significant debt challenges. Projections within the IMF program suggest that the debt-to-GDP ratio could reach 103% by 2028. This indicates that, despite completing debt restructuring and expecting credit rating improvements, Sri Lanka will still face a substantial debt burden by the end of the IMF program. The country's economic future remains uncertain, particularly if anticipated GDP growth is undermined by external shocks, such as disruptions from the ongoing Middle East crisis, which could further increase economic instability. Expanding Sri Lanka's trade relations and enhancing its openness to global markets will also be critical. While Sri Lanka officially adopted an open economy in 1977, its integration into the global market has remained limited. Consequently, advancing free trade agreements (FTAs) has become essential for economic growth. In recent years, Sri Lanka has made strides in this direction, including the reinstatement of the Singapore–Sri Lanka FTA in July 2023 and the signing of an FTA with Thailand in February 2024.

3.5.3 Potential Future Collaboration with BRICS

The future economic direction of Sri Lanka and how it will manage its economic crises is similar to the dialogue between Alice and the cat. The direction of Sri Lanka's economic future depends greatly on where Sri Lanka aims to go. Given the current situation, it is hard to predict because it is highly dynamic and still developing. However, although the future of the Sri Lankan economy is unpredictable, we can assume a few possible scenarios based on ongoing policy changes. Sri Lanka signalled its intention to join the Regional Comprehensive Economic Partnership (RCEP) by submitting a letter of intent in June 2023. However, given that specific terms for new RCEP members remain undisclosed, Sri Lanka's timeline for membership is uncertain, with credit enhancements and negotiations required with all 15 RCEP member states.

Furthermore, Recent foreign policy moves signal Sri Lanka's interest in broader global partnerships, notably with BRICS. Following the presidential election in September 2024, the Sri Lankan government announced its intention to pursue membership in BRICS and the New Development Bank in October 2024, reflecting a potential pivot towards stronger ties with India, China, and Russia. This aligns with a trend among developing countries that view BRICS as an alternative to Western-dominated global institutions, hoping to unlock benefits such as development finance, trade, and investment opportunities. Sri Lanka's Foreign Secretary Aruni Wijewardane is leading the country's delegation to the BRICS Outreach/BRICS Plus Summit in Kazan, Russia, from October 22–24, 2024. BRICS recently expanded to include Iran, Egypt, Ethiopia, and the UAE, enhancing its role in emerging market cooperation. Sri Lanka has formally applied to join BRICS and the New Development Bank, with President Anura Kumara Dissanayake and Foreign Minister Vijitha Herath seeking support from BRICS members. During the summit, Wijewardane will address the gathering, meet senior BRICS representatives to discuss Sri Lanka's application and explore bilateral interests (MFA 2024).

However, Sri Lanka's substantial debt to China necessitates a cautious approach to closer alignment with BRICS. Deepening ties, particularly with China and Russia, could strain relationships with Western countries and organisations wary of BRICS' expanding influence. Sri Lanka must carefully balance these relationships to avoid jeopardising existing partnerships and financial support from Western allies. Managing these interests will be essential for sustainable economic growth and for preserving strategic independence amid a shifting global landscape.

3.5.4 GSP+ and Relationship with EU

Finally, maximising the benefits of the GSP + scheme remains crucial for Sri Lanka's immediate economic recovery. In June 2021, the European Parliament passed a resolution calling for an assessment of the potential withdrawal of Sri Lanka's GSP + status. Considering economic hardships at the time, Sri Lanka was granted continued access to the GSP + scheme until December 2027. Progress in addressing human rights and good governance issues will be crucial for maintaining GSP + eligibility. In the short term, Sri Lanka should prioritise the effective utilisation of GSP + preferences while addressing political issues that could impact eligibility. In the long term, however, as GSP + status is contingent upon income levels, Sri Lanka will eventually exceed the eligibility threshold. Therefore, the country must work proactively toward establishing reciprocal trade agreements with the EU and other high-income markets, such as the United States, to secure stable market access as it advances economically.

All these different activities and scenarios should be carefully screened and analysed by the government of Sri Lanka to avoid the main pitfalls of rush run for quick, sustainable economic development, which, in the case of Sri Lanka, requires its way and partnership alliances with other countries. The authorities also expressed a willingness to introduce reforms in sectors such as tourism to boost foreign exchange earnings for the country. This has also been stressed by many experts (UNDP 2021). These suggestions include promoting sustainable tourism to build a resilient tourism sector. Hence, the future of the Sri Lankan economy, as well as its role in Asia, will also be shaped by how these promises are put into action in real-world contexts.

Therefore, the country's economic journey toward prosperity will be determined by the way responsible parties address these challenges. Furthermore, the entire chapter and its insights can be considered a great socioeconomic learning experience for future economic policy-making for other South and East Asian countries. However, the authors conclude that learning by mistake is an important thing; however, it depends on how the government of Sri Lanka is going to implement its decisions and actions.

Highlights

- Balance IMF (International Monitory Fund) reforms with economic sovereignty.
- Cautiously restructure key state-owned enterprises.
- Expand trade through FTAs (Free Trade Agreements) and global partnerships.
- Address debt challenges and optimise the GSP+ (Generalised Scheme of Preferences Plus) benefits.
- Promote sustainable tourism for resilience.
- Align policies to strengthen Sri Lanka's role in the future Asian economy.
- A lesson for future economic policy-making in other Asian Nations.

References

Abeyagoonasekera A (2023) Sri Lanka's political-economic crisis; corruption, abuse of power, and economic crime. J Financ Crime. https://doi.org/10.1108/JFC-03-2023-0069

Advocata (2019) The state of state enterprises in Sri Lanka-2019. Advocata Institute, Colombo. https://soe.lk/wp-content/uploads/2022/07/THE-STATE-OF-STATE-ENTERPRISES-IN-SRI-LANKA-PS-1.pdf. Accessed 3 Oct 2023

Ahmed A, Duellman S (2007) Accounting conservatism and board of director characteristics: an empirical analysis. J Account Econ 43(2–3):411–437. https://doi.org/10.1016/j.jacceco.2007.01.005

Al-Thaqeb SA, Algharabali BG (2019) Economic policy uncertainty: a literature review. J Econ Asymmetries 20. https://doi.org/10.1016/j.jeca.2019.e00133

Alwis DD (2020) Mass unemployed graduate recruitment program suspended. Univ World News. https://www.universityworldnews.com/post.php?story=20200306124156877. Accessed 3 Oct 2023
Ansell C, Gash A (2008) Collaborative governance in theory and practice. J Public Adm Res Theory 18(4):543–571
Athukorala P, Rajapatirana S (2000) Liberalization and industrial transformation: lessons from the Sri Lankan experience. Econ Dev Cult Change 48(3):543–572. https://doi.org/10.1086/452610
Balasundaram N (2009) An investigation of financial soundness of listed manufacturing companies in Sri Lanka: an application of Altman model. Pet Gas Univ Ploiesti Bull Econ Sci Ser 61(4):19–25. https://ssrn.com/abstract=2118003
Beasley MS (1996) An empirical analysis of the relation between the board of director composition and financial statement fraud. Account Rev 71(4):443–465. http://www.jstor.org/stable/248566
Bhowmick S (2022) Understanding the economic issues in Sri Lanka's current debacle. ORF Occas Pap No: 357, Observer Res Found, pp 1–36. https://www.orfonline.org/research/understanding-the-economic-issues-in-sri-lankas/
Centre for Progressive Network (2022) Big money in politics. Daily FT. https://www.ft.lk/opinion/Big-money-in-politics/14-730566
Cruz MD, Jha CK, Kırşanlı F, Sedai AK (2023) Corruption and FDI in natural resources: the role of economic downturn and crises. Econ Model 119:106122. https://doi.org/10.1016/j.econmod.2022.106122
De Alwis RK (2013) Administrative reforms in Sri Lanka: 1950–2005, 1st edn. Godage International Publishers, Colombo
Ekanayake A, Perera H, Perera S (2010) Contextual relativity of the role of accounting in corporate governance: evidence from the banking industry in Sri Lanka. In: Conference proceedings of the sixth Asia Pacific interdisciplinary research in accounting conference
Fischer F (2012) Participatory governance: from theory to practice. In: The Oxford handbook of governance. Oxford Academic, pp 457–471
Friedman J (2009) A crisis of politics, not economics: complexity, ignorance, and policy failure. Crit Rev 21(2–3):127–183. https://doi.org/10.1080/08913810903030980
Gunasekara T (2019) The unravelling. Daily FT, 16 Dec. https://www.ft.lk/Columnists/The-Unravelling/4-691664
Hapangama A, Mendis J, Kuruppuarachchi KALA (2023) Why are we still living in the past? Sri Lanka needs urgent and timely reforms of its archaic mental health laws. Bjpsych Int 20(1):4–6. https://doi.org/10.1192/bji.2022.26
Holmberg S, Rothstein B (eds) (2012) Good government: the relevance of political science. Edward Elgar, Cheltenham
IMF (2016a) IMF executive board approves three-year US$1.5 billion extended arrangement under EFF for Sri Lanka. Press release. https://www.imf.org/en/News/Articles/2015/09/14/01/49/pr16262. Accessed 3 Oct 2023
IMF (2016b) Sri Lanka to reboot economic policies. IMF Survey. https://www.imf.org/en/News/Articles/2015/09/28/04/53/socar061416a. Accessed 3 Oct 2023
Iveson M (2023) The International Monetary Fund's new extended fund facility arrangement for Sri Lanka. Lakshman Kadirgamar Institute, Colombo. https://lki.lk/explainer-the-international-monetary-funds-new-extended-fund-facility-arrangement-for-sri-lanka/. Accessed 3 Oct 2023
Jayasinghe K (ed) (2015) Public sector accounting, accountability and auditing in emerging economies, 1st edn. Bingley, Emerald
Jiang Q (2022) Debt crisis and national bankruptcy: evidence from Sri Lanka. BCP Bus Manag 29:315–321. https://doi.org/10.54691/bcpbm.v29i.2288. Accessed 20 Apr 2023
Khan NS (2000) Non-economic factors in development. Pak Dev Rev 39(4II):715–725. https://doi.org/10.30541/v39i4IIpp.715-725
Kiel G, Nicholson G (2006) Multiple directorships and corporate performance in Australian listed companies. Corp Gov Int Rev 14(6):530–546

Koperunthevy K, Vijayarani (2014) Corporate governance practices, issues and challenges in Sri Lanka. Globalbizresearch 1–9
KPMG (2007) Corporate governance survey in Sri Lanka, practices and developments of corporate governance
Kumar M, Sharma A (2023) Corruption, education, and media: review of the film Dasvi (2022). Media Asia 50(2):317–323. https://doi.org/10.1080/01296612.2022.2133373
LKI (2018) The International Monetary Fund and Sri Lanka. Lakshman Kadirgamar Institute of International Relations and Strategic Studies. https://lki.lk/publication/the-international-monetary-fund-and-sri-lanka/. Accessed 3 Oct 2023
Madhushani A (2019) Challenges in integrity of sport: current practices and preventive approaches of sport corruption in Sri Lanka. Phys Cult Sport Stud Res 84(1):21–26. https://doi.org/10.2478/pcssr-2019-0023
Mapitiya GS (2015) Degree of compliance with corporate governance best practices and corporate financial performance of public listed companies in Sri Lanka. SSRN Electron J. https://doi.org/10.2139/ssrn.2699719
Marmot M, Bloomer E, Goldblatt P (2013) The role of social determinants in tackling health objectives in a context of economic crisis. Public Health Rev 35(1):9. https://doi.org/10.1007/BF03391694
Mehta S (2023) The economic crisis in Sri Lanka: what led to bankruptcy? In: SAGE business cases. SAGE Publications Ltd. https://doi.org/10.4135/9781529612226
MFA (2024) Foreign Secretary Wijewardane leads the Sri Lanka delegation to BRICS Outreach/BRICS Plus Summit in Kazan, Russian Federation. https://mfa.gov.lk/fs-delegation-to-brics/. Accessed 29 Oct 2024
Mills W (1983) Darwin and the iceberg theory. Notes Rec Roy Soc London 38(1):109–127
MOF (2022) Appointment of international financial and legal advisors for the restructuring of external public debt restructuring of the Government Democratic Socialist Republic of Sri Lanka. Ministry of Finance. https://www.treasury.gov.lk/news/article/112. Accessed 3 Oct 2023
Moravcsik A (2004) Is there a 'democratic deficit' in world politics? A framework for analysis. Gov Oppos 39(2):336–363. https://doi.org/10.1111/j.1477-7053.2004.00126.x
Nadaradjane A (2022) Sri Lanka: democracy in crisis. Center Indep Stud 40:1–7
Niresh A, Pratheepan T (2015) The application of Altman's Z-score model in predicting bankruptcy: evidence from the trading sector in Sri Lanka. Int J Bus Manag 10(12):269–275
Padda I (2011) Fiscal policy of Sri Lanka, past and future. J Econ Econ Educ Res 12(3):73–90
Parathalingam S (2022) How much does money influence Sri Lankan elections? Ceylon Today. https://ceylontoday.lk/2022/07/05/how-much-does-money-influence-sri-lankan-elections/
Rajakarunanayake L (2014) The politics of roofing sheets. Sunday Observer. https://archives.sundayobserver.lk/2004/05/09/fea100.html
Ramasamy R (2020) Governance and administration in Sri Lanka: trends, tensions, and prospects. Public Adm Policy 23(2):187–198. https://doi.org/10.1108/PAP-03-2020-0020
Sadiq A, Priya R (1995) Promoting growth in Sri Lanka: lessons from East Asia. World Bank, South Asia Country Department I, Office of the Director, pp 1–34. https://www.eldis.org/document/A25290. Accessed 1 Mar 2023
Samaratunge R, Pillay S (2011) Governance in developing countries: Sri Lanka and South Africa compared. Int J Public Adm 34(6):389–398. https://doi.org/10.1080/01900692.2011.570003
Sameera TKG (2020) Corporate governance practices and their impacts on corporate risk: evidence from Sri Lanka. Kelaniya J Manag 9(2):21–39. https://doi.org/10.4038/kjm.v9i2.7642
Schneider BR, Evans P, Silva E, Chandler A, Amatori F, Hikino T (1998) Elusive synergy: business-government relations and development. Comp Polit 31(1):101. https://doi.org/10.2307/422108
Seligson MA, Muller EN (1987) Democratic stability and economic crisis: Costa Rica, 1978–1983. Int Stud Q 31(3):301. https://doi.org/10.2307/2600669
Senevirathne K, Kuruppu G (2018) Prediction of bankruptcy risk: evidence from listed manufacturing companies in Sri Lanka. Vidyodaya J Manag 4(1):55–73. https://doi.org/10.31357/vjm.v4i1.3634

Da Silva FC, Clark TN, Vieira MB (2016) Political culture, 1st edn. In: Mazzoleni G (ed) The international encyclopedia of political communication. Wiley, pp 1–10. https://doi.org/10.1002/9781118541555.wbiepc161
Steiner JF, Steiner GA (2016) Business government and society: a managerial perspective, 13th edn. McGraw-Hill India, New Delhi
Stiftung B (2022) Sri Lanka country report 2022 (BTI transformation index). Bertelsmann Stiftung. https://bti-project.org/en/reports/country-report/LKA
Stokke K, Uyangoda J (eds) (2011) Liberal peace in question: politics of state and market reform in Sri Lanka. Anthem Press, London
Suren P (2020) Is board independence a panacea for improved corporate performance? evidence in a developing country setting. Eur J Bus Manag 12(33):49–60. https://doi.org/10.7176/ejbm/12-33-06
Colombo Telegraph (2023) 182 experts call for debt cancellation but private investors play hardball. Colombo Telegraph. https://www.colombotelegraph.com/index.php/182-experts-call-for-debt-cancellation-but-private-investors-play-hardball/
Thalpawila ON (2021) Superstitious beliefs and practices in politics: an aspect of political culture in Sri Lanka. Int J Arts Humanit Soc Stud 3(5):51–57
Transparency International (2022) Corruption perception index. https://www.transparency.org/en/cpi/2022/index/lka
United Nations Development Programme (2021) Putting people first: building a more resilient tourism sector in Sri Lanka. https://www.undp.org/sites/g/files/zskgke326/files/migration/lk/UNDPLKA_Tourism_Roadmap.pdf. Accessed 29 Oct 2024
Wijesooriya N (2017) Government intervention in paddy marketing: issues in purchasing and post-stock management. Hector Kobbekaduwa Agrarian Research and Training Institute, Colombo
Wijewardene W (2020) Conspiracy stories are all over; but why are they so popular and readily accepted? Colombo Telegraph. https://www.colombotelegraph.com/index.php/conspiracy-stories-are-all-over-but-why-are-they-so-popular-readily-accepted/
Zhang T, Cheng S, Luo A (2022) The last resorts to destroy the Democratic Socialist Republic of Sri Lanka. Front Bus Econ Manag 5(3):89–93. https://doi.org/10.54097/fbem.v5i3.1916

Robert Dygas is an assistant professor in the East Asian Economic Studies Department at the SGH Warsaw School of Economics, within The World Economy Research Institute. He earned his Ph.D. in Economics in 2003 from the same institution, where his dissertation focused on strategic management in Poland's mobile telephony industry. In 2011, he completed an Executive MBA at the Warsaw University of Technology. With 23 years of professional experience across IT, telecommunications, and the public sector, Dr. Dygas served as an external expert evaluating Horizon 2020-ECI-FTI proposals for the European Commission from 2015 to 2020

Kelaniyage Shihan Dilruk Fernando is an experienced business professional with thirteen years of experience in banking, finance, and insurance. He has six years of experience in corporate management. Since 2019, he has been teaching at the MBA, master's, and bachelor's levels at the West Pomeranian Business School in Szczecin, Poland. His research interest focuses on sustainable management, and he is an expert in quantitative research design and analysis. Additionally, he is a member of the Winnet Centre of Excellence® and a research team member at the UNESCO Chair for Social Sustainability at the University of Szczecin, Poland. He earned a Diploma in Public Management and a Bachelor of Science in Management (Public) General External Degree from the University of Sri Jayewardenepura. He also holds a Postgraduate Diploma in Business Management and a Master's Degree in Business Management from the University of Colombo. Furthermore, he has obtained a Professional Qualification in Human Resource Management from the Chartered Institute of Personnel Management in Sri Lanka.

Tharindu Udayanga Kamburawala is a lecturer in the Department of Economics at the University of Sri Jayewardenepura in Sri Lanka. He holds a BA Honours degree in Economics from the University of Kelaniya, which he earned in 2019. Currently, he is pursuing an MSc in Economics at the University of Kent in the United Kingdom. Mr. Kamburawala teaches courses on Development Policy and International Economic Policy. Previously, he worked as a Junior Researcher at the Institute of Policy Studies of Sri Lanka, where he focused on macroeconomics, trade, and competitiveness policy. His research interests encompass structural reforms, regional integration, and the political economy of developing countries.

Walisundara Mudiyansela Dhananjaya Madusanka Dissanayaka is a senior lecturer in the Department of Public Administration at the University of Sri Jayewardenepura in Sri Lanka. He is currently pursuing a Ph.D. in Governance and Development at the National Institute of Development Administration in Thailand. He obtained an MBA in International Business from the University of Colombo in 2021 and a B.Sc. in Public Management from the University of Sri Jayewardenepura in 2016. With over seven years of teaching and research experience, his areas of expertise include public administration, public procurement, and public finance.

Chapter 4
Post-pandemic (COVID-19) Economic Dynamics in India and Southeast Asia: Challenges, Opportunities and the Future

Clusters of Future Studies: Past & Futures

Mohor Chakraborty

Abstract Since the onset of the year 2020, countries across the globe have been grappling with the socio-economic, political and humanitarian repercussions of the COVID-19 pandemic. South and Southeast Asia, like other regions of the world, faced the challenges that this crisis unfolded in diverse spheres. India, a highly populous, developing nation in the South Asian region was not immune to the consequences of the pandemic, taking into account its economic potential, performance and attendant impact on the social fabric. Several domestic, regional and international factors like the difficulties in arresting the contagion, the pandemic policy responses of the countries of South and Southeast Asia/Association of Southeast Asian Nations (ASEAN) and underlying vulnerabilities, especially across the financial sector and the overall global slump fuelled the economic distress. It is in this backdrop that the chapter, based on empirical and qualitative analyses, situates the repercussions of the COVID-19 pandemic on India-ASEAN trade relations, while accounting for the compulsions emanating from economic interdependence and regional economic arrangements. It analyses the policies adopted by ASEAN bloc as a whole and India towards addressing the situation and embarking on a resilient economic future. It concludes with a discussion of the challenges, impact and prognosis for unveiling resilient economic future in India-ASEAN relations.

Keywords ASEAN · COVID-19 · Economic · India · Trade

M. Chakraborty (✉)
South Calcutta Girls' College, Kolkata, India
e-mail: mohor_5@yahoo.co.in

E. Lau et al. (eds.), *Future of Economy in Asia*,
https://doi.org/10.1007/978-3-031-95500-6_4

4.1 Introduction

The immediate and long-term socio-economic impact of the COVID-19 pandemic on countries across different regions of the world and their economic policies, endurance and commitment towards resilient, post-pandemic recovery forms a noteworthy subject of enquiry. Like other regions of the world, South and Southeast Asia grappled with the socio-economic, political and humanitarian repercussions of this unprecedented phenomenon since its onset in 2020. India, a highly populous nation in the South Asian region was not immune to the consequences of the pandemic, insofar as its economic potential, policies, guidelines and attendant impact on the politico-economic and social fabric were concerned. The impact of the pandemic was exacerbated by the environmental wrath of cyclone 'Amphan', ravaging the eastern Indian states of West Bengal and Odisha in May 2020. A range of domestic, regional and international factors, stimulated by the pandemic called for rational and timely policy interventions by governments, to protect their respective population from the spiralling consequences, as well as sustaining a functioning economy at the domestic level. Under these circumstances, the policy responses of India and the Association of Southeast Asian Nations (ASEAN),[1] viewed in perspective of the underlying vulnerabilities related to interrelated issues like public health, vaccine rollout and its universal access, economic sustenance, foreign trade and investment, capacity to address the economic impact of pandemic and chart a future course of action etc., emerged as the *sine qua non*, without resort to a universally, standardised approach, since the countries under review not only represent asymmetric levels and standards of economic development, but their diverse demographic strength were also hit to varying degrees by the impact and intensity of the pandemic.

It may be mentioned in this context, that India's policy towards ASEAN is defined by the 'Look East' and 'Act East' policies, the successful execution of which elevated the former from a Full Dialogue Partner, member of ASEAN Regional Forum (ARF), ASEAN+1 Summit-level Partner to a member of East Asia Summit (EAS) and a Comprehensive Strategic Partner, in addition to being ASEAN-India Free Trade Area (FTA) Agreements partner, providing the pedestal for cooperation in the defence, economic, strategic, cultural spheres, etc (ASEAN, India reaffirm Comprehensive Strategic Partnership 2024). Besides, India's Indo-Pacific policy acknowledges the centrality of ASEAN in the Indo-Pacific region and emphasises holistic cooperation to ensure regional peace, stability and prosperity in general and trade and economic relations in particular, through three primary institutional mechanisms like ASEAN Economic Ministers (AEM)-India Consultation, ASEAN-India Framework Agreement on Comprehensive Economic Cooperation and ASEAN-India Business Council (AIBC) (Brief on ASEAN-India Relations 2023).

The economic plight associated with the pandemic was exacerbated by the compelling characteristics of economic interdependence, which forms the hallmark

[1] ASEAN, formed in 1967, consists of 10 member-states, viz. Indonesia, Malaysia, the Philippines, Singapore, Thailand (as founding members) and Brunei, Cambodia, Laos, Myanmar and Vietnam (as subsequent entrants).

of globalisation, fuelled by the global economic slump, with the International Monetary Fund's (IMF) estimated shrinkage of global growth at 4.9% in June 2020 and projected global growth at 5.4% in 2021 (World Economic Outlook Update 2020, 1). It is in this backdrop that the chapter situates the repercussions of the COVID-19 pandemic on India-Southeast Asia trade relations, while accounting for the compulsions emanating from economic interdependence and regional economic arrangements. Based on empirical and qualitative analyses, it highlights the policies adopted by ASEAN bloc as a whole and India towards addressing the situation and embarking on a resilient economic future. It concludes with a discussion of the challenges, impact and prognosis for unveiling resilient economic future in India-ASEAN relations.

4.2 Literature Review

The chapter employs primary data and statistics from official and institutional sources, like official statements, plans of action and reports to situate the context and background of the pandemic and enumerate the economic outlook, dynamics, implications and dimensions of India-ASEAN trade relations as well as its overall impact on the global economic situation. The review of primary sources reveals the fundamental analytical paradigm based on key statistics, comparative analysis of pre-COVID and post-COVID economic data, assumptions, forecasts, projections and pragmatic recommendations for post-pandemic recovery, taking stock of the comprehensive economic impact of the pandemic, both at the global and regional levels and their reflections on India-ASEAN trade (World Economic Outlook Update 2020; Economic Outlook for Southeast Asia, China and India, 2020, 2021; Export–Import Data Bank, 2020–2024).

The review of secondary sources consists of institutional/organisational reports, books, issue briefs and research articles analysing the impact of the pandemic on the economic and trade scenario at the international, regional and ASEAN-India levels, in addition to highlighting its consequences for ASEAN economic integration and community-building (Chakraborty 2023; Ghoshal 2020; Menon and Suvannaphakdy 2021). Besides, to boost the economic, trade and commercial scenario in the region in the post-pandemic situation, the potential of harnessing specific sectors of trade, emerging value chains and economic diplomacy measures has been highlighted (Chanda 2023; Choudhury and Nagda 2020; De and Nida 2021; Gill 2020). The comparison of the nature and volume of India-ASEAN trade in the pre-pandemic and under the immediate circumstances of the pandemic has emphasised the need to identifying the sectors of the economy that have suffered considerably, with attendant recommendations for post-pandemic recovery. In this respect, the imperative of addressing procedural delays in negotiations on ASEAN-India FTA as well as measures to minimise the balance of trade through the implementation of vision-oriented, pragmatic policies, thereby fostering regional economic recovery and trade integration, has been underscored (Chakraborty and Suri 2021; Oikawa et al 2021; Palit 2022; Wignaraja 2022; Zen and Kimura 2023).

Furthermore, within the ambit of secondary sources, newspaper reports from authoritative media sources like *Business Today, CNBC, Deccan Herald, Economic Times, The Financial Express, Hindu Business Line* among others, have provided both the 'real picture' and the analyses of COVID-inflicted circumstances and their impact on national, regional and global economies. These sources have analysed the nature and impact of economic policies followed by India and ASEAN as well as underscored the trends in bilateral trade interactions between the two sides, with emphasis on pragmatic measures for addressing the economic challenges the pandemic entailed. Since the issue under review is a topical and current development, in the context of the pandemic, the paucity of literature concentrating extensively on the economic dynamics and impact on India-ASEAN relations was tangible, thus providing the space for bridging this pedagogical gap in the post-pandemic circumstances.

4.3 Methodology

This case study is based on both quantitative and qualitative/descriptive methodology, using empirical data to contextualise the problem, identify the impact of variables on the global economic situation in general and India-ASEAN trade relations, in particular in the backdrop of the pandemic and derive rational conclusion, considering policy recommendations and findings. Data and statistics from primary sources provide the quantitative framework of analysis, situating the economic reality and challenges associated with the pandemic and accounting for its impact on global economy in general and on India-ASEAN trade relations, in particular. Taking stock of the economic situation, the repercussions for India-ASEAN trade and commercial interactions are supported by comparative trade statistics and export–import data. Statistical reflections of the decline in the overall economic scenario experienced at the global, regional and national levels are revealed from these sources. The challenges and implications of the pandemic are analysed at the levels of India, ASEAN and the international milieu. The quantitative sources provide the analytical frame for qualitative and descriptive methodology, culminating in the interpretation, derivation and logical conclusion, with a prognosis for the future of India-ASEAN trade and commercial relations.

Following a case-study approach, by interpreting quantitative data, the problem is investigated and findings reached by integrating quantitative and qualitative analyses. Besides, the potential and impact of a variety of social, economic and political indicators, acting as variables, attributed at the global, regional and national levels is analysed. In conclusion, recommendations and future prognosis to tide over the crisis unveil the course for post-pandemic economic recovery.

4.4 Findings

Having emerged from the unprecedented impact of the COVID-19 pandemic by navigating the multiple socio-economic and humanitarian challenges it entailed, India and ASEAN find their trade relations recuperated, while acknowledging the imperative of cooperation for sustaining post-pandemic recovery and its impact on the domestic and regional economic scenarios. However, since the future dynamics are unique, yet interrelated for India and ASEAN, the impetus on embarking upon a bright and resilient economic future shall not only be contingent on pragmatic economic and trade policies but also sustained institutional negotiations to overcome the limitations in the domain.

4.4.1 COVID-19 Scourge: Economic Scenario and Implications

The COVID-19 pandemic, *sui generis* in nature, with unique characteristics and unprecedented consequences, pulled the Asia–Pacific region into the vortex of uncertainty and economic volatility. As estimated by the Director of IMF's Asia and Pacific Department, regional economic growth was projected to stall at null in 2020, demonstrating the worst performance in six decades, surpassing the figures of 4.7% and 1.3% recorded during the Global Financial Crisis and the Asian Financial Crisis respectively (Rhee 2020). OECD indicated unprecedented decline in Gross Domestic Product (GDP) and a pronounced recession for the 'emerging Asian economies'[2] while projecting contraction in average annual real GDP growth for ASEAN by 4.3% in 2020. It projected contraction in real GDP growth for India by 10.2% in 2020 (Economic Outlook for Southeast Asia, China and India 2020: ongoing challenges of COVID-19, 2020, 4–5). The OECD projected that regional economic recovery would face significant challenges in 2021, pegging average annual real GDP growth for ASEAN at 5.1%, while India was projected to be the worst hit among the region's emerging economies, accounting for an estimated GDP decline of 9.9% (Economic Outlook for Southeast Asia, China and India 2021, 17). Therefore, the region was poised for a weak and delayed post-pandemic economic recovery and the immediate future portended stunted India-ASEAN trade and commercial interactions, spearheaded by various factors like a decline in demand, supply chain disruption, loss of productivity, impact on financial markets, declining investor confidence, impact on the workforce, restricted travel for tourism, business or medical purposes, among others, thereby stimulating realistic and contingent policy responses by the states concerned.

[2] 'Emerging Asian economies' refer to ASEAN, China and India.

4.4.2 *COVID-19 Scourge: Scenario of Economic Turmoil in India*

COVID-19 struck at a time when India was attempting to recuperate from a domestic economic slowdown. In this situation, the sudden outbreak of the pandemic resulted in a drastic change in the country's policy priorities, from the imperative of reviving economic growth and boosting trade to surviving the calamity, through social welfare measures. The Government of India announced three major relief packages between March-November 2020, targeted at the economically vulnerable sections, committed to food and financial security, provisions for insurance cover, collateral-free loans and employment generation. Furthermore, it focused on Research and Development (R&D) for developing indigenous vaccines, to stimulate self-reliance or 'Atmanirbharta'. Simultaneously, the Government of India, in a bid to arrest the spread of the virus and minimise its effect, introduced stringent measures like imposition of lockdown, social distancing, self-isolation, restrictions on mobility, closure/restricted/online functioning of public and commercial institutions and facilities and stringent travel restrictions for leisure, business or medical purposes, which had severe repercussions for the country's trade and economic potential, connectivity, infrastructure development and maintenance and their contingent impact on the social fabric.

The pandemic-induced effects were palpable on all sectors of the Indian economy, including agriculture, automobiles, aviation and tourism, chemical products, electronics, energy/oil and gas, hospitality, medical travel, pharmaceuticals, telecommunications, etc. The volatility of the financial market, with huge crashes and wealth erosion had a telling effect on consumption levels. The impact of the extended period of lockdown was severe, as stringent restrictions halted most economic activities and caused millions, many of them daily wage earners, to lose their jobs and revenue streams. Furthermore, the migration of labourers exacerbated the challenges faced by the Micro, Small and Medium Enterprises (MSMEs), in addition to having a debilitating impact on their livelihood. A snapshot of the various domains of the Indian economy revealed a slump in industrial production in 2020, resulting in the contraction of production by 55.5% vis-à-vis 2019 figures. The sharpest decline was witnessed by the manufacturing sector, as most industrial establishments posted no production for April 2020. These trends were amply surveyed and reported in the government publications and popular media sources.

The output of eight core infrastructure industries including coal, crude oil, natural gas, refinery products, steel, fertilizers, cement and electricity contracted by 23.4% by May 2020, leading to a dip in the sector's overall output by 30%. Other sectors like travel, tourism and hospitality, aviation, commodity and merchandise trade, construction, foreign trade, hospitality etc. suffered'extreme decline' in activity as well. The services business activity index was rated at 5.4%, far below an industry forecast of around 40% (Lee 2020). This was attributed to the lockdown's impact on reduced consumer footfall in shops and decline in demands, as corroborated by the Reserve Bank of India's Consumer Confidence Survey (telephonic) conducted across 13 major cities. The survey revealed the pessimistic perceptions of the Indian consumer base (Consumer Confidence Survey 2020) further fuelled by the general economic situation, employment scenario and plunging household incomes. The crisis also resulted in retreating foreign exchange reserves. Overall, financial surveys and predictions portrayed a bleak picture as the Indian economy shrank by 23.9% in the first quarter[3] signalling an arduous road to future economic recovery. The situation was compounded by contraction in consumer spending, capital investment, exports and imports by 24.5%, 47.9%, 17.1% and 38.5% respectively under COVID circumstances (Suryamurthy 2020).

4.4.3 Scenario of Economic Turmoil in ASEAN: Impact on India-ASEAN Trade Relations

With respect to the ASEAN economies, in general, the impact of COVID-19 was felt on key sectors, viz. tourism, agro-processing, garments, electronics, manufacturing, etc. ASEAN as an organization, with the objective of strengthening regional cooperation, organizational integration, community-building and balanced recovery, was poised to adopt more targeted policies, taking cognizance of the constraints to finance and resources, while being committed to the need to generate tax revenues for meeting public expenditure, expand the production and consumption base and introduce accommodative policies to strengthen tax administration to create a healthy fiscal space. Besides, ASEAN's imperative of containing the contagion was linked to maintaining the proper functioning of key sectors and services like healthcare, primarily the development and distribution of vaccines, education, employment, social welfare, environmental protection to ensure 'green recovery' and above all, facilitating a sustainable post-pandemic socio-economic recovery and resilience (Supporting post-COVID 19 economic recovery in Southeast Asia 2022).

The sectoral approach important for developing targeted recoveries for the short and medium terms was complemented by ASEAN's focus on 'cross-cutting' policies to support an enabling business environment and enhanced infrastructure. Acknowledging the socio-economic complexities emanating from widespread demand–supply-side shocks and the impact of the pandemic on the population of the constituent member-states of the organisation, ASEAN expressed commitment to strengthen'collective efforts in enhancing resilience to navigate and effectively respond to emerging challenges', while embarking upon policies to address the threats of COVID-19, thus 'minimising its disruption and promoting sustainable recovery and ASEAN's inclusive growth, as well as the need for a holistic and strategic approach to better shield our people against future spillover impacts.' Towards this end, ASEAN was committed to collaborate with its Dialogue Partners like India to 'keep markets open and ensure continued flows of trade and investment, supply chain connectivity, particularly for essential products and strong and resilient post-pandemic recovery' (Chairman's Statement of the 37th ASEAN Summit 2020,

[3] First quarter ended June 30, 2020.

Cl. 4, 13) providing the groundwork for India-ASEAN cooperation and coordination in the backdrop of the pandemic.

As the Indian and ASEAN economies faced severe challenges, arresting the downward spiral was a daunting task and their bleak economic situation linked with the volatility of external demand and supply chain disruptions, did not augur well for trade and commercial interactions. This was reflected in the value of India-ASEAN trade in 2020–2021. During the Financial Year (FY-April to March) 2020–2021, total value of trade between India and ASEAN accounted for US$ 78 billion, as compared to the US$ 86 billion during FY 2019–2020. India-ASEAN trade started improving since FY 2021–2022, recording a total of US$ 110 billion (Export-Import Data Bank 2022) attributed to policies that enabled to arrest the contagion, including the launch of vaccination by January 2021, improvement in public health and emergency services and incremental easing of pandemic-related restrictions.

4.5 Discussion of Implications

The intricate relationship of the domestic-international interface in India-ASEAN economic diplomacy can hardly be exaggerated, with the imperative of stimulating adequate policy responses which would best suit their specific conditions and plight towards addressing the impact and navigating the ccomprehensive challenges posed by the pandemic, including recession, rising protectionism, economic nationalism, scarcity of food and energy resources, disruptions in global value chains, further exacerbated by the United States (US)-China 'trade and technology war' and the unfolding of the Ukraine crisis in the post-pandemic circumstances. As the pandemic progressively intensified through 2020 and its successive resurgent waves in 2021–2022, the IMF recommended prioritisation of post-pandemic recovery policies on 'additional support' in strengthening economic action, as well as successful roll-out of vaccines and their accessibility (Supporting post-COVID 19 economic recovery in Southeast Asia 2022).

4.5.1 Implications of the Pandemic on India's Economic Policies

In keeping with the emerging circumstances, India's policies were focused on bolstering institutional capacity-building and competence across public and private spheres, boosting productivity and growth, promoting linkages with regional economies, like ASEAN, while emphasising domestic welfare measures, viz. employment generation, health and vaccination, food security, education, poverty, digital accessibility and the like. The resumption and gradual boosting of trade and economic relations with other countries and blocs like ASEAN was envisaged as complementary to the domestic economic consolidation and resilience. The Government of India, as an immediate measure to address and mitigate the economic fallout of the pandemic, announced US$ 266 billion support package, corresponding to nearly 10% of GDP and a host of fiscal and monetary measures. Simultaneously, it introduced the 'Atmanirbhar Bharat' (Self-Reliant India mission) programme, as the lynchpin for consolidating the 'Make in India' or indigenisation drive (introduced in September 2014) in conjunction with stimulus measures to address the immediate short-term challenges and a set of reforms in agriculture, labour and industrial sectors for the medium term (Bhalla 2020). This sector-based approach was published in government sources and media reports, which focused on the following measures: First, establishing an Agri-Infrastructure Fund to bridge the gaps along the agricultural value chains; the elimination of the Essential Commodities Act; and the introduction of a new legislation that empowered farmers to sell their produce at remunerative prices by eliminating inter-state barriers to trade and articulating a framework for e-trading of agricultural produce. Therefore, reforms that ensured access of farmers to technology for reducing the impact of natural or climatic disasters like 'Amphan' and neutralise their impact on crop productivity, reduce debt traps and probabilities were embarked upon.

Second, the 'Atmanirbhar Bharat Abhiyan' proposed fast-tracking investments and measures to promote investments, schemes to boost manufacturing, upgrading and augmenting industrial infrastructure and promoting initiatives for improving the quotient of Ease of Doing Business. Although it focused on addressing the concerns of high logistics cost and R&D expenditure, pragmatic measures were required for translating this vision into reality, such that the initiative could accommodate the challenges of employment in general and formal gainful employment in particular. This consideration gained import given the fact that over 90% of the Indian workforce operated in the informal sector and faced the issues of irregular and uncertain earnings, social insecurity and inappropriate working conditions, as per COVID norms (D'souza 2020). Such measures enabled the resumption and increased rate of labour force participation with the gradual abatement of the pandemic.

Third, the 'Atmanirbhar Bharat' package was unveiled to provide relief to MSMEs in the form of an increase in credit guarantees. Besides, collateral free automatic loans for businesses including MSMEs and subordinate debt for stressed MSMEs was envisioned, to extend collateral-free loans to MSMEs with overdue loans (Venkatesh 2020).

Fourth, policies to boost external trade and address the disruptions in global value chains, in tandem with pursuing the vision of transforming India into a global manufacturing hub, while facilitating self-reliance and adopting resilient measures was focused. It was acknowledged that the 'Atmanirbhar Bharat Abhiyan', focused on skill development and mapping employment opportunities to the skill sets of migrant workers, would be effective and complete, following transformation in the educational system, which was poised to enhance employability. Furthermore, the question of generating employment opportunities for the migrant workers according to their skill sets, in keeping with industrial demand, needed to be addressed. Hence, endeavours to make the country a manufacturing hub emerged as a complement to employment generation (D'souza 2020).

4.5.2 *Implications of the Pandemic on ASEAN's Economic Policies*

ASEAN's organisational policies, while being mindful of the inherent asymmetries in the levels of socio-economic development of the member-states, which were bound to have an impact on the respective and cumulative organisational post-pandemic recovery, were generally designed to achieving socio-economic, public health and sustainable environmental goals and harness the potential of the digital medium and bridging the digital divide. It was imperative for ASEAN to pursue policies to facilitate ASEAN integration and community-building along its identified pillars. Besides, post-pandemic recovery was also contingent upon the success of the member-states in containing the infection rates or local transmission and preventing resurgence or new waves, such that national, regional and international trade, investment and movement of goods, services and people could be ensured. Cognisant of the challenges to the worst hit sectors referred to in the previous section, ASEAN embarked upon specific policies related to tourism, agro-processing and garments, while seeking to maximise the benefits of electronic and digital trade.

In the wake of the onset of the pandemic, ASEAN convened the 26th ASEAN Economic Ministers' (AEM) Retreat[4] and issued the seminal document, 'Strengthening ASEAN's economic resilience in response to the outbreak of the Coronavirus disease 2019 (COVID-19)' It subsequently convened the Special ASEAN Summit on the Coronavirus Disease 2019 (COVID-19),[5] which was envisaged to serve as the pedestal for framing the guidelines for addressing the situation and strengthening ASEAN's economic resilience in respect to the outbreak of the pandemic.

In this respect, mention must be made of the 37th ASEAN Summit, held in Vietnam in November 2020, which adopted the 'ASEAN Comprehensive Recovery Framework (ACRF)', affirming that, 'addressing the crisis requires coordinated actions not only within the region but as well as cooperation with its partners.' While the immediate priority and focus for ASEAN was to tide over the challenges entailed by the pandemic, it was mindful of the 'collective and long-term socio-economic recovery strategy.' Towards this end, ACRF served as the consolidated exit strategy from the COVID-19 crisis and articulated ASEAN's response through the different stages of recovery, by focusing on key sectors and segments of society that were the most affected by the pandemic, thereby 'setting broad strategies and identifying measures for recovery in line with sectoral and regional priorities.' ASEAN's recovery efforts envisaged in the ACRF were focused on Five Broad Strategies, viz.

---

[4] 26th ASEAN Economic Ministers' (AEM) Retreatwas held on 10 March 2020.

[5] Special ASEAN Summit on the Coronavirus Disease 2019 (COVID-19) was held on 14 April 2020.

enhancing health systems; strengthening human security; maximising the potential of intra-ASEAN market and broader economic integration; accelerating inclusive digital transformation and advancing towards a more sustainable and resilient future, to be implemented through 'cross-cutting enabling factors' like policy responses and reforms; financing and resource mobilisation; institutions and governance mechanisms; stakeholder engagement and partnership and effective monitoring (ASEAN Comprehensive Recovery Framework 2020: 4–8).

Besides, the 37th ASEAN Summit declared the organisation's commitment to the following objectives: implement the 'Hanoi Plan of Action on strengthening ASEAN economic cooperation and supply chain connectivity in response to the COVID-19 pandemic'; implement the 'Declaration of the Special ASEAN Summit on the Coronavirus disease (COVID-19)' adopted by the ASEAN Heads of State/Government at the Special ASEAN Summit mentioned above; to implement the AEM's Statement on 'Strengthening ASEAN's economic resilience in response to the outbreak of the Coronavirus disease 2019 (COVID-19)'; and explore a temporary arrangement to preserve supply chain connectivity during the pandemic. The ASEAN Summit also adopted an Integrated Food Security Framework to ensure food security, safety and nutrition for all member-states. Furthermore, to facilitate essential intra-ASEAN business travel, it adopted 'Declaration on the ASEAN Travel Corridor Agreement Framework' and undertook strict action to counter the spread of tourism-related fake news and disinformation (Chairman's Statement of the 37th ASEAN Summit 2020, Cl. 18–21). It also provided the pedestal for member-states to fortify collective efforts for enhancing ASEAN's resilience to 'navigate and effectively respond to emerging challenges, address the threats of COVID-19, minimising its disruption and promoting sustainable recovery and ASEAN's inclusive growth.' It charted a blueprint for ASEAN's 2020 'Key Deliverables', consolidating its collective efforts to 'respond to the pandemic and promote regional comprehensive recovery.' With reference to ASEAN's response to the post-pandemic recovery, the blueprint emphasised the need to strengthen the organisational capacity and existing regional mechanisms on public health, emergency response and promote regional and international cooperation with external partners, including its Dialogue Partners and welcoming their contribution to the COVID-19 ASEAN Response Fund (CL. 11–15).

As referred to in the 'Hanoi Plan of Action', the Chairman's Statement alluded to post-pandemic recovery being hinged on reviving trade and investment relations, contingent on advancing sustainable connectivity infrastructure, towards which the finalisation of the 'Framework for improving ASEAN infrastructure productivity' was significant. Through this policy initiative, ASEAN was encouraged to develop a database on trade routes and erect the framework for enhancing supply chain efficiency to minimise the impact of the pandemic. Furthermore, the 'Master Plan on ASEAN Connectivity (MPAC 2025)' focused on synergising relevant connectivity initiatives within the member-states as well as with regional partners (Cl. 36–38).

Demonstrating ASEAN's determination to work closely to identify and address trade disruptions, with ramifications on the flow of essential goods, including food, medicines and medical and other essential supplies in the region, the 'Hanoi Plan of Action' emphasised the imperative of 'mitigating the economic impact of the

COVID-19 pandemic on the region' as well as strengthening 'supply chain connectivity to make supply chains more resilient and less vulnerable to similar changes in the future.' It also underscored the need for facilitating timely information-sharing on trade-related measures and leverage technologies for this purpose. With special reference to strengthening supply chain connectivity in the long term and promote trade and direct investment flows in the region, the document reaffirmed collaboration with 'external and development partners' by building on the existing trade facilitation platforms to 'promote and support supply chain connectivity and ensure unimpeded flow of goods and services in supply chains.'

Finally, it envisaged the promotion of science, technology, innovation and digital economy to foster mechanisms for facilitating customs clearance and processing to allow businesses, specially MSMEs to facilitate 'responsible business practices' among the ASEAN member-states and external partners, thereby empowering businesses to access the regional and global markets thus underscoring the economic revival as the conduit for the conduct of trade (Hanoi Plan of Action on strengthening ASEAN economic cooperation and supply chain connectivity in response to the COVID-19 pandemic 2020, CL. I-II).

4.5.3 Impulses of the Global Economic Scenario

The impact of the pandemic on the global economy was disastrous, with a fall in external demand coupled with a fall in domestic demand, which exacerbated a decline in the manufacturing outputs in several countries. The situation led to the contraction in economic growth in many of the advanced economies like the US, Japan, Britain, Germany, France, Italy and Spain by 5.9%, 5.2%, 6.5%, 7%, 7.2%, 9.1% and 8% respectively in 2020 (Explained: how COVID-19 has affected the global economy 2020). As a part and parcel of the globalisation process and the economic interdependence of nation-states it entails, the spill-over of this overall economic distress was experienced by all regions and countries, not only with respect to the domestic scenario, but also the overall dynamics of bilateral and multilateral trade relations. The situation was aggravated by external impulses, in keeping with the principles of economic interdependence and liberal institutionalism. These principles offer a rationalistic and utilitarian explanation that domestic and international institutions and their manifestation in the form of treaty/institutional arrangements play pivotal roles in determining inter-state interactions, in various dimensions of politics, economics and strategy. It also holds that institutions are rooted in the realities of power and interests, thus facilitating inter-state cooperation among states having common interests (Keohane and Martin 1995, 42). Viewed through this political-economy perspective, the economic impact of factors like the US halting the processing and issuance of non-immigrant work visas; the anti-China rhetoric in India and the sanctioning of Chinese goods, products and rendering the'digital strike' on virtual applications in the post-Galwan (2020) period and the asymmetrical tariff commitments and non-tariff barriers pertaining to the Free Trade

Area (FTA) regimes and their inducement of trade deficits for the Indian economy, was noteworthy. These factors have been discussed below.

First, the United States' proclamation of halting the processing and issuance of non-immigrant work visas of several types, including the Indian H-1B visa for skilled workers, hit the Indian economy very hard. The predicament was worsened by its conjunction with a crucial inflection point for the Indian economy, causing a ripple effect on the Information Technology (IT) sector. India's IT service exports to the US, particularly in telecommunications, computer and information services, R&D and travel sectors, dependent significantly on the H-1B visa, were important constituent elements of Indo-US trade. The visa ban issue also imperilled the employment scenario, with plunging employment rate in the Professional and Business Services super-sector, which included IT services (Lakshman 2020). Washington's decision was a blow to several technological companies in India, which depended on outsourcing of US companies and in turn, had an impact on the remittances.

Second, following the Galwan Valley standoff with China in June 2020, the anti-China rhetoric of sanctioning Chinese goods, products and rendering the'digital strike' on virtual applications had a negative impact on the Indian economy. The move was regarded by industry associations, ranging from pharmaceuticals to telecommunications and automobiles as a'knee-jerk' offensive against China till alternative vendors were finalised or domestic capacities in self-reliance adequately embarked upon. To begin with, imports from China constituted over 14%, accounting for US$ 62 billion of India's total imports in FY 2019–2020, recording the highest share among all nation, while exports to China accounted for over 5.3%, accounting for US$ 15 billion of India's total exports, the third-highest position during the same period (Export–Import Data Bank 2020). China accounted for nearly 12% of India's imports across sectors such as chemicals, automotive components, consumer electronics, intermediate products or components, raw materials and pharmaceuticals. The fact that at least 70% of India's drug intermediary needs were fulfilled by Beijing, while 76.3% of all antibiotics; 45% of total electronics imports; 33% of machinery; 25% of automotive parts and fertilizers; and 40% of organic chemicals were imported from China was significant. Several of these products were used by Indian manufacturers in the production of finished goods, thus thoroughly integrating China in India's manufacturing supply chain. Furthermore, many Indian unicorns or start-ups with over US$ 1 billion valuation had Chinese investors. The Government of India's proposal of banning Chinese telecommunications equipment providers like Huawei and ZTE from participating in the 5G trials also faced mixed reactions on the domestic turf since telecom companies like Vodafone Idea and Bharti Airtel contemplated testing the Chinese equipment along with those from other European and Asian nations. Thus, the outright embargo on Chinese products was expected to hit the Indian economy under the precarious circumstances. However, though the issue was extremely sensitive and emotive, industry experts suggested that the decision of boycotting non-essential Chinese products be left to individual choice, while adopting trade-related measures like raising duties on cheaper raw materials imported from China, rather than embarking on a total embargo. This provided for access to crucial ingredients in the short-term while India could pursue the policy of self-reliance or alternative trading partners like the US, Vietnam, Japan, Mexico and certain European Union countries particularly for critical electronic, vehicular and pharmaceutical components as well (Raghavan 2020; Radhakrishna et al. 2020).

Third, asymmetrical tariff commitments and non-tariff barriers pertaining to the liberal institutional arrangements like FTA regimes and their inducement of trade deficits was another factor. India's cumulative trade with its FTA and bilateral trade agreement partners added up to 15% of its total global trade, with the ASEAN member nations such as Singapore and Malaysia, and other EAS partners like South Korea and Japan, accounting for bulk of the trade. However, asymmetrical tariff commitments and non-tariff barriers in many FTA partner countries were the major contributing factors towards India's rising trade deficit, among which the most significant was the ASEAN bloc. It deserves mention in this context that the ASEAN-India Free Trade Area (AIFTA) constitutes three components, ASEAN-India Trade in Goods Agreement (2010), opening their respective markets by progressively reducing and eliminating duties on goods; ASEAN-India Trade in Services Agreement (2014), with provisions on transparency, domestic regulations, recognition, market access, national treatment and dispute settlement; and ASEAN-India Trade in Investment Agreement (2014), ensuring protection of investors to promote fair and equitable treatment for investments, non-discriminatory treatment in expropriation or nationalisation and fair compensation (Building the ASEAN community: ASEAN-India Free Trade Area 2015).

Besides, the proliferation of non-tariff measures, import controls, import permits, sanitary and phyto-sanitary measures, product standards and technical barriers in the ASEAN markets limited the economic integration of Indian products therein, leading to higher trade deficits faced by India. ASEAN markets had stringent regulations, including different marking/labelling/packaging requirements, testing/quarantine requirements, pre-shipment inspection by authorities, etc. thus increasing the costs borne by exporters. The slow pace of implementation even in cases where Mutual Recognition Agreements (MRA) existed further exacerbated the plight of the Indian exporters. In summary, the utilization rate of FTA provisions, both for India and its partners was moderate (Matthew 2020). Therefore, it was contingent on the Indian policymakers to navigate the liberal institutional issues pertaining to free trading regimes in general and AIFTA, in particular, with pragmatic considerations of whether such agreements generated a complementary and competitive ecosystem, in terms of trade facilitation measures and their conduct. Thus, the international compulsions discussed above had an impact on India's domestic economic scenario, axiomatically having repercussions on its external trade and commercial profile as well.

To sum up, as the world in general and the Asia–Pacific region, in particular, grappled with the multifarious humanitarian and socio-economic challenges of the pandemic, appropriate monetary, fiscal and macroeconomic policy responses embarked upon by India and ASEAN, coupled with the improvement in domestic demand and export performance, paved the way for addressing the economic onslaught, not only at the domestic levels, but also provided the stimulus for rejuvenating trade, commercial and people-to-people interactions, albeit incrementally, to restore the same to pre-COVID levels.

4.6 Future Directions

The positive impact of the policies adopted by India and ASEAN, in its organisational capacity to combat the economic onslaught of the pandemic and create the blueprint for economic revival and future resilience was manifested in the phased and incremental improvement in economic performance and India-ASEAN trade relations, as both sides experienced 'V-shaped' recovery path'.[6] As an indicator of post-pandemic recovery, India's real GDP growth was 8.7% and 6.9% in 2021 and 2022 respectively, while the average of ASEAN's real GDP growth recorded 5.6% in 2022, recovering from 3.2% in 2021 (Economic Outlook for Southeast Asia and India 2023, 1; Gill 2020, 32). Besides, post-pandemic India-ASEAN trade figures improved considerably, accounting for US$ 110 billion and US$ 131 billion in FY 2021–2022 and FY 2022–2023 respectively (Export–Import Data Bank 2023) demonstrating resilience and advantages of cooperation in this sphere, by exploring the right avenues in an optimal manner.

4.6.1 Navigating the Challenges Posed by the Pandemic

Despite these positive trends, the socio-economic vulnerabilities emerging from the pandemic were compounded subsequently by other pressing global issues like the Ukraine War, global economic slowdown, Sino-US trade war, asymmetrical post-pandemic economic recovery, etc. In this broad scenario, specific issues like tariff and non-tariff barriers; inadequacies in the Exclusion List Products; India's increasing trade imbalance vis-à-vis ASEAN, among others called for reviewing the different components of the AIFTA in order to secure greater market access in ASEAN markets, in products offering global value chains connectivity and gaining comparative advantage; harmonising standards, regulations, Rules of Origin, mutual recognition of conformity assessment, relaxation of border procedures, trade diversification, enhanced digital and physical connectivity and infrastructure, particularly underscoring the timely completion of the India-Myanmar Thailand Trilateral Highway (IMT), its proposed extension to the Indochina States as well as the Kaladan Multimodal Transit Transport Project (KMTTP). Of particular concern for India in this regard is the widening disequilibrium in Balance of Trade post-AIFTA and India's widening trade deficit with ASEAN: while India's exports to ASEAN have increased

[6] Characterised by an economic rebound after a sharp decline.

from US$ 25.63 billion in 2010–2011 to US$ 44 billion and US$ 41 billion in 2022–2023 and 2023–2024 respectively, its imports from ASEAN have also increased from US$ 30.61 billion to US$ 87.5 billion and US$ 79 billion in 2022–2023 and 2023–2024 respectively (Export-Import Data Bank, 2011, 2023, 2024).

This imbalance, amounting to more than twice the value since the signing of the FTA (2010), has evolved as a major impediment to harnessing the gains from the FTA process (De 2023) as well as consolidating post-pandemic economic recovery for India. Such a precarious situation has necessitated a review of the AIFTA provisions for analysing the underlying reasons, besides recommending pragmatic policies to contain this trend in India-ASEAN trade. Furthermore, China's economic clout, deeper access, integration of supply chains of ASEAN states and China and predominance of its products in both Indian and ASEAN markets, given its distinction as ASEAN's largest trading partner since 2020, accounts for minimising the positive impact of the FTA, specifically ASEAN-India Trade in Goods Agreement (Marjani 2023). Finally, the emphasis on ecological sustainability through adequate management of urban spaces, rural development, poverty alleviation, labour laws, education and mass awareness, social security, increase in public health expenditure, infrastructure and services, transport infrastructure etc. are common challenges that both sides needed to address to ensure a bright economic future.

4.6.2 Charting the Opportunities and Way Ahead

The challenges discussed above have been addressed with adequate policy responses by India and ASEAN, though more practical, pragmatic and innovative measures are required to sustain post-pandemic recovery and unveil a resilient economic future in India-ASEAN relations, as discussed below:

First, India's emphasis on domestic economic recuperation in tandem with exploring policy options for enhancing trade and commercial interactions with ASEAN to accelerate the post-pandemic recovery process, is significant. India's humane gestures of contributing US$ 1 million to ASEAN's COVID-19 Response Fund and assistance in ASEAN Regional Reserves of Medical Supplies and Public Health Emergencies, to facilitate preparedness, prevention, detection and aid scientific research for vaccine development and distribution, have adequately complemented its pursuit of 'building and maintaining resilient, open and connected supply chains and supporting a sustained economic recovery' (Chairman's Statement of the 17th ASEAN-India Summit 2020, Cl. 3).

Second, the commitment of India and ASEAN to strengthen trade and investment to ensure greater market access and trade facilitation was pivotal to navigating post-pandemic partnership. It was also contingent on ASEAN-India institutional mechanisms like AIBC to promote capacity-building, best practices for MSMEs, start-ups and greater interaction for skills and technological development, for harnessing the advantages of digital connectivity (Cl. 8). In this connection, the implementation of MPAC 2025 and its synergy with India's connectivity policy within the wider 'Act

East' framework to develop and maintain sustainable infrastructure, digital and multi-modal physical connectivity, aims at reviving tourism, trade and commerce. Third, the need to focus on prospective sectors of trade and emphasise trade diversification in keeping with the 'Plan of Action to implement the ASEAN-India partnership for peace, progress and shared prosperity (2021–2025)', can hardly be exaggerated. The Plan of Action identifies pharmaceuticals, healthcare, traditional medicines, cyber, digital and financial connectivity, infrastructure etc. as prospective sectors for future cooperation, providing impetus to innovation and efficiency in digital transformation, R&D support and cooperation in food security and nutrition through information-sharing across food, agriculture, forestry, fisheries and the like (Plan of Action to implement the ASEAN-India partnership for peace, progress and shared prosperity 2020).

Fourth, in order to minimise India-ASEAN trade deficit and contain the trade disequilibrium, periodic review and negotiations on the AFTA provisions, particularly related to different policy areas like market access, Rules of Origin (ROO) and standards, technical regulations, conformity assessment procedures, discrepancies in import duties, as in products like ferro-alloys, metals and products, textiles, fibres, chemicals, etc. deserves emphasis. Simultaneously, both sides should stimulate local/indigenous manufacturing, by promoting measures like production-linked incentives, imports monitoring, introduction of fast-track trade facilitation measures and liberal Product Specific Rules, initiate improvement in the quality of trade integration and trade diversification, from agricultural raw materials and food to manufacturing goods, minerals, mineral fuels and oils, to fulfil the principal objectives of the FTA agreements (De and Rahman 2021; Seshadri 2022). In this context, at the India-ASEAN FTA Review held in New Delhi in March 2024, India has urged the ASEAN to facilitate greater market access for its goods, more flexibility in determining origin of products through product-specific rules and redressal of non-tariff barriers. However, ASEAN's discordance with the Customs Administration of Rules of Origin under Trade Agreements, 2020 CAROTAR, 2020), according to which, rules introduced in India for supplementing the operational certification procedures related to implementing the ROO, emerge as a major stumbling block in the process (Sen 2024). With both sides committing to yield the benefits of the Review, eight sub-committees[7] have been constituted under AIFTA Joint Committee for undertaking negotiations on different areas and pursue the objective of enhancing India's export competitiveness, diversifying export baskets, addressing structural constraints in key sectors and fostering innovation and technological exchanges with ASEAN (India's trade deficit with ASEAN expands since FTA of 2010, 2024; India, ASEAN discuss review of trade agreement 2024).

Finally, it is contingent on both sides to exercise political will and facilitate negotiations at the institutional levels, like AEM + India meetings (business and trade);

[7] The Review covers eight areas, with separate sub-committees for each. The areas under review are national treatment, market access, rules of origin; standards, technical regulations and conformity assessment procedures; sanitary and phytosanitary; legal-institutional issues; customs procedures and trade facilitation; trade remedies and technical cooperation.

ASEAN Digital Ministers (ADGMIN) + India meetings (ICT and digital connectivity); ASEAN-India Ministerial Meeting on Agriculture and Forestry (AIMMAF); ASEAN-India Maritime Transport Working Group (AIMTWG); ASEAN Highways Sub Working Group (AHSWG); ASEAN-India Working Group on Science and Technology (AIWGST), etc. so as to envisage holistic and balanced relations, based on cooperation and resilience between them.

4.7 Conclusion

The COVID-19 pandemic posed socio-economic and humanitarian challenges to the national, regional and global economic scenarios and India and ASEAN member-states were no exceptions to this general experience. Addressing these challenges and prioritising post-pandemic economic recovery was not only imperative for individual states, but also for the socio-economic sustainability, development and prosperity of the Asia–Pacific region, as a whole. This calls for sustained and fortified economic, trade and commercial cooperation between India and ASEAN, by adhering to pragmatic economic and trade policies, poised to boost the domestic and regional economic performances and undertake sustained institutional negotiations to overcome the challenges in the trajectory.

In order to address the impact of the pandemic on India-ASEAN trade relations and pave the way towards the future, it is imperative for both sides to embark upon a pragmatic and pro-active course of implementation and execution of policies. In this context, trade and economic relations can be mitigated by conducting a time-bound review of FTA (in Goods) which assumes importance. Insofar as India is concerned, reducing the trade imbalance with ASEAN is a priority in the future, for which the review mechanism could facilitate an interim mechanism to seek protection to items like palm oil, coal, computers, telecommunication equipment, integrated circuits etc., as these are affected by a surge in imports from ASEAN countries (ASEAN FTA review: India to seek interim relief from import surge 2024). Furthermore, since it is indubitable that the future is contingent upon harnessing and adapting technology, Artificial Intelligence (AI) in a sustainable and ethical manner, the need to focus on digital transformation, new domains and platforms in digital public infrastructure for secure transactions, financial technology ('FinTech') and other innovative methods of economic transactions and relations, can hardly be exaggerated. These initiatives, pivotal for supplementing physical connectivity with digital connectivity, should be facilitated by collaboration in technological capacity-building and knowledge exchange, with a view to achieving the future vision. Towards this end, both sides should exercise pragmatic political will and adept leadership to address the existing challenges, and explore avenues of mitigating them through timely, focused and transparent negotiations, not only to sustain the momentum of economic ties and trade facilitation, but also to chart a resilient future to propagate their time-tested partnership.

Highlights

- Scourge of COVID-19 pandemic felt on India-ASEAN trade relations and the global economy
- Impact and spillover on regional economic integration and ASEAN community-building in future
- Post-pandemic economic recovery contingent on pragmatic national and regional economic and trade policies
- Imperative of institutional negotiations and promoting regional economic linkages and narrow the developmental gap among regional member-states
- Sustained and pragmatic review of FTA negotiations
- Political will and adept leadership to chart a resilient economic future

References

Agreement on Trade in Services under the Framework Agreement on Comprehensive Economic Cooperation between Association of Southeast Asian Nations and the Republic of India , ASEAN secretariat, Jakarta (2014). https://www.asean.org/wpcontent/uploads/2016/04/AI-TISA-Agreement-ASEAN-FINAL.pdf. Accessed 18 May 2024

Atmanirbhar Bharat, Ministry of Finance, New Delhi (2020). https://www.indiabudget.gov.in/anbp/. Accessed 24 May 2024

'Atmanirbhar Bharat' package: Sitharaman announces 1 lakh crore Agricultural Infrastructure fund, plan for farmers' welfare (2020) The Statesman

A crisis like no other, an uncertain recovery (2020) World economic outlook update, international monetary fund (IMF): 1–20. https://www.imf.org/en/Publications/WEO/Issues/2020/06/24/WEOUpdateJune2020. Accessed 18 Apr 2024

ASEAN Comprehensive Recovery Framework (2020) ASEAN Secretariat, Jakarta. https://asean.org/wp-content/uploads/2020/11/2-FINAL-ACRF_adopted-37th-ASEAN-Summit_12112020.pdf. Accessed 20 May 2024

ASEAN FTA review: India to seek interim relief from import surge (2024) The Financial Express. https://www.financialexpress.com/business/industry-asean-fta-review-india-to-seek-interim-relief-from-import-surge-3664393/. Accessed 14 Nov 2024

ASEAN, India reaffirm Comprehensive Strategic Partnership (2024) ASEAN Secretariat, Jakarta. https://asean.org/24th-meeting-of-asean-india-joint-cooperation-committee-convenes/. Accessed 18 Apr 2024

Bhalla S (2020) India's covid package combines short-term macro policy with long-term structural reforms. The Financial Express. https://www.financialexpress.com/opinion/indias-covid-package-combines-short-term-macro-policy-with-long-term-structural-reforms/1976977/. Accessed 15 May 2024

Brief on ASEAN-India relations (2023) Indian mission to ASEAN, Jakarta. https://www.indmissionasean.gov.in/pages/MjU. Accessed 18 Apr 2024

Budget (2020) COVID-19 stimulus: how much did government spend in FY 2021. Business Today. https://www.businesstoday.in/business/expectations/story/budget-2020-covid-19-stimulus-how-much-did-govt-spend-in-fy21-286267-2021-02-01. Accessed 18 Apr 2024

Building the ASEAN Community: ASEAN-India Free Trade Area (2015) ASEAN Secretariat, Jakarta. https://www.asean.org/wp-content/uploads/images/2015/October/outreach-document/Edited%20AIFTA.pdf. Accessed 18 Apr 2024

Chairman's Statement of the 17th ASEAN-India Summit (2020) ASEAN secretariat, Jakarta. https://asean.org/chairmans-statement-of-the-17th-asean-india-summit/. Accessed 12 May 2024

Chairman's Statement of the 37th ASEAN Summit (2020) ASEAN secretariat, Jakarta. https://asean.org/chairmans-statement-of-the-37th-asean-summit/. Accessed 12 May 2024

Chakraborty M, Suri N (eds) (2021) A 2030 vision for India's economic diplomacy. Observer Research Foundation, New Delhi. https://www.orfonline.org/search?q=a-2030-vision-for-india-s-economic-diplomacy. Accessed 20 May 2024

Chakraborty M (2023) ASEAN's tryst with community-building: towards comprehensive dispute settlement. In: Menghour L et al (eds) ASEAN's critical assessment and practical reforms, Special Report, Konrad Adenauer Stiftung-Asian Vision Institute. https://www.kas.de/en/web/kambodscha/laenderberichte/detail/-/content/special-report-asean-s-critical-assessment-and-practical-reforms. Accessed 20 May 2024

Chanda R (2023) Emerging value chain opportunities post-COVID-19 and India-ASEAN relations. In: De P (ed) Thirty years of ASEAN-India relations towards Indo-Pacific. Routledge, London, pp 248–265

Choudhury A, Nagda A (2020) India's COVID-19 opportunity in Southeast Asia. The Diplomat. https://thediplomat.com/2020/06/indias-covid-19-opportunity-in-southeast-asia/. Accessed 15 May 2024

Consumer Confidence Survey (2020) Reserve Bank of India, Mumbai. https://m.rbi.org.in/Scripts/PublicationsView.aspx?id=19455. Accessed 18 Apr 2024

D'souza R (2020) 'Atmanirbhar Bharat' Abhiyan: a true commitment or a cosmetic gimmick? https://www.orfonline.org/expert-speak/atmanirbhar-bharat-abhiyan-68686/. Accessed 15 May 2024

De P, Nida R (2021) Review of ASEAN-India FTA in goods: broad contours and the next steps. ASEAN-India Centre at RIS, New Delhi. https://aseanindiacentre.org.in/sites/default/files/2022-02/AIC%20Commentary%20No%2024%20December2021.pdf. Accessed 15 May 2024

De P (2023) Reactivating the ASEAN-India partnership through an FTA. Hindustan Times

De P, Rahman N (2021) Review of ASEAN-India FTA in goods: broad contours and the next steps. https://aseanindiacentre.org.in. Accessed 18 May 2024

Economic Outlook for Southeast Asia and India (2023) Organisation of Economic Cooperation and Development (OECD) Publishing, Paris. https://www.oecd.org/dev/asia-pacific/economic-outlook/Overview-Economic-Outlook-Southeast-Asia-China-India.pdf. Accessed 18 Apr 2024

Economic Outlook for Southeast Asia, China and India, 2020: ongoing challenges of COVID-19 (2020) OECD Publishing, Paris. https://www.oecd-ilibrary.org/development/economic-outlook-for-southeast-asia-china-and-india/volume-2020/issue-2_5bdef80e-en. Accessed 18 Apr 2024

Economic Outlook for Southeast Asia, China and India, 2021 (2021) OECD Publishing, Paris. https://www.oecd-ilibrary.org/development/economic-outlook-for-southeast-asia-china-and-india/volume-2021/issue-1_711629f8-en. Accessed 18 Apr 2024

Explained: how COVID-19 has affected the global economy (2020) The Indian Express

Export-Import Data Bank (2011) Ministry of Commerce and Industry, New Delhi. https://tradestat.commerce.gov.in/eidb/default.asp. Accessed 15 May 2024

Export-Import Data Bank (2020) Ministry of Commerce and Industry, New Delhi. https://commerce-app.gov.in/eidb/ecnt.asp. Accessed 15 May 2024

Export-Import Data Bank (2022) Ministry of Commerce and Industry, New Delhi. https://tradestat.commerce.gov.in/eidb/irgn.asp. Accessed 15 May 2024

Export-Import Data Bank (2023) Ministry of Commerce and Industry, New Delhi. https://tradestat.commerce.gov.in/eidb/default.asp. Accessed 15 May 2024

Export-Import Data Bank (2024) Ministry of commerce and industry, New Delhi. https://tradestat.commerce.gov.in/eidb/irgnq.asp. Accessed 14 Nov 2024
Ghoshal B (2020) India-ASEAN engagement in the COVID-19 period. ASEAN India Centre at RIS, New Delhi. https://www.aseanindiacentre.org.in/sites/default/files/2021-09/AIC%20commentary%20No8%20August%202020.pdf. Accessed 15 May 2024
Gill S (2020) COVID-19 challenges for the Indian economy: trade and foreign policy effects. ASEAN India Centre-Engineering Export Promotion Council of India, New Delhi. https://aic.ris.org.in/sites/aic.ris.org.in/files/Publication/COVID-19%20Report%202020%20Final%2019_section-min_compressed.pdf. Accessed 20 May 2024
Hanoi Plan of Action on strengthening ASEAN economic cooperation and supply chain connectivity in response to the COVID-19 pandemic (2020) ASEAN secretariat, Jakarta. https://asean.org/hanoi-plan-of-action-on-strengthening-asean-economic-cooperation-and-supply-chain-connectivity-in-response-to-the-covid-19-pandemic/. Accessed 18 May 2024
India, ASEAN discuss review of trade agreement (2024) The Economic Times
India's trade deficit with ASEAN expands since FTA of 2010 (2024) The Economic Times
Keohane R, Martin L (1995) The promise of institutionalist theory. Int Secur 20(1):39–51. https://doi.org/10.2307/2539214
Lakshman N (2020) How will the US visa ban impact India? The Hindu
Lee YN (2020) India's services activity collapses as country shuts down to battle the Coronavirus, a private survey shows. https://www.cnbc.com/2020/05/06/indias-services-activity-collapses-during-coronavirus-lockdown-ihs-survey.html. Accessed 12 May 2024
Lockdown extension to have deep impact on Indian economy (2020) Business Insider. https://www.businessinsider.in/business/news/lockdown-extension-to-have-deep-impact-on-indian-economy-report/articleshow/76132158.cms. Accessed 12 May 2024
Marjani N (2023) Up the economic game with ASEAN. Deccan Herald. https://www.deccanherald.com/opinion/up-the-economic-game-with-asean-1243328.html. Accessed 1 Oct 2023
Master Plan on ASEAN Connectivity 2025, ASEAN secretariat, Jakarta (2016). https://asean.org/wpcontent/ uploads/2021/08/8_compressed.pdf. Accessed 18 May 2024
Matthew J (2020) How to make FTAs work. Business Today
Menon J, Suvannaphakdy S (2021) COVID-19 economic recovery: ASEAN's mixed pattern. Institute of Southeast Asian Studies, Singapore. https://www.iseas.edu.sg/articles-commentaries/iseas-perspective/2021-26-covid-19-economic-recovery-aseans-mixed-pattern-by-jayant-menon-and-sithanonxay-suvannaphakdy/. Accessed 20 May 2024
Oikawa K, Todo Y et al (2021) The impact of COVID-19 on business activities and supply chain in the ASEAN member states and India. Economic Research Institute for ASEAN and East Asia, Jakarta. https://www.eria.org/uploads/media/discussion-papers/FY21/The-Impact-of-COVID-19-on-Business-Activities-and-Supply-Chains-in-the-ASEAN-Member-States-and-India.pdf. Accessed 20 May 2024
Palit A (2022) COVID-19 and its impact on India's FTA strategy. Institute of South Asian Studies, Singapore. https://www.isas.nus.edu.sg/papers/covid-19-and-its-impact-on-indias-fta-strategy/. Accessed 20 May 2024
Plan of Action to implement the ASEAN-India partnership for peace, progress and shared prosperity (2020) ASEAN Secretariat, Jakarta. https://asean.org/speechandstatement/plan-of-action-to-implement-the-asean-india-partnership-for-peace-progress-and-shared-prosperity-2021-2025/. Accessed 20 May 2024
Radhakrishna V et al (2020) Is an economic boycott feasible for India? The Hindu
Raghavan P (2020) Explained: will banning Chinese imports hurt India's exports? The Indian Express
Rhee CY (2020) COVID-19: Asia-Pacific region to expect its lowest growth since 1960s. World Economic Forum. https://www.weforum.org/agenda/2020/04/covid19-pandemic-asia-growth/. Accessed 18 Apr 2024

Sen A (2024) India seeks greater market access, flexibilities in rules of origin in FTA review with ASEAN. Hindu Business Line. https://www.thehindubusinessline.com/economy/india-seeks-greater-market-access-flexibilities-in-rules-of-origin-in-fta-review-with-asean/article67925058.ece. Accessed 1 Apr 2024

Seshadri VS (2022) Preparing for a review of the ASEAN-India Trade in Goods Agreement. Delhi Policy Group Brief. VII (48). https://www.delhipolicygroup.org/publication/policy-briefs/preparing-for-a-review-of-the-asean-india-trade-in-goods-agreement-aitiga.html. Accessed 15 May 2024

Supporting post-COVID 19 economic recovery in Southeast Asia (2022) Asian Development Bank, The Philippines. https://www.adb.org/publications/post-covid19-economic-recovery-southeast-asia. Accessed 12 May 2024

Suryamurthy R (2020) Worst slump on record. The Telegraph

The economic impact of COVID 19 with special reference to India (2020) Lok Sabha Secretariat, New Delhi. https://loksabhadocs.nic.in/Refinput/New_Reference_Notes/English/06072020_125312_1021205239.pdf. Accessed 12 May 2024

Union Budget, Ministry of Finance, New Delhi (2020). https://www.indiabudget.gov.in/budget2020-21/. Accessed 24 May 2024

Venkatesh L (2020) 'Atmanirbhar Bharat': Rs 3.75 crore package won't cost the Government much. https://www.cnbctv18.com/economy/atmanirbhar-bharat-rs-375-lakh-crore-package-for-msmes-wont-cost-the-govt-much-5916311.htm. Accessed 20 May 2024

Wignaraja G (2022) Fostering regional trade integration between South Asia and East Asia for COVID-19 recovery. https://www.elibrary.imf.org/display/book/9781513587219/CH010.xml. Accessed 20 May 2024

Zen F, Kimura F (2023) Reconnecting ASEAN: the impact of the COVID-19 pandemic on trade, transport, e-commerce and mobility of people. Economic Research Institute for ASEAN and East Asia, Jakarta. https://www.eria.org/research/reconnecting-asean-the-impact-of-the-covid-19-pandemic-on-trade-transport-e-commerce-and-mobility-of-people/. Accessed 20 May 2024

Mohor Chakraborty is Assistant Professor in Political Science at South Calcutta Girls' College (aff. To University of Calcutta), Kolkata, India. Her research focus is International Relations and Area Studies (South Asia, Southeast Asia and the Indo-Pacific) and Peace and Conflict Studies. She has three co-authored books to her credit and has contributed several articles and chapters in referred journals and edited books by Springer, Palgrave Macmillan, Routledge, Wiley, IGI Global, Konrad Adenauer Stiftung-Asian Vision Institute and Indian publishers like Indian Council of World Affairs, Pentagon, Gyan, Readers' Service, etc.

Part II
Technological and Transformative Future

Chapter 5
Disruptive Vision and Asia's Future Economy: Insights from Malaysian Firms

Cluster of Future Studies: Corporate Foresight

Rayenda Khresna Brahmana

Abstract Disruptive vision represents a transformative approach to shaping the future economy, characterised by innovative strategies and foresight. In the context of Asia's rapidly evolving economies, where technological advancements and shifting market dynamics redefine competitive landscapes, such vision becomes critical for firms to remain resilient and relevant. This study investigates the impact of disruptive vision on firm performance by analysing the vision statements of 392 Malaysian listed firms through text analysis. Employing robust OLS regression, the findings reveal that firms with disruptive visions outperform their counterparts, demonstrating a significant positive relationship between disruptive vision and performance. These results highlight the potential of disruptive vision to drive competitive advantage and sustainable growth in an increasingly dynamic and uncertain economic landscape. Drawing on impression management and organisational orientation theory, this research shows how crafting and effectively communicating a disruptive vision can enhance organisational outcomes.

Keywords Disruptive vision · Firm performance · Corporate vision · Impression management

5.1 Introduction

Despite the term "Disruption" being a hype word in both academia and industry, it is rarely found empirical research that examines the impact of disruption as a corporate vision on firms' dynamics. Defined as a fundamental change due to technology innovation (Van Balen et al. 2019; Hinterhuber 2022), disruption as part of the corporate vision statements might signal the firm's position within the industry and its utilisation of innovation in maximising the shareholders' wealth (Oliver 2018; Van Balen

R. K. Brahmana (✉)
School of Business, Deakin University Lancaster University Indonesia, Jawa Barat, Indonesia
e-mail: rayenda.brahmana@dli.ac.id

E. Lau et al. (eds.), *Future of Economy in Asia*,
https://doi.org/10.1007/978-3-031-95500-6_5

et al. 2019). Disruptive vision implies the power to be a new industry leader and shift overall market demand from existing business models to new ones. A successful disruption is perceived as the creation of a competitive advantage (Argyres et al. 2015).

Firms can communicate their disruptive plans as an organisation orientation by establishing a formal vision statement, and this is what we called as "Disruptive Vision". It will notify the stakeholders about the radical strategic changes by utilising radical innovation (Ansari et al. 2016). It also serves as a guideline of business operations in collaborating with suppliers for more accessible and inexpensive raw materials, strengthening the market share position for new innovative products, and attracting investment into the firms (Cenamor et al. 2019). The literature has documented that disruptive vision portrays supreme affiliation (Burton et al. 2002), leadership in the industry (Martens et al. 2007), and reputation (Cornelissen 2020). In other words, firms use disruptive vision statements to hint about "who and what the firms are" and "how the firms are different from their competitors."

Our notion aligns with the impression management theory, which posits that an organisation can influence stakeholders' perceptions by communicating the firm's vision (Ansari et al. 2016; Spear 2017; Van Balen et al. 2019). The tenet is that the corporate vision is an excellent organisational communication channel, where it relays the uniqueness of the company and its unique strategic plans to the stakeholders (Ansari et al. 2016; Spear 2017). This theory argues that a powerful vision enables a firm to attract ideas, people, and other resources (Highhouse et al. 2009; Bolino et al. 2016; Van Balen et al. 2019). The literature reveals that a good vision statement positively affects the firm's funding (Van Balen et al. 2019), internationalisation strategy (Gentile-Lüdecke et al. 2019), innovation (Alsaaty et al. 2021), and social responsibility (Peterson and Jun 2009).

Intriguingly, few studies empirically investigate how the disruptive vision affects the organisation dynamics, notwithstanding a few recent studies such as Oliver (2018, disruptive vision effect on firm strategy), Swann and Gill (1993, on technology change), and Van Balen et al. (2019) (on venture funding). Therefore, we focus on filling the lacuna by addressing the impact of the firm's disruptive vision on performance. To do so, we capture the disruptive vision by text-mining from vision statements cross-sectionally, which research suggests is an unobtrusive and reliable measure of corporate vision (e.g., Baum et al. 1998; Baum and Locke 2004; Berson et al. 2001; Slack et al. 2010; Van Balen et al. 2019; Reader et al. 2020). Previous research argues that the motive of communication through vision is to convey a narrative and a picture of the firm's future (Berson et al. 2001). It serves the organisation with innovation orientation. Simply put, disruptive vision is the guideline that may improve firms' performance.

Based on that logic, firms with disruptive vision statements will have radical changes, leading to business transformation that is more effective, efficient, productive, and higher profitability (Westerman et al. 2014; Hitt et al. 2016; Llopis-Albert et al. 2021). Considering the disruptive vision is a guideline or organisational orientation, it is crucial to acquire insight into how firms communicate as it provides a better understanding of the relationship between disruptive and firms' dynamics in the business literature. Specifically, do firms with disruptive vision outperform firms without it?

This research proposes disruptive vision as a type of impression management in the disruption process, with a thematic connotation of communication of vision, defining the goal of disrupting companies, markets, and ecosystems. According to the definition of disruptive innovation, which is an innovation that can transform incumbent goods (Christensen 1997; Oliver 2018; Llopis-Albert et al. 2021), this research uses the term "disruptive vision" to refer to a firm's ambition to alter the market's structure. In relation to this study, the concepts of disruption and disruptive vision demonstrate how organisations and investors adopt disruption practice into their understanding (Cosper 2015; Rachleff 2013).

Our research investigates how disruptive vision communication influences firm performance by using Malaysian listed firms as the research context. Malaysia presents an ideal case to test the hypothesis for several reasons. First, the Malaysian government sets a new economic policy for 2030 digital transformation through disruptive organisations. The government aims to "revolutionise" its economy from a manufacturing-based to a service-based nation. Second, emerging countries like Malaysia present a corporate landscape in which corporate statements are a crucial guideline for business operations (Abdullah et al. 2022). Third, Malaysia offers an interesting research context that represents a dynamic organisational culture and fast-growing industries to advance the understanding of how the corporate vision is used to achieve competitive advantage. Additionally, the complete vision statement dataset for all publicly listed firms is publicly available in companies' annual reports, allowing us to test the hypothesis.

Our study makes several contributions. First, we test one of the most frequently held but understudied assumptions in the business literature: that corporate vision serves as solid guidance for organisational alignment in attaining firm success. Our findings support the theory by revealing that disruptive vision has a positive relationship with profitability. Second, this research contributes by burgeoning access to impression management in terms of literature whereby firms frame communications to build stakeholder trust, loyalty, and commitment (Melewar et al. 2017). To date, there is limited research on the relative influence of future-oriented communications on firm performance. Therefore, this research provides the efficacy of future-oriented communications on results at the firm level with the introduction of corporate financial strategies: the communication of disruptive vision. Finally, our analysis contributes to the growing literature on digital finance. Since radical innovation is embedded in disruptive vision, we reveal that disruptive vision is a crucial guideline for firms in pursuing digitalisation for better performance.

Practically, this study significantly contributes to the future of Malaysia's economy by highlighting the impact of disruptive visions within Malaysian listed firms. The findings align directly with Malaysia's strategic frameworks like the Malaysia Shared Prosperity Vision 2030. Embracing disruptive visions not only propels economic progress and technological advancement toward high-income status and digital maturity but also supports Malaysia's sustainability goals. This emphasises the need for policies fostering disruptive innovation, ensuring ongoing industry transformation, and solidifying Malaysia's position as a forward-thinking, sustainable, and globally competitive economy.

This research is outlined as follows. Section 2 reviews the literature and theoretical argument of the corporate vision. The hypothesis development is also provided in the section. Section 3 provides the methodology and data description. It is followed by Sect. 4, which presents the results and analysis. Section 5 concludes and proposes the implication of the study. Finally, it also indicates the limitation of the study and suggestions for future research.

5.2 Literature Review

5.2.1 Disruptive Vision

Disruptive vision represents a transformative approach to organisational strategy, emphasising bold, innovative shifts that challenge conventional norms and redefine industries (Van Balen et al. 2019). Unlike traditional visions that focus on incremental growth, a disruptive vision identifies deficiencies in current markets and proposes radical change, often heralding entirely new business models, products, or services. This visionary approach communicates a compelling narrative that not only inspires internal stakeholders but also persuades external ones of the need for a paradigm shift.

At its core, disruptive vision is a strategic tool that positions organisations to drive significant innovation and change. According to Cornelissen (2013), a disruptive vision promises a "considerable difference or break from the previous business models and products in an industry or market." It highlights opportunities for improvement, framing disruption as a path to progress and competitive advantage. The agents (managers) play a pivotal role in articulating this vision, leveraging communication to rally support from diverse stakeholders within their ecosystem (Ansari et al. 2016).

By spotlighting innovations that deliver novel functionalities and reimagine market objectives, disruptive visions serve as catalysts for market transformation. These visions align with the concept that innovation promotes fundamental changes, as outlined by Ireland et al. (2003). A successful disruptive vision not only redefines market dynamics but also positions the organisation as a pioneer of the "dominant design," capable of reshaping demand and creating extraordinary returns (Argyres et al. 2015).

Firms that embrace disruptive visions often pioneer new market opportunities (Rasool et al. 2018; Van Balen et al. 2019). It allows them to capture untapped customer segments and redefine value propositions (Westerman et al. 2014; Hitt et al. 2016; Llopis-Albert et al. 2021). This strategic foresight can lead to increased market share (Gentile-Lüdecke et al. 2019), enhanced brand reputation (Cornelissen 2020), and stronger competitive positioning (Argyres et al. 2015). Additionally, by creating a culture of continuous innovation and adaptability, disruptive vision enables firms to navigate uncertainty and respond effectively to external challenges (Martens et al. 2007).

5.2.1.1 Hypothesis Development

We frame our research on two theories, namely, stakeholder and impression management theories, to explain how a disruptive vision affects a firm's performance. Stakeholder theory posits that firms communicate with stakeholders via vision statements (Byerly 2016; Valentinov et al. 2019). Firms will implement their vision statement by altering their operation to fulfill stakeholders' satisfaction and requirement (Ansari et al. 2016; Spear 2017; Van Balen et al. 2019). Meanwhile, impression management theory addresses a vision statement as a communication of symbolic cues and narratives to stakeholders that, in turn, influence how the stakeholders make sense of the firm's operation (Cenamor et al. 2019; Abdullah et al. 2022).

The literature in this area conceptualises and exploits the impact of vision statements on a firm's dynamic by using those two theoretical arguments. For example, the seminal work from Baum et al. (1998) surmises that corporate vision is important for firm growth. Meanwhile, Kantabutra and Avery (2010) reveal that powerful visions significantly impact customer and employee satisfaction. Additionally, Gentile-Lüdecke et al. (2019) find visions as the key factors in explaining the internationalisation process for African banks.

To explain how a disruptive vision affects firm performance, we elaborate on the rational expectation of organisation orientation. Disruption can create new market leaders and transform the current business models into new ones (Argyres et al. 2015; Schoemaker et al. 2020). A successful disruption yields extraordinary performance for the organisation. Thus, firms can create the expectation of those extraordinary performances by communicating a vision of disruption. It formalises the organisation's disruptive orientation into corporate culture, values, strategy, and view of the future (Argyres et al. 2015; Ansari et al. 2016; Spear 2017). It also addresses the firm's commitment to achieving disruptive business for its key stakeholders. Therefore, a disruptive vision is a playbook for firms to ensure a disruptive business model.

Prior studies of disruptive visions have documented how organisations benefit from them. For instance, Alsaaty et al. (2021) address disruptive vision as an essential guideline for the survival of innovative firms. Van Balen et al. (2019) surmise that startups with disruptive vision will have better funding than their peers. Meanwhile, Oliver (2018) uses the case of Sky Plc and Pearson Plc to reveal how disruptive strategic vision is good for firm performance; therefore, we hypothesise:

H_1: *Disruptive vision has positive relationship with firm's performance.*

5.3 Methodology

5.3.1 Model Specification

Economics and business literature has extensively explored the function of performance. It is common to specify the performance model by taking the firm characteristics as the determining factors. Aligning with the hypothesis, we introduce the main effect of this research, which is disruptive vision into the model. The firm characteristics become the control variables to isolate the effect of VISION on firm performance. The specification of the model is as follow.

$$\begin{aligned} Performance_i = {} & \alpha_1 + \beta_1 VISION_i + \beta_2 SIZE_i + \beta_3 LIQUIDITY_i \\ & + \beta_4 AGE_i + \beta_5 LEVERAGE_i + \beta_6 GROWTH_i + \varepsilon_i \ldots \end{aligned} \quad (5.1)$$

Overall, performance is the function of disruptive vision (VISION), and several control variables such as firm size (SIZE), liquidity (LIQUIDITY), firm age (AGE), firm leverage (LEVERAGE), and firm growth (GROWTH). The model is run under robust cross-sectional OLS regression. We run diagnostic test such as Normality, Autocorrelation, Heteroscedasticity, and Multicollinearity before estimating the model. At the end, our estimation model is run with robust standard error OLS.

5.3.2 Measures

Because the key objective of this study is to investigate the relationship between disruptive vision and firm performance, we provide a brief discussion of all the variables. Aforementioned, our dependent variable is firm performance. It can be measured in two ways: accounting book performance and market-based performance. For the accounting book performance, we use return on assets (ROA) and return on equity (ROE) as the proxy. The market-based performance is measured by Tobin's Q. Meanwhile, the main effect of this research is VISION. It is constructed using text-analysis method.

5.3.2.1 Measuring Disruptive Vision

Disruptive vision is measured by text analysis, a standard practice in this area (e.g., Baum et al. 1998; Baum and Locke 2004; Berson et al. 2001). Vision statements are displayed in the annual report, and we collect them cross-sectionally. The process is as follow: (1) write down all Vision statement, (2) recheck the annual report from five years prior to ensure whether the vision statements were changed, (3) operationalise disruptive vision, (4) code the items, and (5) construct the construct.

Table 5.1 Sample for disruptive measurements

Firm name	Vision statements	Fit/suit with item classification number	Score
Fraser and Neave Corporation	To become the leading total beverage company in Malaysia and the region	None	0
Mesinaga Corporation	To be the Malaysian it partner of choice	None	0
IJM Plantation	To be a leading regional plantation group	None	0
Rhone Ma Holding	To be an innovative leader in the provision of products, services and solutions	3 and 4	1
Grand-Flo Berhad	To enhance community and business with innovation, solution and technology	2 and 4	1

A disruptive vision conveys a drastic change in a firm's business operation, showcasing a significant change in the firm's orientation toward its products, service, and business models (Cornelissen 2013). Given that disruptive vision covers radical and fundamental change, it needs to cater the innovation as part of the items. Before operationalising the disruptive vision, we did focus group discussion with the professionals (middle and top management level) to ensure the items practically fit in defining disruptive vision. In another session, we also discuss it with scholars on strategic management, organisational behaviour, accounting and finance, and industry relation.

The items operationalised in constructing disruptive vision are on binary measurement (1 if yes, 0 otherwise). The text analysis has to spot items related to (1) promotes drastic, fundamental, or radical change in the future, (2) features a future that contrast with the status quo, (3) includes radical or innovative ideas, plans, or business operation that contradicts the conventional market, (4) promotes innovation and digital activities. If the firm is diagnosed to have text related to those items, it means it has a disruptive vision, then coded as "1".

For example, Fraser & Neave Corporation has a vision statement: "To become the leading total beverage firm in Malaysia and the region." According to those four items, the vision does not contain anything related to innovation or radical changes. Therefore, we gave a "0" to this firm. Meanwhile, the vision statement from Rhone Ma Holding is: "To be an innovative leader in the provision of products, services, and solutions," which is aligned with items 3 and 4. We gave a score of "1" for Rhone Ma Holdings. Other examples are reported in Table 1.

5.3.2.2 Control Variables

In order to isolate the main independent effect (disruptive vision) on firm performance, we thus control for the model specification. The empirical economics and business literature has long explored the effects of firm's size, liquidity, firm's age, and leverage.

Table 5.2 Variable definitions for the disruptive vision model specification

Variables	Measures	Definition
Dependent variables		
Performance	Returns on Assets (ROA)	Net Income to total assets .
	Returns on Equity (ROE)	Net Income to total equity
	Tobin's Q (TOBIN)	The sum of market value equity and market value of liability divided by the sum of book value equity and market value liability
Independent variables		
Main effect		
Disruptive vision	Vision	Text analysis of vision statements
Control variables		
Firm size	Size	Lognormal of total assets
Liquidity	Current ratio	Current assets to current liabilities
Firm age	Age	Lognormal of establishment's years
Leverage	Debt ratio	Total debt to total assets
Firm growth	Growth	The increase of revenue

The first control variable is firm size, and we use the normal logarithm of total assets to measure it. The liquidity is constructed using current ratio, which is the ratio of current assets to current liabilities. For the firm age, we use the lognormal of total establishment year as the measure. Meanwhile, debt ratio is used to measure firm leverage. Lastly, the growth of revenue is the measure of firm growth. A complete list of all variables used in this study are provided in Table 5.2.

5.3.3 Data and Samples

The samples of this research are all listed companies in Bursa Malaysia (Malaysia Stock Exchange) for the period 2019. The data was obtained from the annual report by downloading the 2019 report. The samples are selected systematically by excluding the "Dead companies" and "Illiquid stocks." We also exclude financial and utility firms because these two industries are governed under a different regulatory and supervisory framework set up by the Malaysian government. Companies that do not provide the 2019 annual report or without a Vision and Mission Statement in the annual report are also excluded. The final list covers 392 Malaysian listed firms from 929 total firms.

5.4 Findings

5.4.1 Descriptive Statistics

Table 5.3 presents the summary statistics for all variables in the model specification. Focusing on the key variables—performance and disruptive vision—aligns with previous research. For example, the mean value of the ROA is 3.58%, which is slightly lower than the reported value by Brahmana et al. (2020), with a value of 5.41% for 319 Malaysian listed companies in the same period samples. For the ROE, our mean value is 5.75%, slightly lower than Alarussi and Alhaderi (2018), who reported a mean value of 6.7% for 120 Malaysian listed companies in 2012–2014. In addition, the mean value of Tobin's Q is 1.34, implying that half of the listed companies are overvalued (Q value > 1). Our mean value is slightly higher than Brahmana et al. (2020, Q mean value = 1.14) and Chia et al. (2020, Q mean value = 1.13). For the disruptive vision, the descriptive statistics reveal that 33% of Malaysian listed companies have the disruptive vision and that one-third of Malaysia's companies are heading toward a disruptive orientation.

Table 5.3 also presents the correlation matrix for all variables, showing a preliminary view of their univariate relationship. The high correlation between ROA and ROE implies that these two variables measure the same dimension: accounting-book performance. Their correlation with Tobin's Q has a relatively high coefficient correlation of 0.49 and 0.48 for ROA and ROE, respectively. Furthermore, it indicates that

Table 5.3 Summary statistics from the disruptive vision's model

Variables	Correlation								Descriptive	
	Roa	Roe	Tobin	Vision	Size	Liquidity	Age	Leverage	Mean	Std. dev.
Roa	1								3.58	5.68
Roe	0.94	1							5.75	9.44
Tobin	0.49	0.48	1						1.34	1.28
Vision	0.48	0.46	0.2	1					0.33	0.47
Size	0.07	0.15	0.05	0.04	1				8.82	0.68
Liquidity	0.19	0.08	0.05	0.08	−0.2	1			2.64	2.79
Age	−0	0	−0.2	−0.12	0.15	−0.08	1		1.36	0.28
Leverage	−0.2	−0	−0	−0.1	0.41	−0.61	0	1	38.4	18.58
Growth	0.19	0.16	0.06	0.12	0.11	−0.06	0	0.06	5.3	42.96
Mean VIF	1.3									

Note Roa denotes Returns on Assets; roe denotes Returns on Equities; tobin denotes Tobin's Q; Vision denotes the disruptive vision; size denotes firm's size; liquidity denotes firm's liquidity; age denotes firm's age; leverage denotes firm's leverage; growth denotes firm's growth. Mean VIF (Variance Inflation Factor) is the estimated average VIF score

Tobin's Q is as good and suitable performance measure as the ROA and ROE. The correlation coefficients of those three dependent variables with disruptive vision are relatively high. Disruptive vision has a correlation of 0.48, 0.46, and 0.20 with ROA, ROE, and Tobin's Q, implying that companies with disruptive vision are positively associated with good performance.

For the control variables, all have the expected signs, with the sole exception of AGE and ROA (negative association). The correlation between performance and all control variables indicates that larger, more liquid, younger, less leverage, and high growth companies are associated with better performance. It is noteworthy that correlation is an indicator (not the test) for multicollinearity. We run the multicollinearity under the Variance Inflation Factor, and find out that our VIF mean value is 1.3. The VIF value is under the threshold of 10, meaning that there is no multicollinearity in our variables.

We plot Fig. 5.1 to portray the performance comparison between disruptive and non-disruptive vision firms. It shows that the firms with disruptive vision have a significant gap in ROA and ROE compared to non-disruptive vision firms. The figure shows that disruptive vision firms' ROA and ROE are more than double those of non-disruptive vision firms. Meanwhile, the two classes' gap (differences between disruptive and non-disruptive) between Tobin's Q is not significantly different. It implies that the market does not consider the vision statement a premium factor.

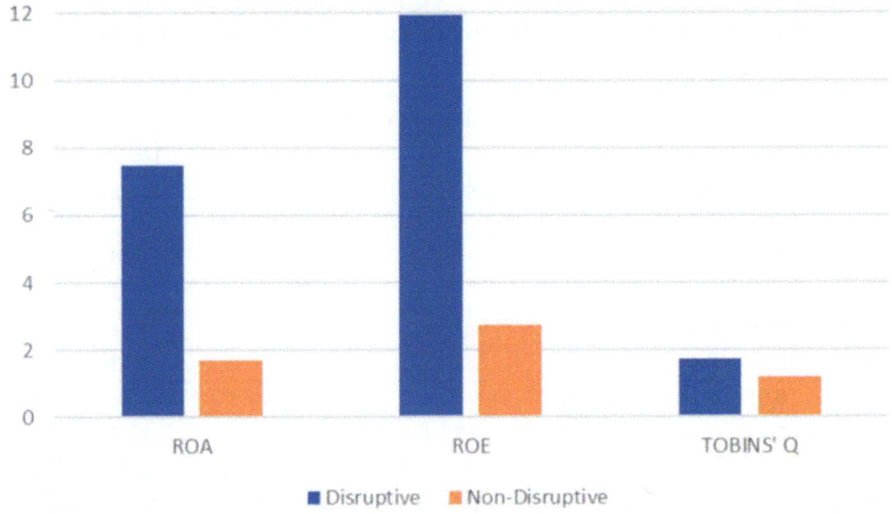

Fig. 5.1 Performance comparison among disruptive and non-disruptive firms. *Note* Roa denotes Returns on Assets; roe denotes Returns on Equities

5.4.2 Regression Results

Table 5.4 provides the statistical inference from the cross-section OLS regression. Before estimating the model, diagnostic tests were conducted. The normality, heteroscedasticity, and autocorrelation of the Classical Linear Regression Model were employed, resulting in the OLS being run under robust standard error.

Our hypothesis states a positive relationship between disruptive vision and performance (H_1). The results in Table 5.4 have two different dependent variables: ROA and ROE. For the ROA, it shows that disruptive vision has a positive relationship with firm performance ($\beta = 5.26$, SE $= 0.60$). This suggests that firms with disruptive vision significantly outperform the ROA of those without disruptive vision. Practically, the results suggest that the ROA difference between disruptive and non-disruptive vision companies is 5.25%.

The results for ROE as a dependent variable remain the same. It shows that disruptive vision has a positive relationship with ROE ($\beta = 8.78$, SE $= 0.98$). It suggests that firms with disruptive vision would perform better ROE than those without disruptive vision. In economic terms, the ROE difference between disruptive and non-disruptive vision companies is 8.78%.

Table 5.4 Regression results of baseline model of disruptive vision

	ROA	ROE
Vision	5.26***	8.78***
	(0.60)	(0.98)
Size	1.09**	1.77**
	(0.46)	(0.84)
Liquidity	0.18	0.32*
	(0.12)	(0.19)
Age	0.2	1.15
	(1.03)	(1.88)
Leverage	−0.05**	0.01
	(0.02)	(0.04)
Growth	0.02**	0.02
	(0.01)	(0.02)
Constant	−6.52*	−15.58**
	(3.88)	(6.56)
Adj. R2	0.28	0.23

Note Presented figures are coefficient values, except the one inside parenthesis, which is standard error. *, **, and *** denote the significant level at 10%, 5%, and 1%, respectively. ROA denotes Returns on Assets; ROE denotes Returns on Equities

Our findings tally with previous research, such as Van Balen et al. (2019) and Oliver (2018). We also confirm the impression management and organisation orientation theory, where the vision statement, especially the disruptive one, is the key factor for firm performance. Having a disruptive vision means the firm has radical changes in the business operation, forcing the organisation to adopt a new "culture" or new "orientation." All business operations are inspired by the vision, and having a disruptive culture may lead companies to work effectively and efficiently. As a result, the companies have better performance.

5.4.3 Regression Results: Alternative Measurement—Tobin's Q

The proxy for the firm performance in this research is accounting-book performance. ROA and ROE are selected as the measurement for the accounting-book performance. Another dimension widely considered in the literature is the market book performance, in which Tobin's Q is accepted as the measure of market book performance. Hence, this research takes Tobin's Q as the alternative measure of performance. We argue that the effect of disruptive performance should remain the same when it tests on Tobin's Q. Shareholders should consider disruptive vision as a signal that the firm will perform better than the others.

We then re-estimate the estimation model using Tobin's Q as the dependent value. The results are presented in Table 5.5. It shows that disruptive vision significantly affects Tobin's Q ($\beta = 0.47$, $SE = 0.18$). It indicates that firms with disruptive vision will have higher Tobin's Q than those without it. We also can interpret that Tobin's Q of disruptive vision firms is higher than the non-disruptive vision. The Tobin's Q difference between the two groups is 0.47.

The consistent results across the two different performance measurements suggest that disruptive vision is crucial in increasing the firm performance. Firms will enjoy better performance when their vision is oriented towards disruption. Overall, the hypothesis is accepted, and our findings support the theoretical argument.

5.5 Discussion of Implication

5.5.1 Implications for Future of Malaysia Economy

The study's findings on the power of disruptive visions in Malaysian listed firms offer a crucial link to Malaysia's overarching strategic frameworks, such as the Malaysia Shared Prosperity Vision 2030 and the Malaysia Digital Economy Blueprint. Embracing disruptive visions within companies aligns directly with the nation's vision for shared prosperity and economic advancement. Companies that integrate

Table 5.5 Regression results with alternative measure of economic performance

	Tobin's Q
Vision	0.47***
	(0.18)
Size	0.14
	(0.10)
Liquidity	0.01
	(0.05)
Age	−0.63**
	(0.25)
Leverage	0.01
	(0.00)
Growth	0.01
	(0.00)
Constant	0.86
	(0.89)
Adj. R2	0.14

Note Presented figures are coefficient values, except the one inside parenthesis, which is standard error. *, **, and *** denote the significant level at 10%, 5%, and 1%, respectively

disruptive visions into their strategies are likely to spearhead innovation and technological advancements, in line with the objectives outlined in these national frameworks. As a result, this alignment could significantly accelerate Malaysia's progress towards becoming a high-income and digitally advanced nation by 2030.

The implications of prioritising disruptive visions also intersect with Malaysia's commitment to sustainability as outlined in its various policies, including the Malaysia Shared Prosperity Vision 2030 and Malaysia New Economy Policy. Companies fostering disruptive visions often prioritise sustainability initiatives, aligning with Malaysia's goals for environmentally friendly and sustainable development. This synergy could amplify Malaysia's efforts to position itself as a sustainable and responsible player in the global market, attracting investments aligned with green and ethical principles. Furthermore, this alignment supports the country's aspirations to develop a diversified and inclusive economy, as emphasised in its strategic policy frameworks.

The implications extend to policy considerations, urging policymakers to create an enabling environment for fostering disruptive visions within the Malaysian business landscape. Crafting policies that incentivise innovation, entrepreneurship, and research and development could fortify Malaysia's position as a hub for forward-thinking enterprises. Furthermore, the potential industry evolution prompted by disruptive visions may necessitate adaptive policies to accommodate the changing economic landscape, ensuring that Malaysia remains at the forefront of innovation and industry transformation.

5.5.2 *Implications for the Future of Asia's Economy*

While this evidence originates from Malaysia, it underscores a broader principle that is highly relevant to Asia's diverse economies: disruptive visions enable firms to innovate, adapt, and excel in dynamic market conditions. In a region characterised by rapid technological advancements and shifting competitive landscapes, adopting such visions can empower firms to lead industrial transformation and seize opportunities in emerging sectors like digital technology, renewable energy, and advanced manufacturing.

This connection between disruptive vision and firm performance has critical implications for Asia's economic trajectory. Firms that prioritise disruptive visions are better positioned to enhance efficiency, embrace new business models, and achieve superior financial outcomes. Governments and policymakers across Asia could capitalise on these insights by encouraging organisations to align their strategic visions with innovation-driven growth. This could include providing targeted incentives for visionary leadership, promoting regulatory frameworks that support disruptive business practices, and cultivating regional networks for knowledge sharing. By embedding disruptive visions into their economic strategies, Asia's firms and economies can unlock sustained growth and secure a leading role in the global economic landscape.

5.6 Future Directions

This study offers significant insights into the relationship between disruptive vision and firm performance but is not without its limitations, which create opportunities for future research. The findings are based on Malaysian listed firms, leaving the broader applicability of disruptive vision to different economic contexts unexplored. Further research is needed to validate these findings and address the limitations, including the context-specific nature of the study. While this research focuses on the effect of disruptive vision on firm performance, it does not account for potential variations across industries or regions, nor does it consider the long-term impact of such visions in dynamically evolving markets. These gaps highlight the need for expanded investigations to deepen our understanding of the role of disruptive vision in driving organisational success.

For instance, the disruptive vision is measured using text analysis from the vision statements. The tenet is that the disruptive vision is a "playbook" of a firm's business operation. A few extensions can be further built upon this approach. For example, future studies may use Management Discussion and Analysis report as the proxy of organisation orientation, and the approach can also be modified using the "tone-at-top" approach.

Another limitation of this research is that the managerial characteristics attributes are not part of the research. According to upper-echelons theory, the orientation and strategy of a firm depend on managerial behaviour and characteristics. Future studies may elaborate on these upper-echelons variables in exploring the disruptive vision-performance nexus. In addition, some corporate governance variables such as board structures, board size, board diversity, and the role of controlling shareholders can be another exciting study extension for this research.

5.7 Conclusion

This study addresses the intriguing topic of organisation study: the effect of disruptive vision on firm performance. Our study is mainly motivated by the lack of attention given to this research area despite the theoretical importance of vision statements for firms. Impression management and organisation orientation theories posit that a vision statement is a guideline for firms' business operation and drive their financial performance. This paper, by all means, lays the foundation for further research on this topic, especially for researchers focusing on digital economics or organisational behaviour.

We adopt the text analysis of the vision statement of Baum et al. (1998), Baum and Locke (2004), Van Balen et al. (2019) with slight modification in its variable operationalisation. Our results bring implications for theoretical framework and empirical findings related to two major theories in economics and business: impression management and organisational orientation. We provide empirical evidence that disruptive vision leads to better performance. Another contributing aspect of our study is that we use text analysis to operationalise the vision statements, which may be common in accounting and finance research but is rarely done in management studies. This research provides a base for future research that Management Discussion and Analysis report from the annual report can be used as unobtrusive data for economics and business research. Practically, this research informs the industry that vision statements may seem simple and a formality; thus, it gives a good orientation for the firms to earn better performance. Moreover, a disruptive vision will benefit the firms with better profitability.

Highlights

- Disruptive visions are essential for the future of Asian corporations, enabling them to anticipate market shifts, embrace new business models, and maintain a competitive edge in the global economy.
- Firms with disruptive visions achieve superior financial outcomes, driving growth and profitability. This is crucial for the future of Asia's economic leadership, ensuring firms remain competitive and resilient in dynamic markets.

- Embracing disruptive visions fosters innovation, positioning firms at the forefront of industry transformation. This is vital for Asia to lead in sectors like digital technology and advanced manufacturing.
- Governments should incentivise visionary strategies to bolster economic leadership and adaptability. Creating regulatory frameworks that support disruptive business practices ensures Asia's competitive edge in the global market.

References

Abdullah Z, Anumudu CE, Raza SH (2022) Examining the digital organizational identity through content analysis of missions and vision statements of Malaysian and Singaporean SME company websites. Bottom Line 35(2/3):137–158

Alarussi AS, Alhaderi SM (2018) Factors affecting profitability in Malaysia. J Econ Stud 45(3):442–458

Alsaaty FM, Kapur K, Ntembe A (2021) Is disruptive innovation your entrepreneurial vision? J Small Bus Entrep 9(2):16–22

Ansari S, Garud R, Kumaraswamy A (2016) The disruptor's dilemma: TiVo and the US television ecosystem. Strateg Manag J 37(9):1829–1853

Argyres N, Bigelow L, Nickerson JA (2015) Dominant designs, innovation shocks, and the follower's dilemma. Strateg Manag J 36(2):216–234

Baum JR, Locke EA (2004) The relationship of entrepreneurial traits, skill, and motivation to subsequent venture growth. J Appl Psychol 89(4):587

Baum JR, Locke EA, Kirkpatrick SA (1998) A longitudinal study of the relation of vision and vision communication to venture growth in entrepreneurial firms. J Appl Psychol 83(1):43

Berson Y, Shamir B, Avolio BJ, Popper M (2001) The relationship between vision strength, leadership style, and context. Leadersh Q 12(1):53–73

Bolino M, Long D, Turnley W (2016) Impression management in organizations: critical questions, answers, and areas for future research. Annu Rev Organ Psych Organ Behav 3(1):377–406

Brahmana RK, You HW, Yong XR (2020) Divestiture strategy, CEO power and firm performance. Manag Res Rev 44(3):418–436

Burton MD, Sørensen JB, Beckman CM (2002) 7. Coming from good stock: career histories and new venture formation. In: Social structure and organizations revisited. Emerald Group Publishing Limited

Byerly RT (2016) A new Institutionalism approach to stakeholder theory. In: A stakeholder approach to corporate social responsibility. Routledge, London, pp 371–386

Cenamor J, Parida V, Oghazi P, Pesämaa O, Wincent J (2019) Addressing dual embeddedness: the roles of absorptive capacity and appropriability mechanisms in subsidiary performance. Ind Mark Manag 78:239–249

Chia YE, Lim KP, Goh KL (2020) Liquidity and firm value in an emerging market: nonlinearity, political connections and corporate ownership. N Amn J Econ Financ 52:101169

Christensen CM (1997) The Innovator's dilemma: when new technologies cause great firms to fail. Harvard Business School Press, Boston, MA

Cornelissen J (2013) Portrait of an entrepreneur: Vincent van Gogh, Steve jobs, and the entrepreneurial imagination. Acad Manag Rev 38(4):700–709

Cornelissen JP (2020) Corporate communication: a guide to theory and practice. Sage, London

Cosper A (2015) Disruption is more than the buzzword it's become. Entrepreneur. https://www.entrepreneur.com/article/245794
Gentile-Lüdecke S, Halaszovich T, Lundan S (2019) What role does CEO vision play in the internationalization process of firms? Evidence from the banking sector in Africa. Thunderbird Int Bus Rev 61(1):13–27
Highhouse S, Brooks ME, Gregarus G (2009) An organizational impression management perspective on the formation of corporate reputations. J Manag 35(6):1481–1493
Hinterhuber A (2022) Digital transformation, the Holy Grail, and the disruption of business models: an interview with Michael Nilles. Bus Horiz 65(3):261–265
Hitt MA, Ireland RD, Hoskisson RE (2016) Strategic management: concepts and cases: competitiveness and globalization. Cengage Learning, Boston
Ireland RD, Hitt MA, Sirmon DG (2003) A model of strategic entrepreneurship: the construct and its dimensions. J Manag 29(6):963–989
Kantabutra S, Avery GC (2010) The power of vision: statements that resonate. J Bus Strateg 31(1):37–45
Llopis-Albert C, Rubio F, Valero F (2021) Impact of digital transformation on the automotive industry. Technol Forecast Soc Chang 162:120343
Martens ML, Jennings JE, Jennings PD (2007) Do the stories they tell get them the money they need? The role of entrepreneurial narratives in resource acquisition. Acad Manag J 50(5):1107–1132
Melewar TC, Foroudi P, Gupta S, Kitchen PJ, Foroudi MM (2017) Integrating identity, strategy and communications for trust, loyalty and commitment. Eur J Mark 51(3):572–604
Oliver J (2018) Strategic transformations in a disruptive digital environment. Strateg Dir 34(5):5–8
Peterson RT, Jun M (2009) Perceptions on social responsibility: the entrepreneurial vision. Bus Soc 48(3):385–405
Rachleff A (2013) 'What "disrupt" really means'. TechCrunch. https://techcrunch.com/2013/02/16/the-truth-about-disruption
Rasool F, Koomsap P, Afsar B, Panezai BA (2018) A framework for disruptive innovation. Foresight 20(3):252–270
Reader TW, Gillespie A, Hald J, Patterson M (2020) Unobtrusive indicators of culture for organizations: a systematic review. Eur J Work Organ Psy 29(5):633–649
Schoemaker PJ, Day GS, Rao G (2020) Converting strategic ambiguity to competitive advantage: how Philips lighting solved the challenge of LED technology disruption. Strat Leadsh 48(2):10–17
Slack FJ, Orife JN, Anderson FP (2010) Effects of commitment to corporate vision on employee satisfaction with their organization: an empirical study in the United States. Int J Manag 27(3):421
Spear S (2017) Impression management activity in vision, mission, and values statements: a comparison of commercial and charitable organizations. Int Stud Manag Organ 47(2):159–175
Swann P, Gill J (1993) Corporate vision and rapid technological change: the evolution of market structure. Routledge, London
Valentinov V, Roth S, Will MG (2019) Stakeholder theory: a luhmannian perspective. Adm Soc 51(5):826–849
Van Balen T, Tarakci M, Sood A (2019) Do disruptive visions pay off? The impact of disruptive entrepreneurial visions on venture funding. J Manag Stud 56(2):303–342
Westerman G, Bonnet D, McAfee A (2014) Leading digital: turning technology into business transformation. Harvard Business Press, Boston

Rayenda Khresna Brahmana is currently academic team for Deakin University Lancaster University Indonesia. Previously, he worked at Coventry University, University of Bahrain and Universiti Malaysia Sarawak. He earned his PhD in Finance from Universiti Sains Malaysia and is a Fellow of the UK Advance Higher Education. Rayenda has taught a wide range of finance courses across undergraduate, postgraduate, and doctoral levels, showcasing his dedication as an educator. His research focuses on corporate strategy and behavioural corporate finance, with an extensive record of publications in leading academic journals. Outside academia, he serves as a senior advisor to several startups and delivers finance and strategy workshops to professionals around the world. Rayenda is also an active member of the academic community, frequently reviewing for high-impact journals and contributing to the advancement of knowledge in the field of finance.

Chapter 6
Embracing Cross-Border Digital Payment in Southeast Asia: Current Development and Future Outlook

Clusters of Future Studies: Corporate Foresight; Past and Futures; Technological Trends

Abu Amar Fauzi, Lufi Yuwana Mursita, Nazlida Muhamad, and Margaret L. Sheng

Abstract The adoption of digital payments within individual Southeast Asia (SEA) countries has been progressive. However, the action toward cross-border expansion remains in its early stages. Using a narrative review approach, which includes an analysis of peer-reviewed journal articles, industry reports, government news and press releases, and reputable press publications, this chapter examines the current state of cross-border digital payment developments in the region and offers a forward-looking perspective. Nowadays, cross-border digital payments in the SEA region have been achieved through bilateral partnerships and regional-level initiatives led by the private sector; the progress marks a significant milestone in extending services to more regional countries. On the one hand, government efforts, market growth, technology penetration, and advancement of the financial industry have been the

A. A. Fauzi (✉)
National Research and Innovation Agency, Jakarta Selatan, Indonesia
e-mail: abua001@brin.go.id

L. Y. Mursita
Universitas Hayam Wuruk Perbanas, Surabaya, Indonesia
e-mail: lufi.yuwana@perbanas.ac.id

N. Muhamad
Prince Sultan University, Riyadh, Saudi Arabia

UBD School of Business and Economics, Universiti Brunei Darussalam, Bandar Seri Begawan, Brunei Darussalam

N. Muhamad
e-mail: nmuhamad@psu.edu.sa

M. L. Sheng
National Taiwan University of Science and Technology, Taipei, Taiwan
e-mail: msheng@mail.ntust.edu.tw

E. Lau et al. (eds.), *Future of Economy in Asia*,
https://doi.org/10.1007/978-3-031-95500-6_6

enabling factors of cross-border digital payment. On the other hand, several challenges are identified, including interoperability, regulatory frameworks, fraudulent activities, currency, speed, and digital divide issues. In practical terms, investment in technology, regulatory compliance and harmonisation, user education and support, cooperation and collaboration, localisation, personalisation, continuous improvement, building trust, and consumer protection are essential future endeavours for advancing cross-border digital payment in the SEA region.

Keywords Cross-border digital payment · Digital economy · Financial services · Regional payment connectivity · Southeast Asia

6.1 Introduction

The financial landscape of Southeast Asia (SEA) is undergoing a significant transformation with the emergence of digital payment systems. Over the past decade, the region has notably increased its adoption of digital payment methods (Morgan 2022). According to Hingel et al. (2023), individual account holdings linked to mobile money have surged substantially across SEA countries, surpassing those of nine countries covered by the World Bank Findex by 60%. Additionally, there has been a marked rise in e-wallet usage in the SEA region. Notably, half of the top ten countries globally in terms of e-wallet penetration rates are in SEA, including Thailand (92%), Vietnam (91%), Philippines (88%), Malaysia (86%), and Indonesia (82%).

Consequently, the gross value of digital payment transactions in the six largest economies within the SEA region—Indonesia, Philippines, Vietnam, Thailand, Malaysia, and Singapore—reached $959 billion in 2022, marking a 13% year-on-year increase. Projections from extensive analysis of Google, Temasek, and Bain & Company in the E-Conomy SEA 2023 Report (Google et al. 2023) suggest that this figure will double to approximately $2.1 trillion by 2030. On the other hand, Google et al. (2023) observed a declining trend in cash (banknotes) usage relative to the total gross transaction value (GTV) of digital payments in the region. The cash-to-total GTV ratio currently stands at 50%, which is projected to decrease to 45% by 2025 and further decline to 34% by 2030. In short, these data indicate that the SEA region is progressing towards a cashless economy, which is already underway and expected to become more pronounced in the coming years as cash gradually diminishes in prominence in consumers' financial transactions.

This payment transformation is propelled by the expanding digital economy within the SEA region. The primary sectors leading the formation of the digital economy ecosystem in the SEA region are e-commerce, transportation and food, online travel, online media, and financial services, including digital payment systems. In this case, digital payments play a crucial role in facilitating individuals and businesses with convenient and efficient transactional solutions, making digital payments an essential component of the digital economy ecosystem (Mützel 2021; Sheng and Fauzi 2023). Unlike traditional payment methods, which often involve complex processes and longer settlement times, digital payments leverage technology to facilitate faster, more secure, and cost-effective transactions (Zetzsche et al. 2021). The definition is not limited to a specific technology. Still, it encompasses a spectrum of digital payment methods, including credit cards, debit cards, prepaid cards, account-to-account (A2A), mobile banking, mobile wallet, QR code, internet banking, mobile payment, and blockchain-based solutions in the form of cryptocurrency (Alkhowaiter 2020; Google et al. 2023;Hashemi Joo et al. 2019).

Given the preceding discussion, the adoption of digital payment systems within individual SEA countries has been notable. However, the expansion of these capabilities across borders has not kept pace. As SEA nations pursue economic integration and regional collaboration (Ishikawa 2021), the necessity for cross-border solutions becomes increasingly evident. Unfortunately, the discourse on cross-border digital payment utilisation in SEA remains nascent, underscoring the need to explore this vital area comprehensively. At this point, a thought-provoking question has been raised concerning the extent of the development of cross-border digital payments in the region thus far. Hence, this chapter overviews the most recent advancements concerning implementing cross-border payments in the SEA region. Through this endeavour, we aim to provide valuable insights to stakeholders, including governments, digital payment service providers, and retailers in the SEA region. These insights are valuable for enhancing cross-border digital payments throughout the area, fostering mutual economic progress, and promoting financial inclusivity in the SEA region.

6.2 Review Methodology

To address the central question of this chapter, we adopted a narrative review approach. This method is particularly effective for evaluating existing research or developing new theories and conceptual frameworks by synthesising studies with diverse topics or methodologies (Baumeister and Leary 1997). To reduce the risk of bias, the narrative review incorporates several steps akin to those used in systematic reviews, such as conducting searches across at least two databases and employing a combination of search strings (Siddaway et al. 2019). Due to the limited availability of scientific literature on cross-border digital payments in SEA, the scope of the inclusion criteria for the reviewed documents was expanded. This broader approach encompassed not only journal articles but also industry reports, government announcements, press releases, and articles from reputable media sources. A similar approach was taken by Pansari and Kumar (2017) in their conceptualisation of customer engagement, where they reviewed marketing literature alongside analyses of articles from popular media sources. To identify relevant literature, we used various combinations of search strings, including: "cross-border digital payment" AND "Southeast Asia"; "regional payment connectivity" AND "Southeast Asia"; and "digital payment" AND "Southeast Asia". These searches were conducted on the

Google Search and Google Scholar platforms between December 2023 and February 2024.

6.3 Findings

6.3.1 Cross-Border Digital Payments in SEA Region

Currently, the SEA region encompasses eleven countries: Brunei Darussalam, Cambodia, Indonesia, Laos, Malaysia, Myanmar, the Philippines, Singapore, Timor-Leste, Thailand, and Vietnam (Ha and Chuah 2023). Among these nations, Singapore, Malaysia, Thailand, Vietnam, the Philippines, and Indonesia have emerged as pivotal figures in the digital payment sphere, boasting robust infrastructures that support an array of electronic payment options (Google et al. 2023). As presented in Table 6.1, payment models in these countries exhibit significant diversity. However, according to Yeo (2023), despite the broad spectrum of digital payment options available in the SEA region, their predominant usage remains for domestic purposes.

Significant strides in developing cross-border digital payments within the SEA region have been achieved through various bilateral initiatives (refer to Table 6.2). According to the Bank of Thailand (2023), bilateral cross-border payment schemes commenced in 2020 with Thailand and Cambodia, facilitating seamless QR-based transactions between the two nations utilising Thailand's Thai QR and Cambodia's KHQR systems. Expanding on this progress, Thailand further bolstered its cross-border digital payment infrastructure in 2021 by integrating its Fast Payment System (FPS) with Singapore's PayNow and Thailand's PromptPay (Banchongduang 2021; Hingel et al. 2023; PwC 2024). During the same period, four additional bilateral cross-border digital payment schemes were launched, connecting Malaysia with Thailand, Malaysia with Indonesia, Thailand with Vietnam, and Thailand with Indonesia (Bank of Thailand 2023; Yulius et al. 2023). In 2022, the Philippines entered a Memorandum of Understanding (MoU) regarding implementing bilateral cross-border digital payments with Indonesia and Malaysia (Yulius et al. 2023). In 2023, Singapore announced the inauguration of a bilateral cross-border digital payment scheme with Malaysia and Indonesia (Bank Indonesia 2023; Hamzah and Yuit 2023). This scheme enables person-to-person money transfers among Singapore's NETS, Malaysia's DuitNow, and Indonesia's QRIS users. Most recently, the National Bank of Cambodia (2024) signed an MoU with Bank Negara Malaysia (BNM) to link domestic payment systems between Cambodia and Malaysia to enable cross-border QR payments.

Furthermore, a significant regional milestone occurred on November 14, 2022, in Bali, with the signing of the MoU on Regional Payment Connectivity (RPC) among five central banks from the SEA region. The participating institutions include Bank Indonesia (BI), Bank Negara Malaysia (BNM), Bangko Sentral ng Pilipinas (BSP), Monetary Authority of Singapore (MAS), and Bank of Thailand (BOT) (Bank

Table 6.1 Payment types and platforms across Southeast Asia

Country	Card	Domestic payment	Mobile wallet	BNPL	Others
Indonesia	Visa, Mastercard	BI-Fast, GPN, ATM Bersama, Prima	GoPay, OVO, Dana, ShoopePay	Akulaku, Kredivo, Indodana, PayLater	Cash on delivery, counter payments
Malaysia	Visa, Mastercard	FPX, DuitNow, Interbank Funds Transfer (IBFT)	Touch 'n Go, GrabPay, Boost	Atome, PayLater, Shopback PayLater	Cash on delivery, counter payments
Philippines	Visa, Mastercard	PesoNet, Instapay	Gcash, PayMaya	Billease, TendoPay	Cash on delivery, counter payments
Singapore	Visa, Mastercard	NETS, FAST, PayNow	GrabPay, DBS Paylah, Singtel Dash	Atome, GrabPayLater	Cash on delivery
Thailand	Visa, Mastercard	PromptPay	TrueMoney, LinePay, ShopeePay	Atome, Traveloka	Cash on delivery, counter payments
Vietnam	Visa, Mastercard	NAPAS Quick Money Transfer Service	MoMo, Zalopay, ViettelPay	WowMelo, PayLater by Lotte	Cash on delivery, counter payments

Compiled from Hingel et al. (2023) and Yeo (2023)

Table 6.2 Bilateral cross-border digital payments in Southeast Asia

Participating country	Eligible payment	Progress
Thailand-Cambodia	Thai QR-KHQR	2020 (launched)
Thailand-Singapore	PromptPay-PayNow	2021 (launched)
Malaysia-Thailand	DuitNow-PromptPay	2021 (launched)
Malaysia-Indonesia	DuitNow-QRIS	2021 (launched)
Thailand-Vietnam	Thai QR-VietQR	2021 (launched)
Thailand-Indonesia	Thai QR-QRIS	2021 (launched)
Indonesia-Philippines	QRIS-InstaPay	2022 (MoU signed)
Malaysia-Philippines	DuitNow-InstaPay	2022 (MoU signed)
Malaysia-Singapore	DuitNow-NETS	2023 (launched)
Indonesia-Singapore	QRIS-NETS	2023 (launched)
Cambodia-Malaysia	Undetailed	2024 (MoU signed)

Compiled from various sources: Banchongduang (2021), Bank Indonesia (2023), Bank of Thailand (2023), Hamzah and Yuit (2023), Hingel et al. (2023), National Bank of Cambodia (2024), PayNet (2024), PwC (2024), and Yulius et al. (2023)

Indonesia 2022). This MoU establishes ASEAN cross-border payment cooperation utilising QR codes or e-wallets, enabling consumers from the five countries to conduct transactions using their local currencies, which will be converted based on prevailing exchange rates in each respective country (Kominfo 2023).

In addition to government-led initiatives, the private sector significantly enhances cross-border payment connectivity in the SEA region, exemplified by efforts undertaken by companies like Alipay+. According to Yulius et al. (2023), Alipay+ has capitalised on Alibaba's extensive asset network and cultivated robust partnerships in Asia to expand its connectivity with various mobile wallets across several countries. These partnerships include DANA in Indonesia, Touch 'n Go in Malaysia, GCash in the Philippines, EZ-Link in Singapore, TrueMoney in Thailand, and KakaoPay in South Korea. As a result, users of Alipay+ and affiliated mobile wallets can seamlessly conduct transactions using their mobile wallets in eight countries, including Indonesia, Singapore, Malaysia, Thailand, the Philippines, South Korea, and Japan.

6.3.2 Regional Benefits

Cross-border digital payments play an essential role in enhancing financial inclusion, offering underserved populations access to formal financial services (Shen et al. 2021). Mobile money platforms, digital wallets, and A2A payment services enable individuals without traditional bank accounts to securely send, receive, and store money, empowering them to save, invest, and build assets. Data on financial inclusion through cross-border digital payments indicates a significant societal penetration of digital payments. Approximately 49% of urban commercial bank customers have utilised e-wallets for transactions, with the unbanked population projected to increase to 58% by 2025 (De et al. 2020). E-wallets have witnessed a surge in adoption alongside daily technologies like online transportation, food delivery, and grocery shopping (Asian Business Review 2021). In addition to e-wallets, the current progress of cross-border digital payments among countries within the SEA region has established numerous payment methods (see Table 6.2). This transition from cash-based transactions to digital financial services enhances financial well-being and enables effective economic participation for all (He 2021).

Cross-border digital payments have the potential to stimulate public consumption and regional economic growth (Liu et al. 2021; Patra and Sethi 2023). By facilitating seamless transactions across domestic and international markets, digital payment methods offer convenience and security to e-commerce platforms, online marketplaces, and digital storefronts, driving increased sales, job opportunities, and revenue. According to Google et al. (2023), the digital economy landscape in SEA has yielded three primary benefits. Firstly, it has fostered business growth, exemplified by a transaction surge involving up to 25 unique merchants on e-commerce platforms and seven million Small and Medium-sized Enterprises (SMEs) restaurants on food delivery platforms. Secondly, it has significantly contributed to job creation, with approximately 160 thousand direct jobs and up to 30 million indirect jobs established. Thirdly, the economic impact is evident in the SEA gross domestic product (GDP) penetration, estimated at around 5–10%, and the anticipated doubling of regional digital economy growth by 2025. These findings underscore the pivotal role of cross-border digital payments in fostering and amplifying economic development across the region.

Cross-border digital payments facilitate economic integration and prosperity in the SEA region by bridging different currencies and national payment systems (Elms 2021). These payments streamline cross-border trade, offering businesses a fast, secure, cost-effective transaction method with overseas partners. With the flourishing e-commerce scene in SEA and notable increases in cross-border transactions, digital payments are becoming increasingly relevant (Google et al. 2022). Furthermore, digital payments will foster collaboration among businesses, governments, and stakeholders, driving initiatives for economic development and infrastructure investment. They reduce transaction costs, increase payment speed, facilitate economic cooperation, eliminate the need for traditional banking intermediaries and physical cash, and allow direct transactions between businesses and individuals regardless of location; thus, simplifying international business operations and encouraging collaboration between companies from different countries. This cooperation extends beyond national borders, enhancing equal economic growth prospects.

6.3.3 Key Enablers

The significant advantages of advancing cross-border digital payments in the SEA region must be further promoted to ensure equitable benefits across all regional countries. Achieving this goal is feasible due to four key enablers: government endeavours, market growth, technology penetration, and advancement of the financial industry.

A. *Government Endeavour*

Governments across SEA have implemented various measures, including establishing regulatory frameworks, supporting infrastructure development, and fostering collaboration with other nations to promote digital payments. As detailed in the preceding section (refer to Table 6.2), eleven bilateral initiatives among different SEA countries are currently aimed at implementing cross-border digital payments, showcasing tangible efforts to facilitate broader adoption in Brunei Darussalam, Laos, Myanmar, and Timor-Leste.

B. *Market Growth*

The e-commerce market in SEA has experienced significant growth. According to Arora et al. (2022), SEA's average e-commerce penetration rate (excluding food and beverage) is 20%. Projections suggest this figure will triple, reaching approximately $230 billion in gross merchandise volume by 2026, with a compound growth rate of 22%. The COVID-19 pandemic has expedited the shift towards online shopping, further fuelling the expansion of the e-commerce sector and positioning the digital payment landscape in SEA for continued growth.

C. *Technology Penetration*

Mobile penetration has rapidly increased in SEA, with over 90% of the population accessing mobile phones (Google et al. 2022). This widespread access to mobile technology has created a fertile market for digital payments and familiarised people with various digital technologies.

D. *Advancement of the Financial Industry*

The rise of fintech has revolutionised the financial industry and significantly impacted cross-border payments (Anshari et al. 2020). Technological developments in infrastructure, big data, and algorithms have greatly facilitated and improved cross-border payment services (Lee and Shin 2018), making them faster, cheaper, and more accessible for individuals and businesses.

6.3.4 Key Challenges

While the factors driving progress in cross-border digital payments are compelling, stakeholders must also recognise the challenges and implement appropriate measures. Those problems include interoperability, diverse regulatory frameworks, potentially fraudulent activities, currency issues, speed challenges, and the digital divide.

A. *Interoperability*

Firstly, adopting cross-border digital payments needs more uniformity among different payment systems and currencies across SEA countries (He 2021; Zhang 2020), resulting in a complex environment for embracing such payments in the region. Addressing interoperability becomes crucial to enable seamless cross-border money transfers and exchanges. The QR code system, currently utilised in some parts of SEA for cross-border digital payments, could serve as a fundamental building block for enhancing the capacity of its operating system (Alkhowaiter 2020; Ng et al. 2021).

B. *Diverse Regulatory Frameworks*

Regulatory frameworks across countries in the region vary (Ha and Chuah 2023), potentially leading to confusion and compliance challenges for companies intending to offer cross-border payment services. Regulators must harmonise these regulations to ensure legal certainty for service providers and address consumer protection concerns (Arner et al. 2022).

C. *Potentially Fraudulent Activities*

Acknowledging that digital payments are susceptible to fraudulent activities like hacking, identity theft, and data breaches is crucial, which is vital chess. These activities erode trust in digital payment systems and impede public adoption and usage of these services (Ambore et al. 2017; Hassan et al. 2020; Sahi et al. 2022). This issue can only be mitigated by requiring national policy to start an intercountry agreement on cross-border security.

D. *Currency Issues*

The SEA comprises diverse currencies, complicating cross-border digital payments due to issues such as determining exchange rates and hidden fees, which can escalate transaction costs and complexities (He 2021; Turban et al. 2018). The currency issues concern the different values between local and foreign currencies (Arauz 2021). Although cross-border payments can be processed, people may feel the exchange rate needs to be revised, especially for those with less strong currency than other currencies. Despite the facility provided to proceed with intercountry transactions, the other consideration will be the value of the money exchanged. Local transactions will be more valuable compared to cross-border transactions. The currency problem can be less than a hiccough if the transaction is a comparable standard of money value or if some countries agree to use the same currency in cross-border transactions. In sum, currency issues show the high cost of transactions, which necessarily becomes the first consideration when using cross-border digital payment.

E. *Speed Challenge*

Speed is one of the essential features of financial transactions. However, this aspect still needs to be solved in cross-border digital payments. Due to the vulnerability to money laundering or other financial crimes, the low speed of payment processing time is subject to multiple cheques and reconciliations in both receiving and sending countries (Financial Stability Board 2021). Thus, the transactions may not be completed in real-time. A possible solution for this challenge is cooperation among financial services in regional countries to achieve the same goal of providing faster transaction facilities. This will involve agreements related to infrastructure, software, security design, and other related aspects. The main challenge to furthering this cooperation is each country's financial services regulation.

F. *Digital Divide*

Although this current era has been called the digital era, the digital divide is still a problem that cannot be ignored. This problem primarily features low-income families, older generations, rural populations, and those who are illiterate (Hung 2024). Despite widespread internet and mobile phone penetration in the region, a significant portion of SEA's population remains unbanked, posing challenges to offering e-payment services (Jahan et al. 2019; Koh et al. 2018). Due to the different levels of knowledge of digital systems, only those equipped with the proximity to the infrastructure and literacy can benefit from cross-border digital payment. The digital divide is not only a challenge for cross-border digital payment. More broadly, it has become the national agenda to improve the equality of rights to enjoy digital facilities and possess the same knowledge about taking advantage of them. Education is the key to determining the success of digital equality, and this role is attached to the government, service providers, and educational institutions.

6.4 Future Directions

6.4.1 The Path Forward

Digital payment systems have become integral to the global economy, providing convenient and efficient means for cross-border transactions. Many countries in the SEA region have adopted this technology, enabling citizens to conduct faster transactions across borders. While this represents significant progress, covering approximately 63% of SEA countries, efforts to enhance cross-border payment inclusivity are necessary to fully realise the vision of economic integration and regional collaboration for all stakeholders in the SEA region.

A comprehensive approach encompassing various institutional initiatives is necessary to build more inclusive cross-border digital payment systems in the SEA region. In the e-commerce context, Chen and Ruddy (2020) highlighted the importance of physical and cyberspace infrastructure, regulations to govern the dynamic online marketplace, connectivity services to enhance value, and collaborative efforts in digital economy governance. Additionally, Chen (2020) proposes a framework for digital connectivity among SEA countries to address cross-border transaction needs. This framework emphasises three critical factors for successful digital connectedness, particularly in e-commerce: logistics connectivity, free trusted data flow, and financial connectivity to facilitate cash flow. These aspects require cohesive policies to facilitate implementation and realisation. Inspired by the framework, we recommend several pertinent strategies to advance cross-border digital payment inclusivity in the SEA region (see Fig. 6.1). These factors can be effectively implemented and utilised by fostering agreement among SEA countries regarding cross-border digital payments. Configuration and linkage are essential for establishing a robust cross-border digital payments ecosystem in the SEA region, addressing challenges to enhance its success.

A. *Investment in Technology*

ASEAN Member States (AMS) exhibit varying degrees of access to financial services, including digital payments. This discrepancy arises primarily from differences in the sophistication of financial ICT infrastructure among the AMS. As a result, adoption rates of digital payment platform development vary across these states (Sinay et al. 2021). Governments should prioritise investment in robust infrastructure, secure payment gateways, and reliable point-of-sale (POS) systems to address this disparity (Ha and Chuah 2023). These measures are crucial to ensure seamless transaction processing and maintain data security standards across the region.

B. *Regulatory Compliance and Harmonisation*

In the SEA region, regulatory frameworks governing cross-border digital payments vary considerably from country to country, reflecting individual nations' diverse regulatory environments and policy priorities. While some countries have adopted

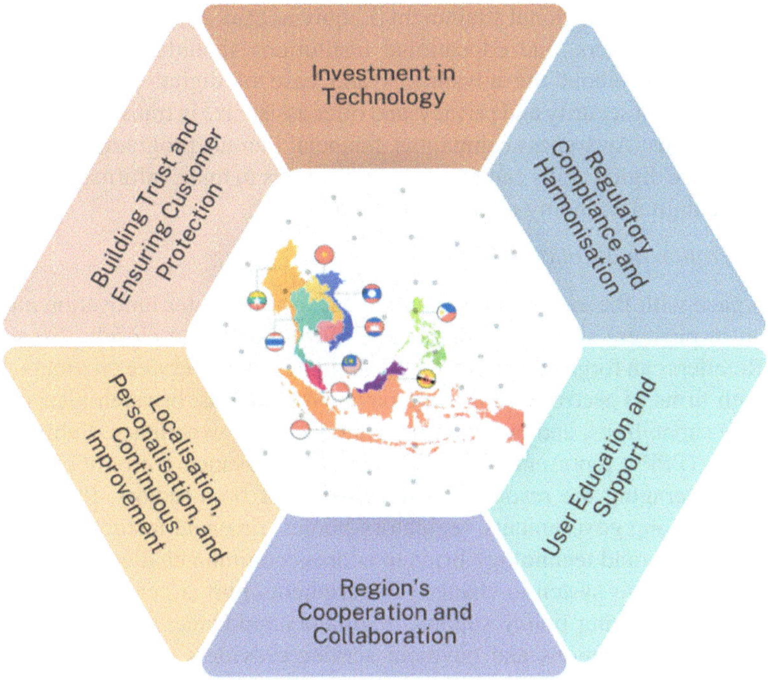

Fig. 6.1 Strategies for advancing cross-border digital payment inclusivity in the Southeast Asia region (The SEA map on this figure was designed by Freepik)

progressive regulations to encourage innovation and facilitate cross-border transactions (e.g., Singapore, Malaysia, Indonesia, Thailand, Vietnam, The Philippines, and Cambodia), others maintain more stringent approaches to probably safeguard financial stability and consumer protection (e.g., Brunei Darussalam, Laos, Myanmar, and Timor-Leste). Retailers must remain updated on regulatory requirements governing cross-border payments to ensure compliance. Regulatory rules must be harmonised, given the diversity of standards and policies across countries. Regulatory authorities are responsible for overseeing providers' compliance with key regulations, including Anti-Money Laundering (AML) and Counter-Terrorism Financing (CTF) measures, Know-Your-Customer (KYC) requirements, data protection laws, and consumer protection measures. Additionally, they must uphold the integrity and security of cross-border payment transactions and safeguard the rights and interests of consumers.

C. *User Education and Support*

In addition to the challenges above, previous research conducted in Vietnam, Indonesia, and the Philippines identified key barriers to financial inclusion, including limited funds, high monetary costs, and the perception that financial services are unnecessary. These obstacles are particularly pronounced among individuals with lower incomes and educational attainment (Esquivias et al. 2021). Hence, the government, service providers, and educational institutions should undertake efforts to educate customers about the advantages and ease of digital payments, address concerns regarding security and privacy, and offer assistance in transitioning to digital payment methods. Awareness campaigns, financial literacy programmes, and initiatives to enhance digital skills can empower consumers to make informed choices and navigate the digital economy effectively.

D. *Cooperation and Collaboration*

To keep pace with the evolving payment landscape and foster innovation and digitalisation in payment systems, stakeholders must collaborate closely (Sinay et al. 2021). Retailers can forge partnerships with digital payment service providers, banks, and fintech firms to access cutting-edge payment solutions, broaden their payment acceptance capabilities, and enrich the overall customer experience. Public–private partnerships (PPPs) represent a pivotal avenue for advancing cross-border digital payments, leveraging the resources and expertise of both sectors. PPPs facilitate cooperation among governments, regulatory bodies, financial institutions, payment service providers, and technology firms to address common challenges, drive innovation, and instigate systemic change. Governments play a central role in facilitating PPPs by offering policy support, regulatory guidance, and financial incentives. Financial institutions and payment service providers contribute by sharing data, knowledge, and resources to develop interoperable payment systems, bolster security measures, and extend access to digital financial services. Technology firms, including fintech startups and digital platform providers, bring inventive solutions, expertise, and agility to the partnership, fostering technological advancement and market disruption.

E. *Localisation, Personalisation, and Continuous Improvement*

To address the challenge of cross-border digital payments, retailers must customise their payment offerings to align with the preferences and requirements of local markets (Choi et al. 2020). This strategy includes providing language support, offering various currency options, and incorporating localised payment methods to boost acceptance and adoption. Additionally, retailers should consistently monitor and assess their cross-border payment procedures, gather customer feedback, and implement enhancements to refine the payment experience and bolster customer satisfaction. Moreover, cross-border digital payments offer substantial opportunities for businesses and retailers to broaden their reach, streamline operations, and enrich customer experiences, thereby driving growth and competitiveness in the global arena.

F. *Building Trust and Ensuring Consumer Protection*

Establishing trust and safeguarding consumer protection is critical in the cross-border digital payments ecosystem, where fraudulent activities, data breaches, and unauthorised transactions pose risks to consumer confidence and trust in digital payment solutions (Daza Jaller et al. 2020). Regulatory authorities, industry stakeholders, and consumer advocacy groups play pivotal roles in safeguarding consumer interests and promoting market transparency and accountability. Regulatory oversight and enforcement are necessary to ensure consumer protection, with regulatory authorities setting standards and overseeing the conduct of digital payment service providers, financial institutions, and other market participants. Additionally, businesses can bolster consumer trust by implementing robust security measures, including encryption, multi-factor authentication, and tokenization, and staying abreast of new technologies (Dinh et al. 2023) to safeguard sensitive financial information and prevent unauthorised access to payment systems. Service providers also play a crucial role in consumer education and awareness, amplifying efforts to protect consumer interests.

6.4.2 The Future Outlook

The array of advancements in cross-border digital payments across the SEA region is remarkable. Progress stems from both governmental and private-sector initiatives. Both initiatives are striving to cultivate a "cashless society" and bolster trust in electronic payments across the region. Despite the cultural sensitivities and diverse economic landscapes in SEA, challenges persist for the widespread adoption of digital payments. However, ongoing efforts to address security concerns and enhance legislative control for e-commerce are underway (Ha and Chuah 2023), indicating significant growth potential for digital payments in the region. Integrating secure and privacy-sensitive technologies like blockchain (Islam et al. 2023; Zetzsche et al. 2021) can further enhance SEA's digital payment infrastructure, addressing security threats and vulnerabilities while fostering greater trust and confidence in the system.

The promising figures in SEA indicate that future growth opportunities are rarely observed in other economic regions (Majumdar et al. 2020). The rise in smartphone usage and internet accessibility in SEA (Kemp 2023) further bolsters the potential for the expansion of digital payments. With more individuals gaining access to smartphones and the internet, the infrastructure supporting the widespread adoption of digital payments is becoming increasingly robust. Furthermore, ongoing collaborations between financial institutions and financial technology (fintech) companies in SEA are driving the development of innovative digital payment solutions. These collaborations are tailored to address the region's unique needs and challenges, creating user-friendly and secure platforms for digital transactions. As consumer behaviour progressively shifts towards digital channels for everyday activities (Sheng and Fauzi 2023), there lies an opportunity for businesses in SEA to leverage this trend by integrating digital payment options into their services. This enhances customer convenience and promotes financial inclusion by granting access to formal financial services for individuals previously underserved by traditional banking systems. In short, there is a promising outlook for cross-border digital payments in SEA as governments and businesses recognise the advantages and take steps to promote their adoption. The region stands ready to undergo a notable transformation in financial transactions, heralding a path toward a more efficient and inclusive digital economy.

6.5 Conclusion

This chapter reveals that the SEA region has made impressive strides in implementing cross-border digital payments. Despite facing various challenges, successfully introducing an interoperable QR code payment system could significantly enhance the feasibility of cross-border digital payments throughout the region. Since SEA is experiencing the fastest economic and digital market growth globally, effective digital payment systems could further strengthen the region's digital economy and facilitate its integration into the global digital economy. In practical terms, stakeholders in SEA are encouraged to consider the framework proposed in this study, which emphasises six key pillars, as outlined in Fig. 6.1. These pillars include investment in technology, regulatory compliance and harmonisation, user education and support, localisation and personalisation with continuous improvement, and building trust while ensuring customer protection. These six pillars can guide stakeholders directly involved in enhancing the technology and utility of cross-border digital payments in the SEA region. From a theoretical perspective, the study contributes to the academic literature by deepening the understanding of the development patterns of cross-border digital payments in SEA, which may also serve as a reference for similar advancements in other global regions.

Although this study offers valuable insights into the implementation and development of cross-border digital payments in SEA, it is important to acknowledge certain limitations. Methodologically, the study employs a narrative review approach, which some scholars consider less rigorous than systematic literature reviews (Farrington and Jolliffe 2017; Greenhalgh et al. 2018). Consequently, future research could adopt a more comprehensive and detailed approach to this topic, yielding more robust and reliable findings. Contextually, this chapter focuses on the macro-level development of cross-border digital payments, without sufficiently examining the adoption of these technological advancements within SEA. Specifically, it does not explore whether the technology has been effectively utilised by the population to drive an increase in cross-border transactions. Future research could expand on this by investigating how communities in SEA adopt and integrate cross-border digital payment technologies. Additionally, researchers could provide recommendations on strategies to encourage the use of these technologies in cross-border transactions across the region.

Highlights

- The Southeast Asia region has achieved significant progress in the adoption and implementation of cross-border digital payments.
- The advancement of cross-border digital payments relies heavily on initiatives from both governments and the private sector.
- The future of cross-border digital payments in the SEA region appears highly promising.

- Enhancing inclusivity is essential for achieving comprehensive economic integration and fostering collaboration within the region.

References

Alkhowaiter WA (2020) Digital payment and banking adoption research in Gulf countries: a systematic literature review. Int J Inf Manag 53(February):102102. Elsevier

Ambore S, Richardson C, Dogan H et al (2017) A resilient cybersecurity framework for Mobile Financial Services (MFS). J Cyber Secur Technol 1(3–4):202–224. Taylor & Francis

Anshari M, Almunawar MN, Masri M (2020) Financial technology and disruptive innovation in business. Int J Asian Bus Inf Manag 11(4):29–43

Arauz A (2021) The international hierarchy of money in cross-border payment systems: developing countries' regulation for central bank digital currencies and facebook's stablecoin. Int J Polit Econ 50(3):226–243. Routledge

Arner DW, Buckley RP, Lammer T et al (2022) Building regional payment areas: the single rule book approach. SSRN Electron J (1016). Epub ahead of print 2022. https://doi.org/10.2139/ssrn.4115851

Arora S, Chu F, Melkinov S et al (2022) E-commerce is entering a new phase in Southeast Asia. Are logistics players prepared? McKinsey & Company (December):1–9

Asian Business Review (9 Jul 2021) SEA named world's fastest-growing mobile wallet region. Asian Bus Rev. https://asianbusinessreview.com/indonesian/node/514515790

Banchongduang S (19 Jun 2021) PromptPay now linked to DuitNow. Bangkok Post. Bangkok. https://www.bangkokpost.com/business/2134811/promptpay-now-linked-to-duitnow

Bank Indonesia (2022) Central banks of Indonesia, Malaysia, Philippines, Singapore and Thailand seal cooperation in regional payment connectivity. https://www.bi.go.id/en/publikasi/ruang-media/news-release/Pages/sp_2430822.aspx. Accessed 31 Mar 2024

Bank Indonesia (2023) Launch of cross-border QR payments linkage between Indonesia and Singapore. https://www.bi.go.id/en/publikasi/ruang-media/news-release/Pages/sp_2531023.aspx. Accessed 31 Mar 2024

Bank of Thailand (2023) Cross-border payment. https://www.bot.or.th/en/financial-innovation/digital-finance/digital-payment/cross-border-payment.html#connectivites-item-e6dea9c0d5. Accessed 31 Mar 2024

Baumeister RF, Leary MR (1997) Writing narrative literature reviews. Rev Gen Psychol 1(3):311–320

Chen L (2020) Improving digital connectivity for e-commerce: a policy framework and empirical note for ASEAN. ERIA Discussion Paper Series (March):7–30

Chen L, Ruddy L (2020) Improving digital connectivity : policy priority for ASEAN digital transformation. Econ Res Inst ASEAN East Asia Key 07(2020):1–4

Choi H, Park J, Kim J et al (2020) Consumer preferences of attributes of mobile payment services in South Korea. Telematics Inform 51:101397

Daza Jaller L, Gaillard S, Molinuevo M (2020) The regulation of digital trade: key policies and international trends. https://openknowledge.worldbank.org/server/api/core/bitstreams/20cc166f-1ef1-516a-be18-3eaec28c3be1/content

Dinh V, Le D-V, Duong D et al (2023) Determinants affecting digital financial consumer protection: evidence from 135 countries. J Econ Asymmetries 27:e00301

Elms D (2021) Increasing access and interoperability of cross-border e-payments in Asia. Hinrich Foundation Report. https://asiantradecentre.org/s/E-payments-in-Asia_Hinrich-Foundation.pdf

Esquivias MA, Sethi N, Ramandha MD et al (2021) Financial inclusion dynamics in Southeast Asia: an empirical investigation on three countries. Bus Strat Dev 4(2):203–215
Farrington DP, Jolliffe D (2017) Special issue on systematic reviews in criminology. Aggress Violent Behav 33:1–3. Elsevier Ltd.
Financial Stability Board (2021) Targets for addressing the four challenges of cross-border payments: final report. https://www.fsb.org/wp-content/uploads/P131021-2.pdf
Google, Temasek and Bain & Company (2022) e-Conomy SEA 2022: through the waves, towards a sea of opportunity. https://services.google.com/fh/files/misc/e_conomy_sea_2022_report.pdf
Google, Temasek and Bain & Company (2023) e-Conomy SEA 2023 report. https://services.google.com/fh/files/misc/e_conomy_sea_2023_report.pdf
Greenhalgh T, Thorne S, Malterud K (2018) Time to challenge the spurious hierarchy of systematic over narrative reviews? Eur J Clin Invest 48(6):1–6
Ha H, Chuah CKP (2023) Digital economy in Southeast Asia: challenges, opportunities and future development. Southeast Asia: Multidiscip J 23(1):19–35
Hamzah A, Yuit CK (31 Mar 2023) Singapore and Malaysia users of Nets, DuitNow can now pay with new QR code link. Straits Times. https://www.straitstimes.com/singapore/qr-code-payment-link-between-s-pore-and-malaysia-launched-person-to-person-money-transfers-by-end-2023
Hashemi Joo M, Nishikawa Y, Dandapani K (2019) Cryptocurrency, a successful application of blockchain technology. Manag Financ 46(6):715–733
Hassan MA, Shukur Z, Hasan MK et al (2020) A review on electronic payments security. Symmetry 12(8):1344
He D (2021) Digitalization of cross-border payments. China Econ J 14(1):26–38. Routledge
Hingel G, Kanithasen P, Varootbangkul W (2023) Shaping the future of cross-border fast payment systems: revolutionizing transactions in South-East Asia. https://www3.weforum.org/docs/WEF_Shapingthe_Future_of_Cross-Border_Fast_Payment_Systems_2023.pdf
Hung J (2024) How southeast Asia can better arrange and deliver internet policies so as to defy the digital divide. In: Radovanović D (ed) Digital literacy and inclusion. Springer International Publishing, Cham, pp 61–77. https://doi.org/10.1007/978-3-031-30808-6_5
Ishikawa K (2021) The ASEAN economic community and ASEAN economic integration. J Contemp East Asia Stud 10(1):24–41. Routledge
Islam MM, Islam MK, Shahjalal M et al (2023) A low-cost cross-border payment system based on auditable cryptocurrency with consortium blockchain: joint digital currency. IEEE Trans Serv Comput 16(3):1–14
Jahan S, De J, Jamaludin F et al (2019) The financial inclusion landscape in the Asia-Pacific region, vol 19(79). IMF Working Papers
Kemp S (2023) Digital 2023: global overview report. https://datareportal.com/reports/digital-2023-global-overview-report. Accessed 24 May 2023
Koh F, Phoon KF, Ha CD (2018) Digital financial inclusion in south east Asia. In: Lee Kuo Chuen D, Deng Digital Finance, and Inclusion, Volume 2 RBT-H of B (eds) Handbook of blockchain, digital finance, and inclusion, vol 2, pp 387–403. Elsevier. https://www.sciencedirect.com/science/article/pii/B9780128122822000152
Kominfo (2023) QRIS simplifies payments in ASEAN countries. https://asean2023.id/en/news/qris-simplifies-payments-in-asean-countries. Accessed 31 Mar 2024
Lee I, Shin YJ (2018) Fintech: ecosystem, business models, investment decisions, and challenges. Bus Horiz 61(1):35–46
Liu Y, Luan L, Wu W et al (2021) Can digital financial inclusion promote China's economic growth? Int Rev Financ Anal 78:101889
Majumdar SK, Sarma AP, Majumdar S (2020) E-commerce and digital connectivity: unleashing the potential for greater India–ASEAN integration. J Asian Econ Integr 2(1):62–81. SAGE Publications India
Morgan PJ (2022) Fintech and financial inclusion in southeast Asia and India. Asian Econ Policy Rev 17(2):183–208

Mützel S (2021) Unlocking the payment experience: future imaginaries in the case of digital payments. New Media Soc 23(2):284–301

National Bank of Cambodia (2024) Joint Press Release Bank Negara Malaysia and the National Bank of Cambodia to strengthen regional cooperation in financial innovation and payments. https://www.nbc.gov.kh/download_files/news_and_events/press_eng/5455English_PR_Malaysia.pdf. Accessed 31 Mar 2024

Ng D, Kauffman RJ, Griffin P et al (2021) Can we classify cashless payment solution implementations at the country level? Electron Commer Res Appl 46(March):101018. Elsevier B.V.

Pansari A, Kumar V (2017) Customer engagement: the construct, antecedents, and consequences. J Acad Mark Sci 45(3):294–311

Patra B, Sethi N (2023) Does digital payment induce economic growth in emerging economies? The mediating role of institutional quality, consumption expenditure, and bank credit. Inf Technol Dev, 1–19. Routledge

PayNet (2024) Cross-border QR payment. https://www.duitnow.my/Cross-Border/index.html. Accessed 31 Mar 2024

PwC (2024) Payments 2025 and beyond: evolution to revolution. https://www.pwc.com/sg/en/financial-services/fintech/payments-2025-and-beyond.html. Accessed 31 Mar 2024

Sahi AM, Khalid H, Abbas AF et al (2022) The research trend of security and privacy in digital payment. Informatics 9(2):32

De SD, Bharadwaj A, Khan I et al (2020) Southeast Asian consumers are driving a digital payment revolution. BCG. https://web-assets.bcg.com/img-src/BCG-Southeast-Asian-Consumers-Are-Driving-a-Digital-Payment-Revolution-May-2020_tcm9-247804.pdf

Shen Y, Hueng CJ, Hu W (2021) Measurement and spillover effect of digital financial inclusion: a cross-country analysis. Appl Econ Lett 28(20):1738–1743. Routledge

Sheng ML, Fauzi AA (2023) Institutional behavior mechanism: exploring the impacts of macro-environmental stimuli on continued digital payment adoption behavior. Comput Hum Behav 149(July):107923. Elsevier Ltd.

Siddaway AP, Wood AM, Hedges LV (2019) How to do a systematic review: a best practice guide for conducting and reporting narrative reviews, meta-analyses, and meta-syntheses. Annu Rev Psychol 70:747–770

Sinay JB, Tumengkol EA, Zendra O (2021) Payment systems in the digital age: case of ASEAN. ASEAN Policy Brief 4(04):1–9

Turban E, Outland J, King D et al (2018) Electronic commerce payment systems. In: Turban E, Outland J, King D et al (eds) Springer International Publishing, Cham, pp 457–499. https://doi.org/10.1007/978-3-319-58715-8_12

Yeo M (2023) How Asia buys and pays 2023: tapping into Asia's regional commerce opportunities. https://go.2c2p.com/wp-content/uploads/2023/10/2C2P-IDC-InfoBrief_AP241443IB.pdf

Yulius, Tjhin D, Lukiman Y et al (2023) Interoperable QR code payment ecosystem in ASEAN: what it means for the world. ASEAN Business Advisory Council. https://www.bcg.com/publications/2023/southeast-asia-qr-code-payment-ecosystem-in-asean

Zetzsche DA, Anker-Sørensen L, Passador ML et al (2021) DLT-based enhancement of cross-border payment efficiency—a legal and regulatory perspective. Comput Hum Behav 15(1–2):70–115

Zhang Y (2020) Developing cross-border blockchain financial transactions under the belt and road initiative. Chin J Comp Law 8(1):143–176

Abu Amar Fauzi is currently a junior researcher at the National Research and Innovation Agency of the Republic of Indonesia. Previously, he served as an Assistant Professor in the Faculty of Economics and Business at Universitas Hayam Wuruk Perbanas in Surabaya, Indonesia. He earned his Ph.D. from the School of Management, Department of Business Administration, at the National Taiwan University of Science and Technology (NTUST) in Taipei, Taiwan. His research primarily focuses on consumer behaviour, the continuance intention of technology-based services, Islamic marketing, and digital marketing. His work has been published in several reputable peer-reviewed journals, including the Journal of Islamic Marketing, Journal of Relationship Marketing, Journal of Small Business Management, Asia Pacific Journal of Marketing and Logistics, Journal of Marketing for Higher Education, Electronic Commerce Research, Computers in Human Behavior, and Asia Pacific Management Review.

Lufi Yuwana Mursita is an Assistant Professor in the Faculty of Economics and Business at Universitas Hayam Wuruk Perbanas in Surabaya, Indonesia. She obtained her master's degree from the Master of Science in Accounting programme at the Faculty of Economics and Business, Universitas Gadjah Mada, Indonesia. Her research interests lie in management accounting, behavioural accounting, and business. She has published academic papers in both international and nationally accredited journals, including the Journal of Accounting and Organizational Change, International Journal of Ethics and Systems, Asian Academy of Management Journal, Jurnal Akuntansi dan Keuangan Indonesia (Indonesian Journal of Accounting and Finance), The Indonesian Journal of Accounting Research, and Integritas: Jurnal Antikorupsi (Integrity: Anti-Corruption Journal).

Nazlida Muhamad is an Associate Professor at the Prince Sultan University, and an adjunct Associate Professor at UBD School of Business and Economics, Universiti Brunei Darussalam (UBDSBE). She earned her PhD in Marketing from the University of Western Australia in 1999. Before joining UBD, she taught marketing at public universities in Malaysia, Saudi Arabia, and Australia. Her research primarily focuses on consumer behaviour, exploring the complex relationship between Islam and Muslim consumer behaviours, tourism, value co-creation in the service industry, and involve in ASEAN's collaborative efforts at member state level. Her scholarly work has been published in esteemed journals, including Psychology and Marketing, Journal of Consumer Behaviour, Asia Pacific Business Review, International Journal of Consumer Marketing, and Journal of Islamic Marketing.

Margaret L. Sheng is a Professor in the Department of Information Management and the Department of Business Administration at the National Taiwan University of Science and Technology in Taiwan. She earned her PhD from the University of Minnesota, Twin Cities, USA. She was a visiting scholar at the University of California, Berkeley, USA, from August 2014 to July 2015, and at Peking University, China, in 2006. Before joining NTUST, she taught at Hamline University in Minnesota, USA. She has published extensively in high-ranking academic journals, including Technovation, International Journal of Information Management, Journal of Business Research, Journal of International Marketing, Journal of International Management, Industrial Marketing Management, and Journal of Retailing and Consumer Services. Additionally, she has authored a related book titled "Global Integrated Supply Chain Implementation: The Challenges of E-procurement." In 2023 and 2024, she was awarded as the World Top 2% Scientist, recognised by Stanford University.

Chapter 7
The Future of Economy in Japanese Innovation of Pre- and Post-COVID-19: A Systematic Review

Clusters of Future Studies: Past and Futures; Post-normality and Complexity

Chutinon Putthiwanit

Abstract The colossal spread of the COVID-19 pandemic has affected the whole world's economic performance unavoidably. This qualitative research employs a systematic review roofed the literature search, covering pre- and post-COVID-19, from years 1986–2022. In primarily 423 articles were explored, and later 370 articles were examined. Thence, 32 articles were investigated to note the context of Japanese innovation using the quantitative tool—KH-Coder. The results show that, prior to COVID-19, Japan performed well in the past. However, substantial drops were observed during the late 1990s, even before COVID-19, due to having a challenging time with the Galapagos Syndrome. Regarding correspondence analysis and the cluster analysis on Japan's sophisticated organisational strategy, the research demonstrates that Japan handles well with technology, development, research, and even the environment when compared to its counterpart—China. This enlightens us in perceiving about future of the economy in Asia comprehension. Nevertheless, when discussing success in COVID-19 prevention, China tends to perform better than Japan. For future research, the author highly recommends applying the meta-analysis method, to gain even more meaningful, profound, and precise results. It is also interesting to investigate whether different organisational culture also affects the innovative way of COVID-19 prevention in the Asian region.

Keywords Development · Innovation · Keiretsu · Technology · COVID-19

C. Putthiwanit (✉)
Department of Business Administration, Kasetsart University (Chalermphrakiat Sakon Nakhon Province Campus), Sakon Nakhon, Thailand
e-mail: chutinon.p@ku.th

E. Lau et al. (eds.), *Future of Economy in Asia*,
https://doi.org/10.1007/978-3-031-95500-6_7

7.1 Introduction

Globalisation has driven Japan to become one of the most promising innovation-led countries. Technological spillover played an important role in Japan's rise to become a key industrialised state (Yokura et al. 2013). Regarding, the future of the economy of Japan, one key success factor for their success at innovation is the governmental support given to Keiretsu by the Ministry of Economy, Trade, and Industry (METI) (Nishimura and Okamuro 2016; Yokura et al. 2013). This support is beneficial for growing Japan's innovation success when compared to other countries like Singapore, which lacks government support (Carney and Zheng 2009), or even Chile, which lacks economic liberalisation (del Sol 2010). Another factor to note is the Japanese lifetime employment policy (Carney and Zheng 2009), which even supports employees during recessions (Yamagishi et al. 1998), increases the commitment to innovation (Dirks et al. 2000), and diminishes employment turnovers (Yusuf and Othman 2016). Saito and Vuszkun (2010) support this policy in that this practice was part of Japan's research and development (R&D) efforts in the past. The other key factor is the outward Foreign Direct Investment (FDI) abroad, especially in China (Cheng 2008; Xu and Yeh 2013). By combining these groups of amalgams, it is not surprising that Japanese innovation is more advanced than the innovation in most of the rest of the world. Will this positive success continue as it has in the past and still in the present? Will there be any new renaissance efforts made when Japan's innovation is seen as now declining and diminishing?

Japan's innovation is not only accumulated in its capital city but also in the countryside. Ibata-Arens (2008) reported that the ancient capital of Japan (Kyoto) has become the most profitable and innovative city in Japan. This remarkable result is a fine example for other cities in Japan and abroad. Apart from the company and government support, it is believed that there must be certain psychological issues involved as well. For example, Krockow et al. (2018) indicated that Japanese people are cooperative by nature and live their lives based on trust. They tend to favour certainty in society, while avoiding rejection and expressing less about themselves (Hashimoto and Yamagishi 2016). Such trust between people is a required element for positive relationship in business and between employees (Sołoducho-Pelc 2017) and suppliers (Nakayama 2000). As a result, Japan has a history of a collective practice of business conduct called 'Keireitsu'.

There have been some research study on the concept of Keiretsu and its various aspects, such as that of Miwa and Ramseyer (2000), who reimagined the Keireitsu relationship in the automobile business; Mutoh et al. (2020) provided insights about factors affecting Japanese innovation during the rise after World War II and the fall in 1990s. Balderbos and Carree (2002) reported on the effect of Keireitsu on location choice when Japanese businessmen invest in China. Nevertheless, there are not yet enough studies on how Japanese innovation performs vis-à-vis the influence of innovation, when juxtaposing the rise, the fall, and the renaissance of Japanese Innovation. Hence, the author here explores more on this aspect by using a systematic review approach.

7.2 Literature Review

The author reviews the history of Innovation in Japan, as driven by Keiretsu for more systematic new analysis and as structured into four distinct segments. The first segment details the state of Innovation in pre-war and post-war Japan using key historical aspects (the rise). In the second segment, the author describes the innovation in China, Japan's competitor. In the third segment, the author analyses Japan's industrial structure from a contemporary academic perspective (the fall). The final segment hypothesises how new economic and social pressures will affect both the rate and the direction of future innovation in Japan (its renaissance and greater success) going forward.

7.2.1 The Rise of Innovation in Japan (Pre-COVID)

Keiretsu is a close formation of business groups based on family bonds (Krockow et al. 2018). It is collectivistic and reciprocal by nature (Lincoln et al. 1996). It is different from other business networks, which seems to be more individualistic (Morash and Clinton 1997) or heterogeneous (Bell et al. 2017). One example, the bigger Keiretsu companies can collectively control the smaller Keiretsu companies (Miwa and Ramseyer 2000). This relationship-specific investment (Ibid.) is believed to have emerged, pre-war, even perhaps as early as the 1870s (Flath 1993). This innovative formation network in the business world has a unique pattern that has been used to develop Japan's future economy (Yokura et al. 2013) and allowing changes in market mechanism operations (Lincoln et al. 1996). It also creates a strong relationship between sellers and suppliers. Keiretsu was previously called 'Zaibatsu' but changed its name to 'Keiretsu' after World War II (Flath 1993).

Today, Japanese mega enterprises exploit the concept of Keiretsu by transforming the supplier relationship into an instrument that help them drastically cut business costs and innovate quickly (Aoki and Lennerfors 2013). Therefore, we cannot deny the fact that Keiretsu represents and offers a major picture of and for innovation. In addition, the rise of Japan in monopolistic innovation let her gain key strategic bargaining power (Nishimura and Okamuro 2016). Also, Japanese employees tend to have a great relationship with their respected employers (Cheng and Kalleberg 1996); and, given the lifetime employment (LTE) policy in the past, this aspect also produces innovative Japanese employees with children becoming even more desirable for LTE (Saito and Vuszkun 2010).

7.2.2 China's Innovation: Japan's Counterpart

Most of the Japanese enterprises have operations abroad (especially in China), but smaller enterprises in the Keiretsu networks tend to operate locally (Tanaka 2005), or closer to Japan (Balderbos and Carree 2002) due to the economic turmoil during the 1990s, which coerced Japanese enterprises to operate abroad (Mutoh et al. 2020). Due to that oriented proximity in location (Balderbos and Carree 2002), by operating in a neighbouring country like China, there is no wonder that Japanese enterprises also brought innovation along with them to these venues. Apart from that circumstances, China's success in innovation was heavily driven by the support of the Chinese government, indeed one of the first priorities that was able to make that nation become a world-class industrializing powerhouse (Simon 2007; Xie and Li-Hua 2009; Fu and Mu 2014; Lauer and Liefner 2019; Xu et al. 2023) and boost its entrepreneurship research during 1980s (Ju et al. 2023).

Like Japan, location also has had its effects on innovation. For example, Lan et al. (2022) mention that innovation is more flourishing more in the Western and Central cities of China than in other regions of the country. Contrary to Japan's innovation, which is more conservative and collective, China's innovation is based on a government-led national innovation system (Cao et al. 2009). This system has allowed her to move forward, even under the pandemic (Li 2020) and rise from being to imitator to become an innovator (Yu et al. 2017).

7.2.3 The Fall of Innovation in Japan

During the 1990s, even though Japanese companies were successful abroad, they faced high pressure from foreign investors to pay those investors higher dividends (Jung and Mun 2016). Yamagishi et al. (1998) provide evidence that Japanese businessmen tend to reciprocate those in inner circle only, not the outsiders. Both that pressure and regulations were limiting the ongoing creativity and innovation in enterprises. Could this circumstance have been a sign of the coming fall of innovation in Japanese enterprises? Packer (1999) and Mutoh et al. (2020) suggest that Japan's economy actually started to fall during the 1990s, which affected to future of the economy of Japan. In terms of the role of Keiretsu, Gul (1999) mentions that Keiretsu has had a negative relationship with a firm's growth opportunities. Bigger Keiretsu companies sometimes have to help the disturbed smaller Keiretsu companies, even to resolve scandals (Lincoln et al. 1996). This situation affects the growth rate and speed of the bigger firms. One bit of evidence from Chang et al. (2014) shows that Japanese companies tend to stay closer in terms of their grouping with Keiretsu companies with the same nationality, which makes them succumbed to Taiwanese companies and those also located in China. Japanese companies also resist to change, especially in postal services (Erickson and Fleisher III 2014).

Interestingly, the Galapagos syndrome also occurred during this period of the fall in Japanese innovation. The Galapagos syndrome, emerged in the late 1990s and is a process, service, or even a product that emphasises a single culture/market and thus fails to be compatible with other nations (Butterman 2021), it is the company headquarters that always who make the key business decisions (Bowen and Leinbach 2003). One example is Japanese mobile phones, which were so unique to each customer group (that is sometimes the markets were too small/niche), due to the different company designs that were then incompatible when used abroad (Deguchi 2016). One clear example is that some Japanese people still use flip phones. This special type of Japanese mobile phones thus cannot be sold abroad. Hence, it is, no wonder that the Japanese mobile phone business is somehow lacking behind that in China and South Korea.

7.2.4 The Renaissance and the Future of Innovation in Japan (Post-COVID)

One key component of the renaissance and the future of Japanese innovation is the public sector and the collaboration of academia (Arita et al 2006; Yokura et al. 2013). The contemporary support of research and development consortia in both small and large enterprises in Japan and that public financial support created an innovation and knowledge spillover in the Japanese region (Nishimura and Okamuro 2016). That technological transfer was beneficial for more business growth (Ford et al. 2008). Another positive component has been the support given to open innovation, open culture, and work-life balance (Tubiera 2019) as well as the support of start-up establishments in the Kyoto area (Ibata-Arens 2008). This kind of localisation will help distribute more small and medium enterprises and let them flourish in the region (Ceptureanu et al. 2017). Japan also has taken advantage of modern Keiretsu linkages by establishing foreign-invested enterprises (FIEs) abroad (Xu and Yeh 2013). As a result of these Keiretsu networks, the following research questions can be asked: 1) What are the key drivers for Japanese innovation in the past and the present, especially during the pre- and post-COVID? and 2) What is Japan's current organisational strategy for greater innovation, especially when compared to its competing country, like China?

7.3 Methodology

This research employs a systematic review to explore the concept of Keiretsu in detail and how it affects Japan's innovation efforts in the past, the present, and the plausible future. Systematic review is a way to summarise key aspects during research to answer the proposed research questions (Clarke 1996). This research method has been popularised nowadays and used in both the social science and science fields (Putthiwanit 2015). Since the current research wants to understand Keiretsu-driven innovation in Japanese global mega enterprises omni-dimensionally, a multilayered and systematic review is highly suggested (Kwon et al. 2019).

The initial search on this topic was based on a full-text journal, rather than a full textbook, where it would be difficult to filter the related information effectively. After an initial search, ten well-known and major databases were found to be available for a successful journal search (*Academic Search Ultimate, Business Source Ultimate, EconLit™ with Full Text, Emerald Management, Sage Journals, ScienceDirect eJournals, SCOPUS, SpringerLink (E-journals), Taylor & Francis Journals, and the Wiley Online Library*). The search term used was 'Keiretsu' (since it was the only keyword used, Boolean operator did not apply here), based on the availability of the full text academic journals in English. The researched publication period ranged from year 1983–2022. The initial identified papers totaled 423 papers (excluding duplicated papers). The quality of papers was also rechecked on the publisher and the journal esteemed names. This screening process responded to Research Question 1 and the inclusion process responded to Research Question 2. Figure 7.1 shows the systematic reviews process, according to recommendation by Kurek et al. (2023), for this research. In terms of data analysis, the author extracted keywords (provided by authors) from each selected paper to perform that analysis. Later, KH-Coder, a program for qualitative analysis, was used to find meaningful results for Keiretsu and Innovation in Japan and its counterpart country, China.

7.4 Findings

The systematic reviews of the selected articles offered key findings. Summaries findings produced the following findings of Japanese innovation prior to COVID-19 (1) Based on an article by Lincoln et al. (1996) in *American Sociological Review*, Keiretsu companies tend to perform less profitably than do those that are non-Keiretsu companies; (2) Centred on a research by Flath (1993) offered in *The Review of Economics and Statistics*, major Keiretsu enterprises using their strategy for business growth based on the number of shareholder and debtholder; (3) As noted in a paper by Mutoh et al. (2020) in *Plos One*, Keiretsu corporations develop an organisational strategy for innovation by expanding their operations abroad, especially when the markets in Japan has started to plummet; (4) Based on a article by Tanaka (2005) in *Harvard Asia Quarterly*, Keiretsu multinational corporations (MNCs) have more power than do their competing Keiretsu companies; and (5) Applying the work by Cheng and Kalleberg (1996) in *Social Forces*, Japanese businessmen in bigger Keiretsu firms incline to change their jobs often more than Japanese businessmen in smaller Keiretsu firms. The years and number of publications reviewed for these findings (32 articles) are also exhibited in Fig. 7.2. The names of selected journals to see the covered area are also exhibited in Table 7.1.

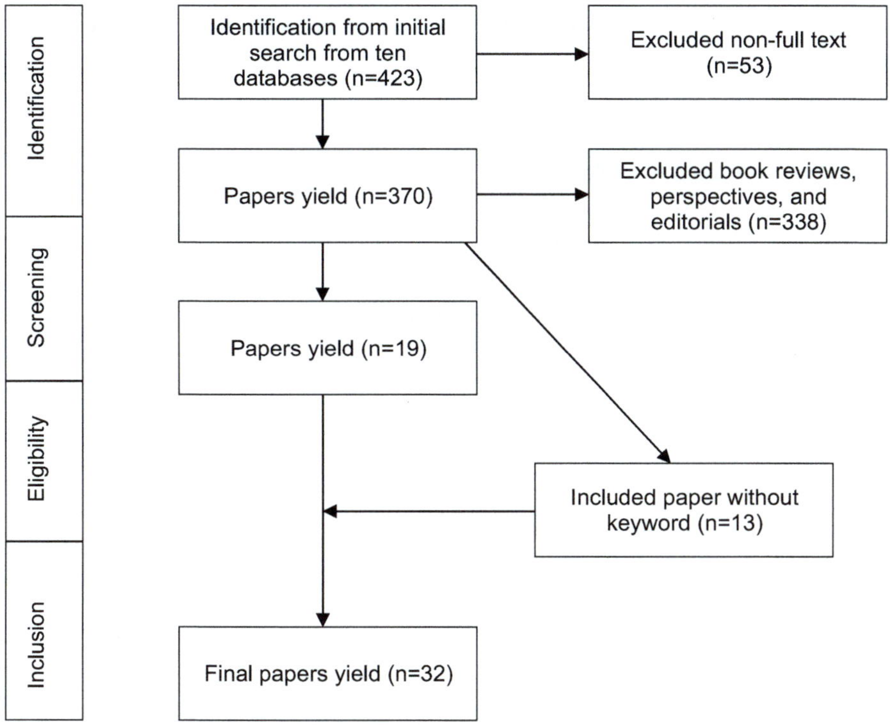

Fig. 7.1 Systematic review process using Keiretsu as a keyword

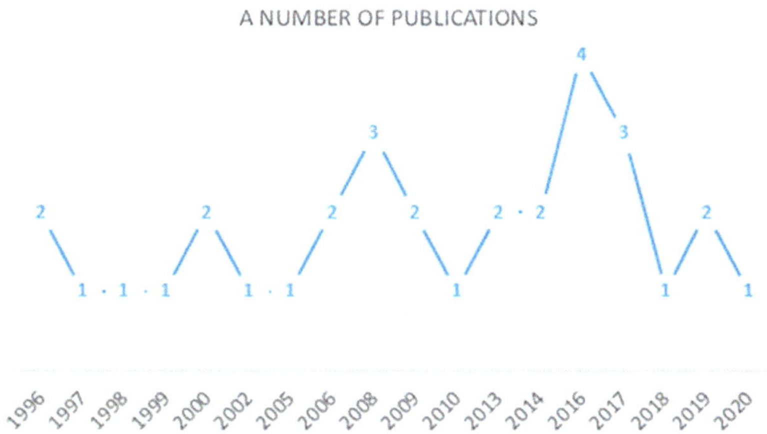

Fig. 7.2 The selected publications and years used for the systematic review

Table 7.1 The name of selected journals for the systematic review of research about Keiretsu

Journal name	Frequencies
American Sociological Review	1
Area	1
Asian Journal of Social Psychology	2
Current Issues in Economics and Finance	1
Entropy	1
Harvard Asia Quarterly	1
International Journal of Contemporary Management	1
International Journal of Japanese Sociology	1
Japan Studies Association Journal	1
Journal of Asia Pacific Studies	1
Journal of East Asian Studies	1
Journal of Information Technology	1
Journal of Open Innovation: Technology, Market, and Complexity	1
Journal of Regional Science	2
Judgment and Decision Making	1
Michigan Law Review	1
Papers in Regional Science	3
Plos One	1
Prometheus	1
Review of Regional Studies	1
RURDS	1
Science and Public Policy	1
Scientific Reports	1
Social Forces	1
Sociological Forum	1
Tijdschrift voor Economische en Sociale Geografie	1
Transportation Journal	1
Universia Business Review	1

Moreover, to discover meaningful context from these systematic reviews, the author also adopted various approaches in KH-Coder such as Correspondence Analysis (CA), Co-occurrence Network of Codes (CNC), and Cluster Analysis (Hierarchical). Figure 7.3 shows the most frequent keywords (top 10) in the Japanese Keiretsu context with Japan and FDI being the most frequent. Further, according to Fig. 7.4, CA is also used to see the relationships between the keywords. Unsurprisingly, given her well-developed organisational strategy in innovation, this review shows that Japan copes well with development, technology, research, and even the environment (when compared to her counterpart—China), for which environment can actually affect technology (Chen 2019). Notedly, the rise of China for FDI since 1978 (Cheng 2006) also has pressures Japan to become more innovative. It is also interesting to explore more about Chaebol concept in South Korea and see how it affects to Japanese innovation.

Moreover, CNC (see Fig. 7.5) is analysed to see the words that most occurred and their network that resulted from the systematic review. These results show that there are robust interwoven relationships between Japan, Keiretsu, and innovation. In contrast, there are strong interwoven relationships between China, FDI, (location)

1	Japan	ProperNoun	5
2	FDI	ProperNoun	4
3	China	ProperNoun	3
4	choice	Noun	3
5	location	Noun	3
6	research	Noun	3
7	development	Noun	2
8	dividend	Noun	2
└	dividends		2
9	economy	Noun	2
└	economies		2
10	employee	Noun	2

Fig. 7.3 The most frequent keywords gathered from the systematic reviews of Keiretsu

Fig. 7.4 Correspondence analysis of the selected papers on Keiretsu topic

choice and location. We can thus plausibly infer that China is an attractive location for FDI, while Japan is attractive in terms of innovation. Finally, a hierarchical cluster analysis (see Fig. 7.6) was also performed to see the relationships and the categories for the grouped words. These results yield similarly those in Fig. 7.4 but show that Keiretsu somehow has had a stronger impact on Japan than innovation. Indeed, Japan has gained her innovation competitiveness from the technology transfer that has occurred between institutions (Yokura et al. 2013). With the differences in religious and cultural backgrounds found in the industrialised countries in Northern Europe (Krockow et al. 2018), based on these results, the research and the development offer strong evidence that supports Japan as a positive country for future innovation, driven by Keiretsu. No matter the obstacles or caveats (e.g. COVID-19) that Japan has to encounter—though, she has 27% of population infected of COVID-19, when compared to 7% of Chinese population infected (Mathieu et al. 2023)—she still can survive and flourish by using her own innovation style, which is corresponding to United Nations' Sustainable Development Goal (SDG) 9 (industry innovation and infrastructure) (United Nations 2023).

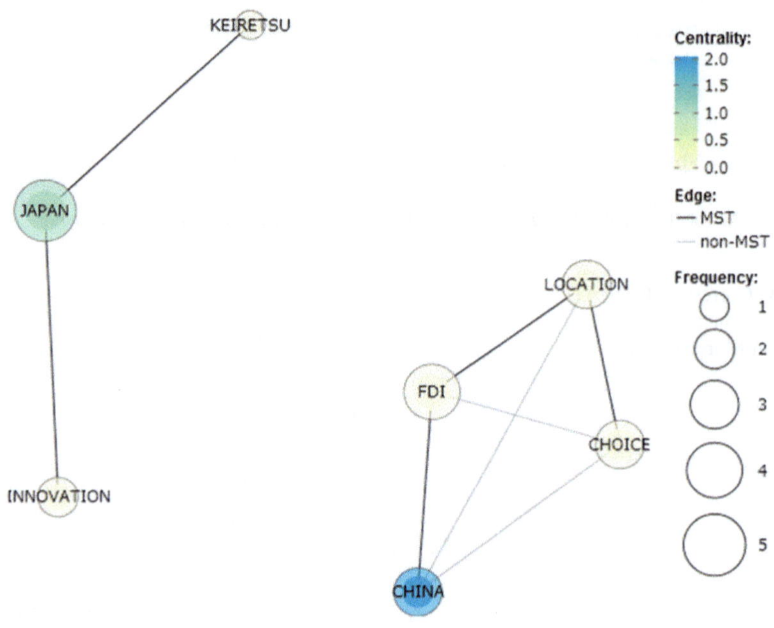

Fig. 7.5 The Co-occurrence network of codes (CNC) based on a systematic review of Keiretsu

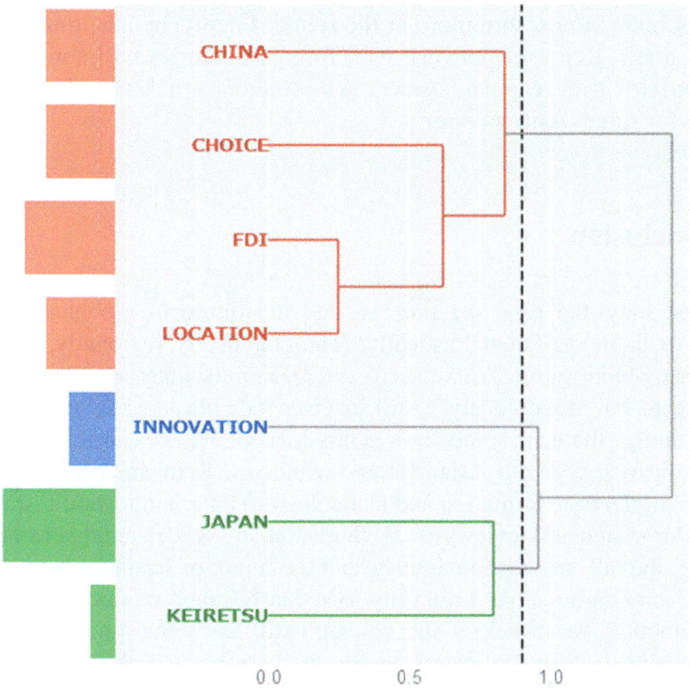

Fig. 7.6 Hierarchical cluster analysis (HCA) on the selected research about Keiretsu

7.5 Future Directions

This current chapter of the book provides the interplay of regional cooperation, digital transformation, and of course innovation, as the interplay of future of economy in Asia. We cannot deny the facts that the advancement of technology in Japan tends to enhance not only her innovation, but also her leadership reinforcement in digital transformation; while maintaining Keiretsu concept. In order for Japan to reach her innovation paragon, it is highly recommended to initiate collaboration with academia and open innovation, this will reinforce Japan's research and development sector. Therefore, these acts of Japan will align her with other Asian countries' economy, which emphasises the enhancement of innovation and digital transformation. Nevertheless, such enhancement and advancement also draws her counterpart, China, to compete against her in being innovative.

In addition, Japan should concentrate on additional plausible regional integration. This is a topic we can discuss more in the future, when talking about economic future of Asian continent. In addition, instead of competing against each other, it is better to foster and nurture each other vis-à-vis economic bloc with ASEAN or even South Korea on innovative digital economy framework. This is a challenging quest for scholars to explore more in the future. Finally, by adapting modified-to-cultural-context 'Keiretsu' network with foreign countries will somehow deepen the relationships with regional business and economics in Asia and consolidate the prosperity for future Asia economy.

7.6 Conclusion

Come what may, the past, the present, and the future of globalisation has and continued to challenge Japan drastically, demographically, regionally, and economically, in term of innovation (Yokura et al. 2013). Keiretsu may seem to be outdated in some respects, but miraculously, it still survives today in Japan (even after COVID-19). Even though there are some flaws in this concept, but we can at least take a good thing from it to apply with Asian future businesses. Keiretsu concept can also go along with digital transformation and technology to enrich innovation strategy post-COVID. Our systematic review reveals that are many hidden amalgams responsible for the rise, the fall, and the renaissance and the future of Japanese innovation. This lesson learned can be applied not only to Asian business practice, but also global business practice. Nevertheless, this research still has some limitations: First, any systematic review is subjective by nature, as the topic and theme can be deviant from any shallow analysis. Secondly, it is very difficult to review Keiretsu systematically in various languages, since it is difficult to code data in KH-Coder. Thirdly, any systematic review is less generalised, when compared to a meta-analysis. For future research, the author highly recommends applying meta-analysis method, to gain even more meaningful, profound, and precise results. Other business networks like the Chinese Guanxi (e.g. Jones (2017) suggests that the domestic economy has a robust effect on FDI) and the Korean Chaebol is also suggested for future studies and applying the qualitative/quantitative methodology in the same fashion.

Highlights

- Keiretsu concept can be applied to other Asian network's conduct of business.
- Sometimes, unforeseen obstacles, no matter what size, cannot always block the advancement/enhancement of innovation, especially in the case of the future Asian economy.

References

Arita T, Fujita M, Kameyama Y (2006) Effects of regional cooperation among small and medium-sized firms on their growth in Japanese industrial clusters. RURDS 18:209–228

Balderbos R, Carree M (2002) The location of Japanese investments in China: agglomeration effects, Keiretsu, and firm heterogeneity. J Jap Int Econ 16:194–319

Bell M, Perera S, Piraveenan M et al (2017) Network growth models: a behavioral basis for attachment proportional to fitness. Sci Rep 7(42431):1–11

Bowen T Jr, Leinbach TR (2003) Air cargo services in Asian industrializing economies: electronics manufacturers and the strategic use of advanced producer services. Pap Reg Sci 82:309–332

Butterman E (9 Jun 2021) Galapagos syndrome: unique innovator or progress prohibitor? Mizuho Americas. https://www.mizuhogroup.com/americas/insights/2021/06/galapagos-syndrome-when-isolation-breeds-deviation.html/

Cao C, Simon DE, Suttmeier RP (2009) China's innovation challenge. Inno Mgt Policy Prac 11(2):253–259

Ceptureanu SI, Ceptureanu EG, Marin I (2017) Assessing the role of strategic choice on organizational performance by Jacquemin-Berry Entropy Index. Entropy 19:1–12

Chang KI, Hayakawa K, Matsuura T (2014) Location choice of multinational enterprises in China: comparison between Japan and Taiwan. Pap Reg Sci 93(3):521–538

Carney RW, Zheng LY (2009) Institutional (dis)incentives to innovate: an explanation for Singapore's innovation gap. J East as Stud 9:291–319

Chen X (2019) Antecedents of technological diversification: a resource dependence logic. J Op Inno Tech Mkt Comp 5(4):1–18

Cheng MM, Kalleberg AL (1996) Labor market structures in Japan: an analysis of organizational and occupational mobility patterns. Soc for 74(4):1235–1260

Cheng S (2006) The role of labor cost in the location choices of Japanese investors in China. Pap Reg Sci 85(1):121–138

Cheng S (2008) Location decision variations of Japanese investors in China. Rev Reg Stud 38(3):395–415

Clarke CL (1996) Keiretsu networks and corporate performance in Japan. Am Soc Rev 61(1):67–88

Deguchi M (2016) Globalization, glocalization, and Galapagos syndrome: public interest corporations in Japan. Int J N-f-pro Law 18(1):249–257

del Sol P (2010) Chilean regional strategies in response to economic liberalization. Uni Bus Rev 25:112–130

Dirks D, Hemmert M, Legewie J et al (2000) The Japanese employment system in transition. Int Bus Rev 9(5):525–553

Erickson EL, Fleisher AA III (2014) The price sensitivity of letter demand in Japan. Jap Stud Asso J 12:83–102

Flath D (1993) Shareholding in the Keiretsu, Japan's financial groups. Rev Econ Stat 75(2):5–14

Ford TC, Rork JC, Elmslie BT (2008) Considering the source: does the country of origin of FDI matter to economic growth? J Reg Sci 48(2):329–357

Fu X, Mu R (2014) Enhancing China's innovation performance: the policy choices. Chi Wo Econ 22:42–60

Gul FA (1999) Growth opportunities, capital structure and dividend policies in Japan. J of Corp Fin 5:141–168

Ibata-Arens K (2008) The Kyoto model of innovation and entrepreneurship: regional innovation systems and cluster culture. Prometheus 26(1):89–109

Jones J (2017) Agglomeration economies and the location of foreign direct investment: a meta-analysis. J of Reg Sci 57:731–757

Ju X, Jiang S, Zhao Q (2023) Innovation effects of academic executives: Evidence from China. Res Pol, 52(3). https://doi.org/10.1016/j.respol.2022.104711

Jung J, Mun E (2016) Bending but not breaking?: Foreign investor pressure and dividend payouts by Japanese firms. Soc for 31(3):663–684

Krockow EM, Takezawa M, Pulford BD et al (2018) Commitment-enhancing tools in Centipede games: evidencing European-Japanese difference in trust and cooperation. Judg Dec Mak 13(1):61–72

Kurek J, Brandli LL, Frandoloso MAL et al (2023) Sustainable business models innovation and design thinking: a bibliometric analysis and systematic review of literature. Sustainability 15(988):1–22

Kwon J, Pohlmann M, Schöttli J (2019) Transnational corporate elites in Japan: international career mobility in East and South Asia. Int J Jap Soc 28:132–147

Lan F, Liu T, Li M (2022) Research on the influence of constructing national innovation demonstration zone on urban innovation level in China. Emer Mkt Fin T 58(14):4162–4171

Lauer J, Liefner I (2019) State-led innovation at the city level: policy measures to promote new energy vehicles in Shenzhen. China. Geo Rev 109(3):436–456

Li P (2020) The outlook for China's innovation-driven development beyond COVID-19. Cult of Sci 3(4):232–244

Lincoln JR, Gerlach M, Ahmadjian CL (1996) Keiretsu networks and corporate performance in Japan. Am Soc Rev 61(1):67–88

Mathieu E, Ritchie H, Rodés-Guirao L et al (2023) Coronavirus pandemic (COVID-19). https://ourworldindata.org/coronavirus/

Miwa Y, Ramseyer JM (2000) Rethinking relationship-specific investments: subcontracting in the Japanese automobile industry. Mich Law Rev 16:2636–2667

Morash EA, Clinton SR (1997) The role of transportation capabilities in international supply chain management. Am Soc Tran Log 36(3):5–17

Mutoh T, Kotani K, Kakinaka M (2020) An oversea business paradox: are Japanese general contractors risk takers? PLoS ONE 15(9):1–26

Nakayama M (2000) E-commerce and firm bargaining power shift in grocery marketing channels: a case of wholesalers' structured document exchanges. J Info Tech 15:195–210

Nishimura J, Okamuro H (2016) Knowledge and rent spillovers through government-sponsored R&D consortia. Sci Pub Pol 43(2):207–225

Packer F (1999) Credit risk in Japan's corporate bond market. Cur is Econ Fin 5(15):1–5

Putthiwanit C (2015) Exploring the impact of organizational culture on employees in multinational enterprises: a qualitative approach. Procedia – Soc Behav Sci, 207:483–491

Simon DF (2007) Whither foreign R&D in China: some concluding thoughts on Chinese innovation. As Pac Bus Rev 13(3):471–480

Saito T, Vuszkun B (2010) What is the 'lifetime employment'? Empirical research from Japan. J as Pac Stud 1(3):591–614

Sołoducho-Pelc L (2017) The importance of trust in the implementation of the strategic management process. Int J Con Mgt 16(4):237–261

Tanaka I (2005) The importance of trust in the implementation of the strategic management process. Harv as Quar 9(3):4–13

Tubiera A (20 Jun 2019) Japan in the face of breaking into new working culture. https://worldfinancialreview.com/japan-in-the-face-of-breaking-into-new-working-culture/

United Nations (2023) The 17 Goals. https://sdgs.un.org/goals/

Xie W, Li-Hua R (2009) What will make China an innovation-oriented country? J Know Inno Chi 1(1):8–15

Xu R, Shan Y, Liu M et al (2023) Can government subsidies improve innovation performance? Evidence from Chinese listed companies. Econ Mod, 120. https://doi.org/10.1016/j.econmod.2022.106151

Xu Z, Yeh A (2013) Origin effects, spatial dynamics and redistribution of FDI in Guangdong. China. Tijd Econ Soc Geo 104(4):439–455

Yamagishi T, Jin N, Miller AS (1998) In-group bias and culture of collectivism. As J Soc Psy 1:315–328

Yokura Y, Matsubara H, Sternberg R (2013) R&D networks and regional innovation: a social network analysis of joint research projects in Japan. Area 45(4):493–503

Yu Y, Yu J, Pan X et al (2017) Governance and the China innovation economy. As Pac J Reg Sci 1:63–84
Yusuf SM, Othman R (2016) Leadership for creativity and innovation: is Japan unique? J Ad Mgt Sci 4(2):176–180

Chutinon Putthiwanit is an Assistant Professor at Kasetsart University, Sakon Nakhon, Thailand. Putthiwanit is a Ph.D. Candidate in Management Science from Turiba University, Latvia; an MBA in International Business from Providence University, Republic of China; and a BS in Applied Statistics from Assumption University, Thailand. He taught partial courses at Providence University, Republic of China; Turiba University, Latvia; and Mykolas Romeris University, Lithuania. He has published research papers in refereed journals and conferences (one paper in the Asian World Trade Organization Network Conference). His case study research was published by the National Institute of Development Administration.

Chapter 8
Future Trends in Digital Banking in Asia: The Role of Reputation, Risk, and Financial Literacy in Malaysia

Cluster of Future Studies: Past and Futures

Kelvin Yong Ming Lee, Pick Soon Ling, Mohamad Jais, Woon Leong Lin, and Irfah Najihah Basir Malan

Abstract On 29 April 2022, the Central Bank of Malaysia (Bank Negara Malaysia) announced the list of five successful digital bank applicants. The approval of these digital bank licenses is also one of the Malaysian government's efforts in transforming Malaysia to cashless society. By extending the technology acceptance model (TAM) to include additional determinants, such as firm reputation and digital financial literacy, this study aimed to fill this gap and provide a deeper understanding of user acceptance in the digital banking context. A total of 238 valid responses were collected and analysed by using partial least square structural equation modelling (PLS-SEM). The results confirmed that perceived ease of use, perceived usefulness, firm reputation, perceived risk, and digital financial literacy positively related to the potential users' intention to use digital bank. The outcome of this study also serves as the reference for policymakers and financial institutions to develop targeted strategies to facilitate the usage of digital banks. As such, Malaysia can pave the way for a

K. Y. M. Lee (✉)
School of Accounting and Finance, Taylor's University, Selangor, Malaysia
e-mail: kelvinyongming.lee@taylors.edu.my

P. S. Ling
School of Business and Management, University of Technology Sarawak, Sarawak, Malaysia
e-mail: ling.pick.soon@uts.edu.my

M. Jais
Faculty of Economics and Business, Universiti Malaysia Sarawak, Sarawak, Malaysia
e-mail: jmohamad@unimas.my

W. L. Lin
School of Management and Marketing, Taylor's University, Selangor, Malaysia
e-mail: woonleong.lin@taylors.edu.my

I. N. B. Malan
Faculty of Business and Management, Universiti Teknologi MARA, Malaka Campus, Malaysia
e-mail: irfah@uitm.edu.my

E. Lau et al. (eds.), *Future of Economy in Asia*,
https://doi.org/10.1007/978-3-031-95500-6_8

more inclusive, resilient, and digitally driven economy that benefits all the segments of society.

Keyword Digital bank · Firm reputation · Perceived risk · Digital financial literacy

8.1 Introduction

In recent years, rapid technological advancements have brought revolution in various industries, including the banking industry. The growth of the Internet has coincided with the widespread invention and use of various devices by businesses, including laptops, tablets, and smartphones (Dasgupta and Gupta 2019). Technological advancement also transformed the ways people access information and digital services (Goyal et al. 2022). Malaysia is no exception to this trend, as the emergence of fintech has contributed to expanding improvements in technology capacity within its banking sector (Kamis et al. 2022). The current banking activities, such as online banking and self-service machines have changed user's perception regarding the need for digital banking (Windasari et al. 2022). In Malaysia, the introduction of digital banks has caught significant attention since the authorisation of five digital banking licenses in April 2022. As of May 2024, there were only three digital banks commenced their operation, which include; (i) GX Bank; (ii) Boost Bank Berhad; and (iii) AEON Bank (M) Berhad (see Table 8.1).

Digital banking refers to financial services conducted through mobile or digital platforms, enabling users to execute transactions, engage in trading, and access transaction history records (Baptista and Oliveira 2015). Digital banking is different from the other technology-enabled banking services, as it is branchless and offering a wider range of services (Sha and Mohammad 2017). The availability of internet connectivity and smartphone usage also facilitates the usage of digital banking in a particular country (Ramayanti et al. 2024). The high internet penetration rate in Malaysia provides a favourable environment for the usage of digital banking. According to

Table 8.1 List of digital banks and their shareholders in Malaysia (as of January 2024)

Name of the digital bank	Shareholders
GX Bank	Grab Holdings Limited, Singapore Telecommunications Ltd., and Kuok Group
Boost Bank Berhad	Axiata Group Berhad, Boost Holdings Sdn Bhd, and RHB Bank Berhad
AEON Bank (M) Berhad	AEON Financial Service Co. Ltd., AEON Credit Service (M) Berhad and MoneyLion Inc.

Sources Skrine Advocates and Solicitors: https://www.skrine.com/insights/alerts/january-2024/digital-banks-commence-operations-in-malaysia

Datareportal's (2024) official website, there were currently 33.59 million internet users, and 44.55 million cellular mobile connections were active in Malaysia.

Generally, factors such lower costs, convenience, and effective banks operation also contributed to the growth of digital banking (Shin et al. 2019). Notably, digital banking operates under the assumption that potential users for digital banking must own the specific digital devices and internet connectivity, together with the necessary digital literacy (Khan et al. 2019). Indirectly, this helps the potential users to ease the transaction processes and efficiently manage financial resources. Aside from the evident advantages of digital banking, people are still underusing the services. The potential users are also looking for technical assistance that could help them perform better (Khan 2022).

To remain competitive in the market, banks must embrace digitisation as customers seek a more tailored, quick and easy banking experience. Innovations in digital banking have given rise to new services and products like robo-advisory, online payments, and mobile banking. While technology innovation offers countless benefits, it also may bring certain risks, such as cybersecurity threats and privacy concerns. On 1 June 2023, Bank Negara Malaysia (BMN) or the Central Bank of Malaysia released *Risk Management in Technology (RMiT) Policy Document* and stated the Bank's requirements for the management of technology risk. BNM also stated that all the Banks must implement strong risk management controls beyond the minimal regulatory requirements to provide secure financial services and prevent the exploitation of weak links in interconnected networks and systems.

Meanwhile, technology also contributes to the well-being of society (Brey 2018). For instance, the smartphone provides the users with opportunities to open bank accounts without physically visiting a bank branch. The emergence of financial technology (fintech) has made digital financial literacy crucial for both existing and potential users (Goldstein et al. 2019). There is also a growing need for educators to equip themselves with sufficient digital and financial literacy to address the needs of 21st-century learners (Golden and Cordie 2022). Besides that, there is also a need for the Malaysian government to educate the Malaysians about the importance of digital financial literacy. The enhancement of digital financial literacy is not only helping the Malaysians to utilise digital banking effectively, but also preventing the misuse of services offered by digital banks.

In the context of Malaysia, there is a notable gap in the existing literature concerning the factors affecting the usage of digital banks. To fill this gap, this study modified the technology acceptance model (TAM) by including digital financial literacy, security risk and firm reputation as the factors affecting potential users' intention to use digital banking. Without a physical branch, a digital bank's reputation may significantly influence potential user's intention to use it. As digital banks have only recently commerce their operations, security risk is one of the major concerns. Somehow become one of the main concerns. Additionally, a sufficient level of digital financial literacy can help users manage their money properly by using digital banks. Hence, these modifications of TAM model are to ensure that the model accurately captures the specific phenomena and challenges associated with the usage of digital banking in Malaysia, providing a comprehensive understanding of users' intentions to use digital banks. Besides that, understanding these factors is vital for the success and sustainability of these newly established digital banks. By identifying the key determinants that shape potential users' intentions to use digital banking services, the banks can tailor their strategies and offers to meet customer expectations and preferences effectively. Thus, this study aims to examine how perceived ease of use, perceived usefulness, firm reputation, perceived risk, and digital financial literacy affect the acceptance of digital banking.

8.2 Literature Review

8.2.1 Technology Acceptance Model (TAM)

The technology acceptance model (TAM) model is widely used theoretical framework in various fields. Davis (1989) proposed the TAM model to explain the factors for the acceptance of a technology product. This model is basically based on the idea that perceived usefulness and perceived ease of use are key determinants of an user's intention to use a particular technology product. However, these two factors are insufficient in explaining the intention to use a particular technological product. Numerous studies have modified the TAM model to address its limitations and enhance its applicability in different contexts. For instance, Bastari et al. (2020) incorporated intrinsic motivation as a predictor in the TAM model and tested its impact on the intention to use the Bank's application. These modifications also acknowledge the complex nature of technology adoption decisions and aim to provide a comprehensive understanding of potential users' behaviours. The TAM model remains a useful framework for researchers to understand potential users' intention to use a particular technological product.

This study extends the TAM model by incorporating digital financial literacy, firm reputation and perceived risk as the potential factors that affect the potential users' intention to use digital banks. Firstly, the inclusion of digital financial literacy (DFL) acknowledges that financial literacy is essential for understanding and leveraging the functionalities of digital banking, reflecting the need for user education and competency in digital financial environments. Secondly, FR addresses the importance of organisational credibility and trust in the financial sector, reflecting how users' perceptions of a bank's reputation can significantly impact their adoption decisions. Lastly, the inclusion of PR in the research framework provides a more balanced view of technology adoption, highlighting that users' concerns about risks need to be addressed to foster trust and encourage the use of digital banking services. This extended model captures the multifaceted nature of digital banking adoption, offering valuable insights for enhancing user acceptance and addressing barriers in this evolving landscape. Figure 8.1 shows the conceptual framework of this study.

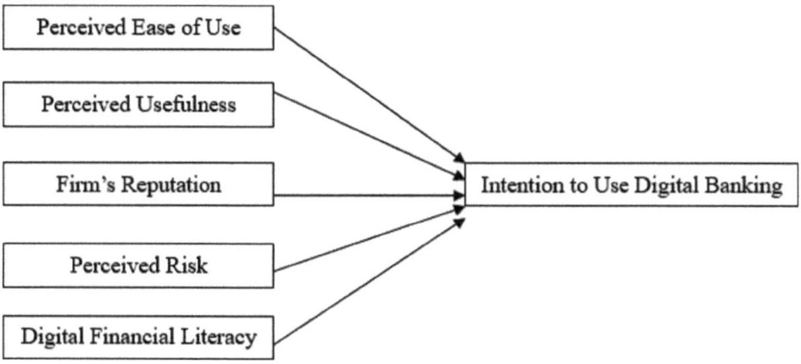

Fig. 8.1 Conceptual framework showing potential factors influencing the intention to use digital banking in Malaysia. *Sources* Author's Computation

8.2.2 Perceived Ease of Use and Perceived Usefulness

Perceived ease of use refers to the extent to which an individual believes that a specific technology product would be free of effort (Davis 1989). Usually, the potential user might be concerned about the difficulty to use a new specific technology product (Malaquias and Hwang 2019). Amongst various technology products available in the market, an individual might prefer to use a less complicated technology product. The perceived ease of use is positively related to the potential users' intention to use technology products (Giovanis et al. 2019; Kar 2020). For instance, Nguyen et al. (2022) found that perceived ease of use, along with other e-service quality dimensions, such as responsiveness and interface quality, positively impacts the attitude towards using video teller machine (VTM) services, and the intention to use them.

Besides that, perceived usefulness refers to the extent to which an individual believes that a specific technological product can enhance their performance (Davis 1989). Previous studies found that perceived usefulness tends to an important factor influencing user's intention to use various fintech products, such as E-wallet (To and Trinh 2021) and P2P lending platform (Hutapea and Andista 2021). This is due to the fact the potential users perceived that technology products tend to improve their performance and career advancement (Alalwan et al. 2018). For instance, Huang and Chueh (2022) revealed that perceived usefulness positively affects usage attitude, which in turn influences the intention to use digital assessment systems. Furthermore, research on online banking in the Indian context highlights that perceived usefulness and perceived ease of use are positively related to consumers' attitude towards using online banking, which subsequently influences adoption intention and customers' satisfaction (Ghani et al. 2022). These findings collectively emphasise the significance of perceived ease of use and perceived usefulness in shaping users' intentions to engage with digital banking services.

In the context of digital banking, perceived ease of use refers to how effortless users find the digital banking platform. When users feel that the platform is easy to navigate and operate, they are more likely to adopt it (Tiong 2020). A straightforward and intuitive interface reduces the effort required to perform banking tasks, making users more comfortable and confident in using the service (Joshi et al. 2019). Recent studies conducted showed that perceived ease of use and perceived usefulness positively affect the intention to use digital banks (Rathnaweera and Karunasena 2020). Moreover, the combined effect of perceived ease of use and perceived usefulness creates a compelling case for the adoption of digital banking. When a digital banking platform is both easy to use and highly functional, users are more likely to have a positive experience, leading to higher adoption rates (Bastari et al. 2020). The digital bank also comes in the form of a mobile application. Therefore, an app that is user-friendly and provides valuable features, such as secure transactions and personalised financial advice is likely to attract more users (Sun et al. 2019). Thus, this study hypothesises that perceived ease of use and perceived usefulness positively related to the usage of digital banking.

Hypothesis 1: Perceived ease of use is positively associated with the intention to use digital banking.

Hypothesis 2: Perceived usefulness is positively associated with the intention to use digital banking.

8.2.3 Firm's Reputation

Reputation is treated as a strategic intangible and invisible asset for banks. Effective reputation management also tends to bring significant benefits, such as improved financial performance, easier access to funding, and creating value for the banks (Li 2010). The recent global financial crisis in 2008 also taught a valuable lesson to stakeholders about the importance of a good bank reputation (McGrath and Walker 2023). In the context of digital banking, a firm's reputation plays a crucial role in influencing customers' intention to use their services (Rathnaweera and Karunasena 2020). Customers who perceive a bank to have a positive reputation are more likely to trust and feel confident in using their services (Tangmanee and Sritadawut 2021), and believe that their transactions are secured (Ahmed et al. 2023). Furthermore, a bank with a strong reputation is more likely to be seen as credible and trustworthy, which can positively influence customers' perception of the quality and reliability of their digital banking services.

Additionally, the reputation of a firm emerges as a critical factor that may significantly influence customer's behaviour and firm's performance (Yi 2023). A good firm reputation is also an essential component for them to gain the consumer's trust and reliability in their financial dealings. As digital banking continues to evolve, maintaining a strong reputation becomes increasingly important in retaining customer loyalty and gaining a competitive edge (Aguayo and Slusarczyk 2020). This is especially relevant as customers have access to numerous alternatives and can easily switch to other service providers if their expectations are not met. A recent study by

Windasari et al. (2022) revealed that Gen X and Gen Y perceived that firm reputation is a significant factor in affecting their intention to use digital banking. Similarly, Nicholas et al. (2023) revealed that reputation tends to affect the potential user's attitude towards the usage of digital banking and the intention to use it. However, the age range for the respondents included in the study of Nicholas et al. (2023) is too wide. Thus, this study hypothesises that firm reputation is positively associated with the intention to use digital banking.

Hypothesis 3: Firm's reputation is positively associated with the intention to use digital banking.

8.2.4 Perceived Risk

The theory of perceived risk (PR) explains that PR can be understood in two ways: (i) as the likelihood of experiencing a loss, and (ii) as the subjective sense of unfavourable consequences (Mitchell 1992). PR can also be defined as the perception of adverse impacts or consequences arising from the utilisation of a particular service (Featherman and Pavlou 2003). For instance, Singh and Sinha (2020) found that perceived risk negatively related to the user's willingness to use mobile wallet. Moreover, Hassan et al. (2023) revealed that perceived risk significantly affects the behavioural intention to use fintech services. Like the other fintech services, digital banking is also presumed to be affected by the perceived risk.

Users might hesitate to adopt digital banking services if they perceive high risks associated with them, such as potential financial loss, security breaches, or privacy concerns (Aguayo and Slusarczyk 2020). Digital banks need to address these concerns by implementing robust security measures, offering guarantees, and educating users about the safety of their platforms (Musyaffi et al. 2024). Effective communication about risk management practices can help alleviate these fears and build trust amongst potential users. Additionally, understanding the specific types of risks that concern different user segments can enable banks to tailor their strategies more effectively, thus encouraging wider adoption of digital banking services (Sundararaj and Priya 2024). Thus, this study hypothesises the perceived risk is negatively associated with the intention to use digital banking.

Hypothesis 4: Perceived risk is negatively associated with the intention to use digital banking.

8.2.5 Digital Financial Literacy

Digital financial literacy (DFL) simply refers to a combination of both financial literacy and digital literacy. The person also needs to have basic financial knowledge and basic digital skills to be qualified as a person with digital financial literacy

(Lyons and Kass-Hanna 2021). A recent study conducted by Yang et al. (2023) found that people with higher financial literacy significantly affect their frequency in using digital finance, such as mobile payment and online borrowing services. However, individuals with high financial literacy and low digital literacy or basic digital skills may refuse to use the digital finance platform (Prete 2022). Individuals who are more digitally and financially literate were more likely to use digital tools to set their financial goals and perform budgeting (Gosal and Nainggolan 2023). Moreover, higher level of DFL enables individuals to make better financial decisions and prevent irresponsible behaviour associated with the usage of digital finance platforms (Andreou and Anyfantaki 2021).

When individuals are knowledgeable about digital financial services, including understanding concepts like online security, digital transactions, and mobile banking, they are more likely to feel confident and comfortable using digital banking platforms (Mahdzan et al. 2022). Additionally, those with higher digital financial literacy are better equipped to make informed decisions about which digital banking products and services best suit their needs. This, in turn, positively impacts their intention to use digital banking as part of their financial management (Kumar et al. 2023). Moreover, digital financial literacy can empower individuals to effectively navigate the complexities of digital banking, leading to a more seamless and efficient user experience. As such, it is evident that an individual's level of digital financial literacy plays a vital role in shaping their intention to use digital banking, indicating the importance of promoting and enhancing digital financial literacy to drive widespread adoption of digital banking services (Widyastuti et al. 2024). DFL is not only important for individuals' confidence and decision-making in using digital banking but also for their overall financial wellbeing (Nawang and Shukor 2023). Based on the reviews, this study hypothesises that DFL positively related to the usage of digital banking.

Hypothesis 5: Digital financial literacy is positively associated with the intention to use digital banking.

8.2.6 Summary

Digital banking is a relatively new concept; it is easier for customers to draw conclusions based on their direct experience with the product or service rather than trusting the reputation or structure of the institution itself. Perceived usefulness and ease of use are straightforward and immediate factors that customers may perceive. If the digital banking service is seen as easy to use and helpful in everyday life, it is more likely to be endorsed and used, regardless of the customer's knowledge of the firm providing the service. Firm reputation and perceived risks are also essential factors. If the firm is well known and has a good reputation, the perceived risk is low, and customers are more likely to trust and use the service. Increasing customers' digital financial literacy enhances their skills to operate digital banking services competently, and directly affects the customer engagement and satisfaction. By identifying such product or service-related elements, this study can identify what aspects specifically attract customers to use digital banks. The identified focus is meant to help create a plan to formulate more accurate measures that drive the change process and the experience creation process. This focus can further boost their use and the capacity to build more trust, thereby supporting the use of these kinds of services across wide populations and demographics.

8.3 Methodology

This section discusses the data and methodology of this study. This study carried out an online survey to collect primary data from the respondents by using Google form. Online sampling and data collection were chosen for several reasons. Firstly, it allows for a broad reach, enabling the collection of data from a diverse group of respondents efficiently. Secondly, it offers convenience for participants, as they can complete the survey at their own pace and from any location with internet access. This method also ensures cost-effectiveness and faster response times as compared to traditional paper-based surveys (Nayak and Narayan 2019). The survey consists of two different sections. The first section of this survey collects the basic information about the respondents. Meanwhile, the second section of this survey collects the respondents' feedback for the statements related to perceived ease of use, perceived usefulness, digital financial literacy, perceived risk, and intention to use digital banking. Table 8.2 lists the statements presented to the respondents. Specifically, the measurement items for each variable were assessed by using a 5-point Likert scale.

The survey design and distribution followed a structured protocol to ensure the reliability and validity of the data collected. Memon et al. (2020) mentioned that the minimum number for the respondents should be five times the independent variables. Since there are five independent variables included in this study, then the minimum number of 25 respondents is required. The sampling strategy used for this study was non-probability sampling, specifically convenience sampling, due to the nature of online surveys and the need for timely data collection. This method was chosen because it allowed for easy access to many respondents quickly and efficiently. This study also distributed the survey form through various social media platforms, such as Facebook, WhatsApp, and Instagram. As a result, a total of 238 valid responses were obtained from the potential users of digital bank over the period December 2023 to January 2024.

It is important to note that at the time this study was conducted, the digital bank had not yet started its operations. This context is crucial as it means that the survey targeted individuals who are potential users, rather than current users, of the digital bank. This study aimed to understand the factors influencing the intention to use digital banking services before they become available. By focusing on potential users, this study captures the anticipatory perceptions and attitudes that could influence the adoption of digital banking once it launches. This approach provides insights into the considerations and expectations of individuals who may adopt digital banking

Table 8.2 Statements presented to respondents to collect their level of agreement using a 5-point Likert scale

Variable	Items	References
Perceived Ease of Use (PE)	PE1: I think that digital bank transactions save me a lot of time and energy	To and Trinh (2021)
	PE2: I think that the use of digital banks is clear and understandable	
	PE3: I think that digital bank apps are easy to use	
Perceived Usefulness (PU)	PU1: I think that using digital bank helps me to accomplish my tasks more efficiently	To and Trinh (2021)
	PU2: I think that work will become easier after using digital bank	
	PU3: I think that digital bank improves the quality of my life	
Firm Reputation (FR)	FR1: The digital bank always fulfils its customer promises	Yi (2023)
	FR2: The owner of a digital bank has a good reputation	
	FR3: I feel positive about the owner of digital bank	
Perceived Risk (PR)	PR1: Using digital bank is associated with a high level of risk	Hassan et al. (2023)
	PR2: There is high potential for loss in using digital bank	
	PR3: There is considerable risk involved in using digital bank	
Digital Financial Literacy (DFL)	DFL1: I have a good understanding of the usage of any finance-related technological product	Author's computation
	DFL2: I have a good understanding of the concept of digital finance	
	DFL3: I have a good understanding of digital payments	
Intention to Use Digital Bank (INT)	INT1: I will use digital bank in the future	To and Trinh (2021)
	INT2: I will always try to use digital bank in my daily life	
	INT3: I plan to use digital bank more frequently	

services in the future, offering valuable information for digital banks to tailor their strategies and address potential barriers to adoption.

8.3.1 Method of Analysis

This study also assesses the reliability and validity of constructs by using the Cronbach's alpha, composite reliability (CR), and average variance extracted (AVE). The Cronbach's alpha that is greater than 0.8 indicates high internal consistency reliability for a set of items (Khairul et al. 2018). On the other hand, Hair et al. (2009) also suggested that AVE and CR should be greater than 0.5 and 0.7, respectively. This study also assesses the discriminant validity by using heterotrait-monotrait (HTMT). Henseler et al. (2015) mentioned that HTMT ratio should be below the threshold of 0.8. Then, this study used the partial least squares structural equation modelling (PLS-SEM) to estimate the path coefficient.

8.4 Findings

This section presents the results of this study. Table 8.3 shows the demographic information of the respondents. Notably, the respondents of this study consisted of the potential users, but not the existing users for digital bank. Majority of the respondents were male (53.8%) and aged between 26 and 30 years old. On the other hand, most of the respondents worked as employees (52.9%) and earned a salary of RM 3,001 to RM 4,000 (31.5%).

Hair et al. (2009) suggested that the factor loading should be more than 0.5. Table 8.4 shows that the factor loadings for all the items were above the threshold of 0.5, indicating the high individual item reliability. The Cronbach's alpha (CA) values ranged between 0.705 and 0.849, indicating that the measurement for the variables were reliable. The composite reliability (CR) and average variance extracted (AVE) values were also greater than the threshold of 0.5 and 0.7, respectively. These results confirmed the convergent validity of the constructs. Table 8.5 shows that all the HTMT ratio were below 0.9. This result confirmed that there was no discriminant validity issue.

Table 8.6 and Fig. 8.2 shows the results of PLS-SEM. PE, BR and DF were significantly and positively related to the INT at the level of 5%. Whereas PU and PR were significantly and positively related to the INT at the level of 10%. These findings also supported Hypothesis 1 to Hypothesis 5. This study supported the findings of Kar (2020) and Yang et al. (2023) that perceived ease of use and perceived usefulness is positively associated with the potential users' intention to use digital bank. PE is a critical factor that influences the adoption of digital banking. When users find a digital banking platform that is easy to navigate and use, they are more likely to develop a positive attitude towards using it. This ease reduces the cognitive load

Table 8.3 demographic information of the respondents participating in the study on digital banking adoption in Malaysia

		Frequency	Percentage (%)
Gender	Male	128	53.8
	Female	110	46.2
Age (years old)	18–25	82	34.4
	26–30	118	49.6
	31–40	30	12.6
	More than 40	8	3.4
Occupation	Employee	126	52.9
	Self-employed	44	18.5
	Student	68	28.6
Income (RM)	Below 1,200	30	12.6
	1,200–2,000	53	22.3
	2,001–3,000	48	20.2
	3,001–4,000	75	31.5
	4,000 and above	32	13.4

Table 8.4 Results for factor loadings, Cronbach's alpha, Average Variance Extracted (AVE), and Composite Reliability (CR) to assess convergent validity for all variables

	Items	Factor loadings	CA	CR	AVE
Perceived Ease of Use (PE)	PE1	0.862	0.849	0.908	0.767
	PE2	0.872			
	PE3	0.894			
Perceived Usefulness (PU)	PU1	0.890	0.738	0.847	0.656
	PU2	0.895			
	PU3	0.610			
Firm Reputation (FR)	FR1	0.871	0.787	0.875	0.701
	FR2	0.845			
	FR3	0.794			
Perceived Risk (PR)	PR1	0.713	0.705	0.835	0.629
	PR2	0.855			
	PR3	0.805			
Digital Financial Literacy (DFL)	DFL1	0.779	0.739	0.852	0.658
	DFL2	0.816			
	DFL3	0.837			
Intention to Use Digital Bank (INT)	INT1	0.856	0.739	0.852	0.658
	INT2	0.888			
	INT3	0.831			

Table 8.5 Results of Heterotrait-Monotrait (HTMT) ratio analysis to evaluate discriminant validity across variables

	PE	PU	FR	PR	DFL	INT
PE						
PU	0.61					
FR	0.831	0.632				
PR	0.77	0.785	0.863			
DFL	0.438	0.408	0.678	0.69		
INT	0.703	0.58	0.854	0.809	0.722	

Notes PE refers to Perceived Ease of Use; PU refers to Perceived Usefulness; FR refers to Firm Reputation; PR refers to Perceived Risk; DFL refers to Digital Financial Literacy; and INT refers to the Intention to Use Digital Bank

Table 8.6 Path analysis results for factors influencing intention to use digital banking

Independent variable	Coefficient	P-values	Decision
PE → INT	0.154	0.041	Supported
PU → INT	0.092	0.095	Supported
FR → INT	0.327	0.000	Supported
PR → INT	0.144	0.064	Supported
DFL → INT	0.242	0.000	Supported

Notes PE refers to Perceived Ease of Use; PU refers to Perceived Usefulness; FR refers to Firm Reputation; PR refers to Perceived Risk; DFL refers to Digital Financial Literacy; and INT refers to the Intention to Use Digital Bank

on users, making their interactions with the platform more seamless and enjoyable. For instance, intuitive design elements, such as clear instructions, simple navigation menus, and quick access to essential features can significantly enhance the user experience.

Besides that, a digital banking platform that is perceived as useful can improve users' financial management by offering features such as real-time account updates and easy fund transfers. These functionalities can save users time and effort, making their banking experience more efficient and effective. These results reinforce the significance users' experience in affecting potential users' intention to use digital bank. It is also important to design user-friendly interfaces to enhance users' acceptance of a particular technological product.

Similar to the previous findings of Windasari et al. (2022) and Nicholas et al. (2023), this study asserted that the firm reputation is crucial in influencing the potential user's intention to use digital banking. In this study, the digital banks under consideration are mostly non-traditional banks, which means they do not have the longstanding history or physical presence of traditional banks. This makes the establishment of a strong and positive firm reputation even more critical for these institutions. Specifically, a positive reputation may encompass trustworthiness

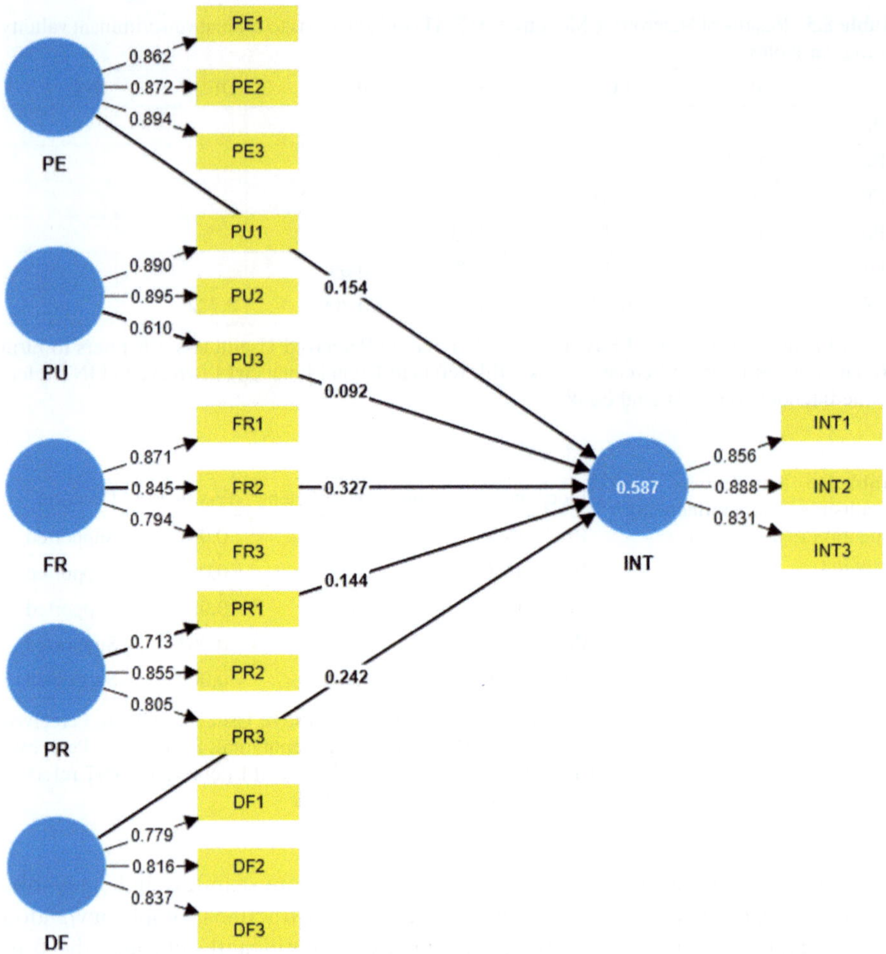

Fig. 8.2 Results of PLS-SEM analysis on factors influencing intention to use digital banking. *Sources* Author's Computation

and customer service. Since these banks were lacking physical branches, providing exceptional online and mobile support has become paramount. Undoubtedly, trust is one of the important components to be considered when dealing with money. Besides that, responsive and effective customer support can address the potential issues faced by potential users. Thus, a better firm reputation may attract more users for digital banking.

Digital bank is a brand-new term for Malaysians, and they may concern about the risk of using digital bank. Contradict with the findings of Singh and Sinha (2020) and Hassan et al. (2023), this study found that the perceived risk is positively associated with the intention to use digital banking. This result suggests that potential users are willing to accept these risks due to the perceived advantages of digital banking, such as convenience, accessibility, and enhanced financial management tools. Additionally, the release of BNM's RMiT policy document may have contributed to boosting the potential user's confidence level in digital bank's risk management capabilities. By adhering to these standards, digital banks can demonstrate their commitment to safeguarding user information and maintaining secure operations, thereby mitigating user concerns.

In line with the findings of Andreou and Anyfantaki (2021) and Gosal and Nainggolan (2023), this study revealed that the potential users with higher level of DFL tend to be use digital bank. This could be explained by the fact that potential users with greater DFL may feel comfortable and confident in adopting the digital platforms, leading to an increase in probability of adopting digital banking. DFL encompasses a range of skills and knowledge that enable individuals to effectively manage their finances using digital tools and platforms. These skills include understanding online financial transactions, recognising digital security measures, and utilising digital financial services to their fullest potential. Moreover, higher DFL may contribute to a better understanding of the benefits associated with digital banking. An individual with higher DFL tends to make better financial decisions and prevent irresponsible behaviour associated with the usage of digital finance platform.

8.5 Discussion of Implications

Using the partial least squares structural equation modelling (PLS-SEM), this study confirmed that all the five potential factors are positively related to the potential users' intention to use digital bank. Firstly, the perceived ease of use and perceived usefulness positively related with the intention to use digital bank. Digital banks should prioritise the development of user-friendly interfaces to enhance potential user's acceptance and trust in digital banking services. On the other hand, this study found that firm reputation tends to attract potential users to use the digital bank. A reputable digital bank is typically perceived as trustworthy, reliable, and customer-focused, thereby enhancing users' confidence in engaging with its services. Thus, digital banks should invest in building and maintaining a positive firm reputation, including efforts to enhance trustworthiness and customer service. Such effort is needed for digital banks to attract and retain more users, especially under the competitive digital banking landscape.

8.6 Future Directions

Digital banks can improve financial inclusion by providing underserved communities with access to financial services. It also reduces operational costs for businesses and enhances their efficiency. To maximise these benefits, policymakers should focus on improving digital financial literacy through nationwide education programs. This will help individuals and businesses better understand and use digital banking platforms effectively. However, building trust in digital banking is equally important. Financial institutions should invest in strong security measures and transparent communication to address user concerns about risk. Simplifying digital banking platforms can further encourage adoption by making them more user-friendly and accessible. Future economic strategies should include robust regulations to manage risks associated with digital banking. Collaboration between financial institutions and regulators is key to creating a secure and supportive environment for digital banking. By taking these steps, Malaysia can position itself as a regional leader in digital finance. This transformation can drive sustainable economic growth while aligning with global sustainability goals.

8.7 Conclusion

This study tested five potential factors that affect the potential users' intention to use digital bank in Malaysia. This study collected 238 valid responses from the potential users. The findings of this study are basically consistent with the previous study, except for the relationship between perceived risk and intention to use digital bank. This result could be explained if the potential users believe digital banks bring more advantages than disadvantages. Noteworthy, Bank Negara Malaysia (BNM) also released the RMiT policy statement may somehow boost up the potential users' confidence on the digital bank's capacity for risk management. Lastly, this study also found that potential users with higher level of digital financial literacy tend to use the digital bank. These potential users are more likely to feel confident in using the digital bank platform. Digital banks or relevant authorities should educate and raise awareness amongst potential users about the benefits and functionalities of digital banks. In summary, the outcome of this study also serves as guidelines for digital banks to implement effective strategies to attract potential users. Hence, the SDG 8 (Decent Work and Economic Growth) and SDG 9 (Industry, Innovation, and Infrastructure) can be achieved by attracting more users for digital bank. As a result, Malaysia has the potential to lead the way towards an economy that is inclusive, resilient, and technologically driven.

In the Asian context, the findings related to perceived ease of use, perceived usefulness, and firm reputation are likely relevant across many Asian countries, where digital banking is rapidly gaining popularity. For instance, countries like China, India, and Singapore have seen significant growth in digital banking adoption due to similar factors. The cultural emphasis on technological advancement and the younger population's familiarity with digital platforms in these countries align well with the findings of this study. Furthermore, the importance of firm reputation is equally critical in the Asian context, where trust and customer service play pivotal roles in financial transactions. The strategies recommended in this study can thus be adapted and applied to other Asian markets, considering local nuances and regulatory environments.

Highlights

- Digital financial literacy and firm reputation are key drivers of digital banking adoption in Malaysia.
- Perceived risk is outweighed by convenience and enhanced financial tools.
- The findings can guide Asian digital banks in designing secure, user-friendly platforms.
- Enhancing trust and digital literacy is crucial for scaling digital banking across Asia.

References

Aguayo FZ, Slusarczyk B (2020) Risks of banking services' digitalization: the practices of diversification and sustainable development goals. Sustainability 12:4040. https://doi.org/10.3390/su12104040

Ahmed E, Enebang H, Hashim S et al (2023) The moderating effect of social media on the relationship between value co-creation and brand equity in the services sector (banking) context. Int J Acad Res Bus Soc Sci 13(8):424–441. https://doi.org/10.6007/IJARBSS/v13-i8/17960

Alalwan AA, Dwivedi YK, Rana NP et al (2018) Examining factors influencing Jordanian customers' intentions and adoption of internet banking: extending UTAUT2 with Risk. J Retail Consum Serv 40:125–138. https://doi.org/10.1016/j.jretconser.2017.08.026

Andreou PC, Anyfantaki S (2021) Financial literacy and its influence on internet banking behavior. Eur Manag J 39:658–674. https://doi.org/10.1016/j.emj.2020.12.001

Baptista G, Oliveira T (2015) Understanding mobile banking: the unified theory of acceptance and use of technology combined with cultural moderators. Comput Human Behav 50:418–430. https://doi.org/10.1016/j.chb.2015.04.024

Bastari A, Eliyana A, Syabarrudin A et al (2020) Digitalization in banking sector: the role of intrinsic motivation. Heliyon 6:e05801. https://doi.org/10.1016/j.heliyon.2020.e05801

Brey P (2018) The strategic role of technology in a good society. Technol Soc 52:39–45. https://doi.org/10.1016/j.techsoc.2017.02.002

Dasgupta S, Gupta B (2019) Espoused organizational culture values as antecedents of internet technology adoption in an emerging economy. Inf Manag 56:103142. https://doi.org/10.1016/j.im.2019.01.004

Datareportal (2024) Digital 2024: Malaysia. https://datareportal.com/reports/digital-2024-malaysia. Accessed 24 Feb 2024

Davis FD (1989) Perceived usefulness, perceived ease of use, and user acceptance of information technology. MIS Q 13(3):319–340. https://doi.org/10.2307/249008

Featherman MS, Pavlou PA (2003) Predicting e-services adoption: a perceived risk facets perspective. Int J Hum Comput Stud 59:451–474. https://doi.org/10.1016/S1071-5819(03)00111-3

Ghani EK, Ali MM, Musa MNR et al (2022) The effect of perceived usefulness, reliability, and COVID-19 pandemic on digital banking effectiveness: analysis using technology acceptance model. Sustainability 14:11248. https://doi.org/10.3390/su141811248

Giovanis A, Athanasopoulou P, Assimakopoulos C et al (2019) Adoption of mobile banking services. Int J Bank Mark 37(5):1165–1189. https://doi.org/10.1108/ijbm-08-2018-0200

Golden W, Cordie L (2022) Digital financial literacy. Adult literacy education. https://doi.org/10.35847/WGolden.LCordie.4.3.20

Goldstein I, Jiang W, Karolyi GA (2019) To FinTech and beyond. Rev Financ Stud 32(5):1647–1661. https://doi.org/10.1093/rfs/hhz025

Gosal GG, Nainggolan R (2023) The influence of digital financial literacy on Indonesian SME's financial behavior and financial well-being. Int J Prof Bus Rev 8(12):1–13. https://doi.org/10.26668/businessreview/2023.v8i12.4164

Goyal DS, Harahap EP, Santoso NA (2022) Analysis of financial technology implementation on the quality of banking services in Indonesia: SWOT analysis. IAIC Trans Sustain Digit Innov 4(1):77–82. https://doi.org/10.34306/itsdi.v4i1.573

Hair JFJ, Black WC, Babin BJ, Anderson, RE (7th eds) (2009) Multivariate data analysis. Pearson Prentice Hall, New Jersey, USA

Hassan MS, Islam MA, Yusof MF et al (2023) Users' fintech services acceptance: a cross-sectional study on Malaysian insurance & Takaful industry. Heliyon 9:e21130. https://doi.org/10.1016/j.heliyon.2023.e21130

Henseler J, Ringle CM, Sarstedt M (2015) A new criterion for assessing discriminant validity in variance-based structural equation modeling. J Acad Mark Sci 43(1):115–135. https://doi.org/10.1007/s11747-014-0403-8

Hutapea RS, Andista DR (2021) The effect of perceived ease of use, perceived usefulness, and risk on user interest in using financial technology peer-to-peer lending. In: Proceedings of 2nd international seminar of science and applied technology, Bandung, Indonesia

Huang DH, Chueh HE (2022) Usage intention model of mobile apps in membership application. J Bus Res 139:1255–1260. https://doi.org/10.1016/j.jbusres.2021.10.062

Joshi R, Goel R, Garg S (2019) A study on customer's perception on adoption of digital banking in Indian banking sector. SSRN. https://doi.org/10.2139/ssrn.3363623

Kamis R, Ismail S, Rahman NHA (2022) Go cashless: a proposed conceptual framework for e-wallet acceptance. Int J Acad Res Bus Soc Sci 12(11):1797–1808. https://doi.org/10.6007/IJARBSS/v12-i11/15264

Kar AK (2020) What affects usage satisfaction in mobile payments? Modelling user-generated content to develop the digital service usage satisfaction model. Inf Syst Front 23(5):1341–1361. https://doi.org/10.1007/s10796-020-10045-0

Khairul MDA, Khidzir NZ, Ismail AR et al (2018) Validity and reliability of instrument to measure social media skills among small and medium entrepreneurs at Pengkalan Datu River. Int J Dev Sustain 7(3):1026–1037

Khan IU (2022) How does culture influence digital banking? A comparative study based on the unified model. Technol Soc 68:101822. https://doi.org/10.1016/j.techsoc.2021.101822

Khan IU, Hameed Z, Hamayun M (2019) Investigating the acceptance of electronic banking in the rural areas of Pakistan: an application of the unified model. Bus Econ Rev 11(3):57–88. https://doi.org/10.22547/BER/11.3.3

Kumar P, Pillai R, Kumar N, Tabash M (2023) The interplay of skills, digital financial literacy, capability, and autonomy in financial decision making and well-being. Borsa Istanbul Rev 23(1):169–183. https://doi.org/10.1016/j.bir.2022.09.012

Li P (2010) Analysis on concept, identification and key management elements of bank reputation. In: 2010 international conference on management and service science, Wuhan, China, August 2010

Lyons AC, Kass-Hanna JA (2021) Methodological overview to defining and measuring "digital" financial literacy. Financ Plan Rev 4:e11113. https://doi.org/10.1002/cfp2.1113

Mahdzan NS, Sabri MF, Rahim HA et al (2022) Digital financial services usage and subjective financial well-being: evidence from low-income households in Malaysia. Int J Bank Mark 41(2). https://doi.org/10.1108/IJBM-06-2022-0226

Malaquias RF, Hwang Y (2019) Mobile banking use: a comparative study with Brazilian and U.S. participants. Int J Inf Manag 44:132–140. https://doi.org/10.1016/j.ijinfomgt.2018.10.004

Memon MA, Ting H, Cheah J-H, Thurasamy R, Chuah F, Cham TH (2020) Sample size for survey research: review and recommendations. J Appl Struct Equat Model 4(2):1–20

McGrath J, Walker C (2023) Regulating ethics in financial services: engaging industry to achieve regulatory objectives. Regul Gov 17:791–809. https://doi.org/10.1111/rego.12482
Mitchell VW (1992) Understanding consumers' behavior: can perceived risk theory help? Manag Decis 30(3):26–31. https://doi.org/10.1108/00251749210013050
Musyaffi AM, Johari RJ, Sobirov B et al (2024) Examining initial trust in adoption of digital banking platform: a personal innovativeness and security perspective. J Syst Manag Sci 14(1):67–86. https://doi.org/10.33168/JSMS.2024.0105
Nawang WRW, Shukor SA (2023) Digital financial literacy among young adults in Malaysia. Int Bus Educ J 16(2):115–126. https://doi.org/10.37134/ibej.Vol16.2.9.2023
Nayak MSDP, Narayan KA (2019) Strengths and weaknesses of online surveys. IOSR J Humanit Soc Sci 24(5):31–38. https://doi.org/10.9790/0837-2405053138
Nicholas BYH, Jaolis F, Wijaya S (2023) The effect of perceived risk, subjective norm, brand reputation on the attitude and the intention to use digital bank services in Indonesia. Int J Financ Invest Stud 4(1):1–8. https://doi.org/10.9744/scriptura.4.1.1-9
Nguyen HV, Vu TD, Nguyen BK et al (2022) Evaluating the impact of e-service quality on customer intention to use video teller machine services. J Open Innov Technol Mark Complex 8:167. https://doi.org/10.3390/joitmc8030167
Prete AL (2022) Digital and financial literacy as determinants of digital payments and personal finance. Econ Lett 213:110378. https://doi.org/10.1016/j.econlet.2022.110378
Rathnaweera L, Karunasena A (2020). Influential factors of adopting digital banking by users in western province of Sri Lanka. In: 2nd international conference on advancements in computing (ICAC), Malabe, Sri Lanka, December 2020
Ramayanti R, Rachmawati NA, Azhar Z et al (2024) Exploring intention and actual use in digital payments: a systematic review and roadmap for future research. Comput Human Behav Rep 13:100348. https://doi.org/10.1016/j.chbr.2023.100348
Sha N, Mohammad S (2017) Virtual banking and online business. Banks Bank Syst 12(1):75–81. https://doi.org/10.21511/bbs.12(1).2017.09
Shin JW, Cho JY, Lee BG (2019) Customer perceptions of Korean digital and traditional banks. Int J Bank Mark 38(2):529–547. https://doi.org/10.1108/IJBM-03-2019-0084
Singh N, Sinha N (2020) How perceived trust mediates merchant's intention to use a mobile wallet technology. J Retail Consum Serv 52:101894. https://doi.org/10.1016/j.jretconser.2019.101894
Sun T, Shi L, Viswanathan S et al (2019) Motivating effective mobile app adoptions: evidence from a large-scale randomized field experiment. Inf Syst Res 30(2):523–539. https://doi.org/10.1287/isre.2018.0815
Sundararaj J, Priya RS (2024) Customer's perception towards digital banking services in Andaman Islands. Educ Adm Theory Pract 30(5):6793–6798. https://doi.org/10.53555/kuey.v30i5.4015
Tangmanee C, Sritadawut N (2021) Effects of perceived risk and bank reputation through online trust on the use of internet banking services. Bus Manag Strat Issue E-Commer 12(2):229–244. https://doi.org/10.7903/ijecs.1927
Tiong WN (2020) Factor influencing behavioral intention towards adoption of digital banking services in Malaysia. Int J Asian Soc Sci 10(8):450–457. https://doi.org/10.18488/journal.1.2020.108.450.457
To AT, Trinh THM (2021) Understanding behavioral intention to use mobile wallets in Vietnam: extending the TAM model with trust and enjoyment. Cogent Bus Manag 8(1):1–14. https://doi.org/10.1080/23311975.2021.1891661
Widyastuti U, Respati DK, Dewi VI et al (2024) The nexus of digital financial inclusion, digital financial literacy and demographic factors: lesson from Indonesia. Cogent Bus Manag 11(1):2322778. https://doi.org/10.1080/23311975.2024.2322778
Windasari NA, Kusumawati N, Larasati N et al (2022) Digital-only banking experience: insights from Gen Y and Gen Z. J Innov Knowl 7:100170. https://doi.org/10.1016/j.jik.2022.100170
Yang J, Wu Y, Huang B (2023) Digital finance and financial literacy: evidence from Chinese households. J Bank Financ 156:107005. https://doi.org/10.1016/j.jbankfin.2023.107005

Yi MR (2023) Corporate reputation and users' behavioral intentions: is reputation the master key that moves consumers? Sage Open 13(1). https://doi.org/10.1177/21582440231154486

Kelvin Yong Ming Lee is a Senior Lecturer and Program Director at Taylor's University, Malaysia. He focuses on finance, sustainability, and climate-related research. He also supports efforts to promote sustainability and responsible education at the university.

Pick Soon Ling is a lecturer at the School of Business and Management, University of Technology Sarawak, Malaysia. He graduated with a Doctor of Philosophy from Universiti Kebangsaan Malaysia. His research interest is focused on the finance and investment field such as efficient market hypothesis, portfolio management, and behavioural finance.

Mohamad Jais is an experienced finance lecturer with a strong record in teaching and academic service. Holding a Ph.D. and Master's in Finance, he excels in simplifying complex financial concepts. His research focuses on corporate finance and investment, showcasing his dedication to advancing knowledge in these fields.

Woon Leong Lin is an Associate Professor and Program Director of the DBA at Taylor's University, Malaysia. Specializing in ESG, CSR, Blue Economy, Circular Economy, Sustainability and Ethical Leadership, he is an accomplished researcher, mentor, and conference organizer. Dr. Lin actively contributes to academic excellence, policy innovation, and environmental stewardship through impactful publications and grants.

Irfah Najihah Basir Malan currently holds the position of Senior Lecturer and Coordinator (Indexed Publications) at the University Teknologi MARA, Malaka campus. With over 7 years of experience in tertiary education, her expertise encompasses teaching, research, and industry practices within the dynamic field of finance.

Chapter 9
Unlocking the Future of Asia and Emerging Economies: The Role of Leadership and Telecommunication Technology in Shaping the Finance-Growth Nexus

Clusters of Future Studies: Humanity at the Limen; Technological Trends

Chor Foon Tang and Mannir Salisu

Abstract This study attempts to empirically examine the role of national leadership quality, telecommunication technology (ICT), and financial development in shaping the future economic growth of over 50 emerging and Asia economies from 1997 to 2017. This study also examines the moderating effects of national leadership quality and ICT infrastructure development on the finance-growth nexus. Knowledge of the moderating effect is fundamental, particularly for uplifting and sustaining the future economic growth of emerging economies. To achieve this goal, we employ advanced non-stationary panel data approaches, namely panel unit root, panel cointegration tests, and the Group-Mean Fully Modified Least Squares (FM-OLS) estimator. Generally, our study finds that financial development, ICT, and the quality of national leadership are positively related to long-term economic growth. Moreover, our findings also show that national leadership quality and the availability of telecommunication technology are found to be important enablers (moderators) of financial developments, which leads to greater effects on economic growth. As such, improving both the quality of national leadership and telecommunication technology is necessary to unlock and shape the future economic development of the emerging and Asia's economies.

Keywords Asia · Economic growth · Financial development · National leadership · ICT · Panel data

---

C. F. Tang (✉)
Centre for Policy Research, Universiti Sains Malaysia, George Town, Malaysia
e-mail: tcfoon@usm.my

M. Salisu
Department of Banking and Finance, Hassan Usman Katsina Polytechnic, Katsina, Nigeria

E. Lau et al. (eds.), *Future of Economy in Asia*,
https://doi.org/10.1007/978-3-031-95500-6_9

9.1 Introduction

Sustainable economic development is crucial for addressing unemployment, poverty, and inequality in emerging countries. The financial sector has been identified as a key driver of growth in recent decades (Levine 1997). It plays an intermediary role by mobilising savings and deploying financial resources to meet the needs of investors and savers (Guru and Yavad 2019). However, there is considerable debate over whether the financial sector contributes to economic advancement. Asterious and Spanos (2022) attested that the positive contribution of financial development on economic growth is no longer valid, especially after the global financial crisis in 2007/2008 that jeopardised the sector's performance. Challenges such as misaligned interest rate policies, bureaucratic limitations, and excessive reserve requirements have hindered financial institutions in developing countries from fulfilling their intermediation role effectively, leading to reduced investment opportunities and economic activity.

The current state of financial growth in emerging nations presents significant challenges for strengthening the financial sector's effectiveness for development purposes. Weak national leadership undermines efforts to reform the financial sector through freedom granted to financial markets and sound monetary policies. This results in socio-economic marginalisation, inequity, corruption, and political instability that discourage investment. The technological gap between emerging nations and industrialised ones further impacts internet penetration rates due to socio-economic imbalance and infrastructure restrictions. This lack of technological advancement and infrastructure will continue affecting financial growth, consequently reducing its contribution towards economic development (Chinn and Fairlie 2010).

The economic growth literature has primarily focused on examining the connection between finance and growth, while overlooking the influence of factors such as national leadership and telecommunications technology. This study aims to observe how national leadership and telecommunications technology affect the relationship between financial development and economic development in 50 developing countries from 1997 to 2017. The year 1997 was chosen as the starting point of this study for two reasons. First, the dataset for some variables and countries under investigation is only available from 1997 onward. Second, the Asian financial crisis of 1997–1998 had profound impacts on the economies of East and Southeast Asia. This crisis significantly influenced Asian economic policies and financial structures, leading to the implementation of strict capital controls and fixed exchange rate regimes. Therefore, our sample represents a significant turning point in Asian economic and financial development.

Furthermore, the findings are expected to provide new insights into the complex link between financial sector growth and economic development, contributing to knowledge in this area and informing policymaking. The results could be used to develop effective long-term economic prosperity policies by understanding how telecommunication technology and national leadership impact financial growth for economic development. Additionally, this research can serve as a foundation for future studies on sustainable economic development concerns, leading to more comprehensive policy measures for promoting effective growth.

The organisation of this article comprises several sections. Specifically, the literature review is presented in Sect. 9.2. The methodologies applied in this research are illustrated in Sect. 9.3, followed by Sect. 9.4 that illustrates the empirical results. Section 9.5 concludes this article and offers policy implications.

9.2 Literature Review

This section reviews the related literature in this research. Specifically, the literature on the role of financial sector development, national leadership and ICT on economic growth will be segregated into three sub-sections as below.

9.2.1 Finance-Growth Nexus

Economic growth theories suggest that financial intermediary services can drive economic development through revolution, capital distribution, and entrepreneurship (Bagehot 1873; Mises and Batson 1953; Gurley and Shaw 1955; Schumpeter 1961; McKinnon 1973; Shaw 1973). This concept was first empirically studied by Goldsmith (1969) and gained attention in the 1990s. Researchers like Levine and Zervos (1998), Rajan and Zingales (2003), and Levine (1997) found that a well-structured financial system greatly benefits development and promotes financial stability. Furthermore, Doarte et al. (2023) attested that the finance-led growth hypothesis still survives in a study of a panel sample of 36 countries over the period from 1983 to 2019. However, Nguyen et al. (2022) added that the economic impact of financial development depends on both ICT diffusion and the institutional quality of a country.

De Gregorio and Guidotti (1995) studied the finance-growth relationship using yearly data from 99 countries (1960–1985) and panel data from 22 Latin American countries (1950–1985). They found that unregulated financial liberalisation in Latin America negatively impacted economic development. Similarly, Kilimani (2007) found that financial sector growth significantly benefited Uganda's economic development (1970–2002). Hassan et al. (2011) also found that financial development positively impacted long-term economic growth across all regions. However, they noted a negative relationship when domestic credit was used as a proxy for financial development, suggesting that banking credit is a better indicator of financial development in emerging nations. Despite a large strand of literature supporting the finance-led growth hypothesis, Levine (1997) cautioned that financial sector development does not always benefit economic development. Supporting this, Ram's (1999) research on 95 economies found a negative relationship between financial sector growth and output development. Dahmani et al. (2023) added that economic growth is not the result of financial development but ICT diffusion.

9.2.2 Leadership and Economic Growth

Leadership can be defined as the practice of authority in managing a company's or nation's political and economic frameworks, and plays a crucial role in overall success (Zolcsak 2015; Northouse 2017). However, the empirical relationship between national leadership and economic growth remains underexplored. Studies by Jones and Olken (2005), Iyoha (2008), and Brown (2020) suggest that leaders significantly influence growth through their policy decisions. Poor leadership can negatively impact a nation's output growth.

Leadership is closely tied to governance, with corruption management and political stability being key indicators of leadership quality. Good national leaders control corruption, provide a stable political environment, and are committed to national goals. They also maintain integrity and oppose corruption to gain credibility, contributing to successful policy execution.

Given that the quality of leadership is closely linked to good governance, the corruption-growth nexus studies were considered in the current review. The corruption-growth nexus studies, such as Mauro (1995), show a negative relationship between development and corruption. However, Quinn and Woolley (2001) found that democratic leaders are less likely to negatively impact development and are more conducive to balanced economic development. Policies in a democratic system are more likely to follow economic rationales, promoting economic development.

Jones and Olken (2005) found that leaders significantly affect a country's economic development, with changes in leadership in democracies having no impact. Demetriades and Law (2006) found that the contribution of finance to real output growth is directly related to institutional quality. Girma and Shortland (2007) found that political stability and democracy are major factors in financial sector growth and economic development. Karikari (2010) found that good governance strengthens the effect of the financial sector on output increase, attributed to the balance offered in democratic actions. Elmi and Ariani (2010) found that good governance positively and significantly affected the financial sector and economic development in MENA countries, with corruption management, political balance, and rule of law being crucial.

Law and Azman-Saini (2012) found that good institutional environments and governance positively contribute to financial systems, particularly in the banking sector. Owoye and Bissessar (2014) found that corruption, exacerbated by inadequate institutional safeguards, continuously harmed African economies. Cherif and Dreger (2016) found that the institutional environment was the primary factor of financial growth in MENA economies, with corruption being the most crucial element in the banking sector. Uddin et al. (2017) found that political stability had a positive and significant impact on economic growth in emerging economies, while political instability posed a risk to their development. Adewosi et al. (2018) revealed that several factors, including law regulation, political stability, government expenditure, foreign direct investment, savings, and inflation, had a positive influence on financial development in West African economies. However, private sector credit, GDP, interest rate, capital formation, and openness negatively impacted financial sector development. They suggested the construction and application of fiscal and monetary policies to boost financial development.

Samarasinghe (2018) conducted a study on the relationship between governance and development in 145 economies from 2002–2014. The research included governance factors such as corruption management, political stability, opinions, and responsibility. The results showed that corruption management was a key factor for output growth, with improved corruption management leading to a 6.9% increase in output. Hence, it was suggested that protecting corruption management and political stability could lead to strong economic development. Recent studies, such as those by Spyromitros and Panagiotidis (2022) and Singha and Singh (2022) consistently emphasised that the failure to achieve high economic performance is largely attributed to poor corruption control and political instability, as investors perceive these as risks. Therefore, they conclude that good governance and political stability are important to promote growth.

9.2.3 Telecommunication Technology and Economic Growth

Numerous studies have highlighted the significant role of Information and Communication Technology (ICT) in economic growth. For instance, Jorgenson et al. (2000) and Jorgenson and Stiroh (2000) emphasised ICT's substantial contribution to GDP growth in the United States. Similar findings were reported in the United Kingdom by Haskel and Heden (1999) and Clayton and Waldron (2003).

In emerging economies, Bahrini and Qaffas (2019) identified telecommunication technology as a primary factor of real GDP. Kumar (2014) and Kumar and Kumar (2012) demonstrated the positive role of ICT in the GDP growth of Vietnam and Fiji, respectively. Lal (2007) highlighted that the globalisation of the Nigerian economy was mainly driven by the implementation of telecommunication technologies. Roller and Waverman (2001) found that outstanding telecommunication infrastructure was positively and significantly correlated with real GDP. Besides, Nguyen et al. (2022) discovered an inverted U-shape relationship between ICT and economic growth in emerging economies. However, the study claims that over-diffusion of ICT may harm rather than help the economy. Recently, Tang and Lim (2022, 2024) consistently attested that the advancement of ICT uplifts private consumption in Malaysia. They further elucidate that this impact is significantly amplified in the presence of exemplary leadership, thereby leading to greater economic performance.

In this literature review, we discover many studies on ICT-growth, finance-growth and institutions-growth relationships, research on the moderating impact of national leadership and ICT on the finance-growth nexus is scarce, especially in emerging nations. Therefore, we attempt to fill in this research gap.

9.3 Methodology

9.3.1 Empirical Model

Solow's (1956) growth framework is used in this study to analyse the impact of national leadership and ICT on the relationship between finance and growth in emerging nations. The model is expressed as follows:

$$\ln y_{it} = \ln A_0 + \theta_i \ln Z_{it} + \frac{\alpha}{1-\alpha}\ln k_{it} - \frac{\alpha}{1-\alpha}\ln(n+g+\delta)_{it} \tag{9.1}$$

where ln represents natural logarithm, $\ln A_0 = \beta_0 + \varepsilon_{it}$, y_{it} denotes the per capita output, k_{it} is the capital stock, n denotes the exogenous population growth rate, g indicates the exogenous technology growth rate, and δ represents the depreciation level of capital stock. Meanwhile, Z_{it} encompasses explanatory factors such as the quality of national leadership (QL), financial development(FD), and telecommunication technology(ICT). Therefore, the growth model is further simplified as:

$$\ln\mathrm{GDP}_{it} = \beta_0 + \beta_1 \ln k_{it} + \beta_2 \ln(n+g+\delta)_{it} + \theta_3 \ln\mathrm{FD}_{it} + \theta_4 \ln\mathrm{ICT}_{it} + \theta_5 \ln\mathrm{QL}_{it} + \varepsilon_{it} \tag{9.2}$$

Furthermore, to estimate the moderating effects of QL and ICT on the effect of FD on economic growth in emerging nations, the interaction terms, $\ln(\mathrm{FD} \times \mathrm{QL})_{it}$ and $\ln(\mathrm{FD} \times \mathrm{ICT})_{it}$ are incorporated into the growth model as below:

$$\begin{aligned}\ln\mathrm{GDP}_{it} = {} & \beta_0 + \beta_1 \ln\mathrm{k}_{it} + \beta_2 \ln(n+g+\delta)_{it} + \theta_3 \ln\mathrm{FD}_{it} \\ & + \theta_4 \ln\mathrm{ICT}_{it} + \theta_5 \ln\mathrm{QL}_{it} + \gamma_1 \ln(\mathrm{FD} \times \mathrm{QL})_{it} + \varepsilon_{it}\end{aligned} \tag{9.3}$$

$$\begin{aligned}\ln\mathrm{GDP}_{it} = {} & \beta_0 + \beta_1 \ln\mathrm{k}_{it} + \beta_2 \ln(n+g+\delta)_{it} + \theta_3 \ln\mathrm{FD}_{it} \\ & + \theta_4 \ln\mathrm{ICT}_{it} + \theta_5 \ln\mathrm{QL}_{it} + \gamma_2 \ln(\mathrm{FD} \times \mathrm{ICT})_{it} + \varepsilon_{it}\end{aligned} \tag{9.4}$$

If γ_1 is statistically significant, it indicates that impact of FD on economic growth is facilitated by the QL. Similarly, the significance of γ_2 demonstrates that the finance-growth nexus is contingent on ICT. The marginal effects of FD on economic development can be computed using the partial derivatives, namely $\partial\ln\mathrm{GDP}_{it}/\partial\ln\mathrm{FD}_{it} = \theta_3 + \gamma_1 \ln\mathrm{QL}_{it}$ for Eq. (9.3) and $\partial\ln\mathrm{GDP}_{it}/\partial\ln\mathrm{FD}_{it} = \theta_3 + \gamma_2 \ln\mathrm{ICT}_{it}$ for Eq. (9.4).

Table 9.1 List of 50 selected Asian and emerging countries

No	Country	No	Country	No	Country
1	Albania	18	Equatorial Guinea	35	Namibia
2	Angola	19	Fiji	36	Nigeria
3	Armenia	20	Gabon	37	Pakistan
4	Bangladesh	21	Georgia	38	Paraguay
5	Belarus	22	Ghana	39	Peru
6	Benin	23	Guatemala	40	Philippines
7	Bolivia	24	Honduras	41	Romania
8	Brazil	25	India	42	Senegal
9	Bulgaria	26	Indonesia	43	South Africa
10	Cambodia	27	Jordan	44	Thailand
11	China	28	Kazakhstan	45	Tunisia
12	Colombia	29	Kenya	46	Turkey
13	Costa Rica	30	Lebanon	47	Uganda
14	Cote D'ivoire	31	Malaysia	48	Ukraine
15	Dominican Republic	32	Mexico	49	Vietnam
16	Ecuador	33	Moldova	50	Zambia
17	Egypt	34	Morocco		

9.3.2 Sources of Data

This study utilised balanced yearly panel data of 50 emerging nations from 1997 to 2017. The choice of countries and the sample period, as listed in Table 9.1, were based on data availability. Data were sourced from the World Bank's *World Governance Indicators* (WGI) and *World Development Indicators* (WDI). The growth rate of the working-age population (n) was measured following Mankiw et al. (1992) and the assumption of $g + \delta = 0.08$ was adopted based on the dataset.[1]

Following the strategy of Tang and Tan (2014), Ang and McKibbin (2007), and Stock and Watson (2002), a financial development index was created using the Principal Component Method (PCM) with four indicators: the ratio of domestic credit to private credit to GDP, the ratio of M2 to GDP, the ratio of M3 to GDP, and the domestic credit provided by the banking sector to GDP. Similarly, the ICT index was formed using PCM based on the number of Internet services, phone line subscribers, and mobile phones for each 1,000 population. Furthermore, a national leadership index was constructed through PCM with a number of national leadership-related indicators: political stability, corruption control, government efficacy, and voice and

[1] Along with the calculation in Mankiw et al. (1992), g and δ were computed based on the development rate of per capita GDP and the capital-output ratio with the study dataset. The average development rate of per capita GDP was approximately 0.03 and the capital-output ratio was approximately 0.05. Therefore, $g + \delta = 0.08$ was determined.

accountability, as recommended by Tan et al. (2010) and widely applied in empirical studies by Tang and Lim (2024), and Tang and Rosidi (2025). All data was converted into a natural logarithm for standardisation.

9.4 Findings

Before conducting panel regression analysis, it is crucial to examine the stationarity of the variables. If any variable in the regression is non-stationary and not cointegrated, the regression analysis could yield false outcomes. Thus, the panel unit root tests were conducted using the Im-Pesaran-Shin (IPS), ADF-Choi, and PP-Choi panel unit root tests.

The results in Table 9.2 showed that all variables were non-stationary at the level but became stationary after the first differencing, demonstrating that the variables were integrated at order one, $I(1)$. Following this, the presence of cointegration or long-term association among the variables was tested through Pedroni's (1999) residuals-based test for panel cointegration.

Table 9.2 Results of panel unit root tests on GDP, capital, population growth, financial development, ICT, and leadership quality

Variables	IPS test		ADF-Choi Z-test		PP-Choi Z-test	
	Constant	Constant and trend	Constant	Constant and trend	Constant	Constant and trend
Level:						
$\ln GDP_{it}$	–0.547	–2.116	7.046	1.980	8.352	1.184
$\ln GFC_{it}$	–1.222	–2.128	2.259	0.823	0.451	–1.729**
$\ln(n+g+\delta)_{it}$	–1.035	–1.700	3.711	3.659	4.045	2.674
$\ln FD_{it}$	–1.300	–1.687	1.608	3.953	1.094	2.727
$\ln ICT_{it}$	–2.563***	–1.277	–7.840***	7.356	–18.218***	5.757
$\ln QL_{it}$	–1.588	–2.234	–0.477	–0.530	–1.528	–1.581**
First difference:						
$\Delta\ln GDP_{it}$	–2.951***	–2.945***	–9.194***	–5.243***	–13.915***	–10.345***
$\Delta\ln GFC_{it}$	–3.208***	–3.571***	–11.291***	–9.499***	–18.408***	–14.171***
$\Delta\ln(n+g+\delta)_{it}$	–2.970***	–3.554***	–9.881***	–9.401***	–15.404***	–12.741***
$\Delta\ln FD_{it}$	–2.744***	–3.505***	–8.434***	–8.963***	–16.493***	–15.184***
$\Delta\ln ICT_{it}$	–1.564	–3.836***	0.295	–11.681***	–7.023***	–13.273***
$\Delta\ln QL_{it}$	–3.442***	–4.066***	–12.668***	12.536***	–19.186***	–16.443***

Note The asterisks *** and ** denote statistical significance at the 1% and 5% levels, respectively. The optimal lag order for ADF-Choi and IPS tests is selected by the Modified Akaike Information Criterion (MAIC), while bandwidth for the PP-Choi test is based on Newey-West using Bartlett Kernel. The critical values for the IPS test refer to Im et al. (2003)

Table 9.3 Results of Pedroni's Panel Cointegration among GDP, capital, population growth, financial development, ICT, and leadership quality

Tests	Statistics	*p*-values
Panel v-statistic	0.010	0.350
Panel rho-statistic	4.135	0.999
Panel PP-statistic	–1.435***	0.000
Panel ADF-statistic	–2.169***	0.000
Group rho-statistic	6.349	1.000
Group PP-statistic	–8.734***	0.000
Group ADF-statistic	–7.433***	0.000

*Note**** denotes statistical significance at the 1% level. Deterministic intercept and trend are involved in the test. The lag length and bandwidth are chosen by AIC while the Bartlett kernel is based on Newey-West

The results in Table 9.3 showed four of the seven statistics exhibited significance at 1%, rejecting the null hypothesis of no cointegration.[2] This indicated a long-term association among the variables. The same outcomes were also recorded in other research works in emerging nations (Ahmed 2010; Asghar and Hussain 2014). After verifying the cointegration between the variables, the long-run coefficients are estimated.

Four long-run growth models were estimated using the panel FMOLS estimator, as shown in Table 9.4. The results were consistent with the economic development theory and several available empirical research works (e.g., Law and Habibullah 2006; Mankiw et al. 1992). The empirical findings from Model 2 showed that financial development positively affected economic development in emerging nations. This result is not in line with Dahmani et al. (2023), but aligned with many other past studies, such as Guru and Yadav (2019), Tang et al. (2020), and Duarte et al. (2023).

A 1% increase in financial development would improve the GDP by approximately 0.064%, holding other factors constant. Similarly, ICT was found to have a significant and positive impact on growth, indicating that ICT progress significantly enhanced the development in emerging nations. This result aligned with the results by Roy et al. (2014), Lee and Brahmasrene (2014), and Kumar (2014) who highlighted that ICT is an essential factor of growth. Apart from that, the quality of national leadership also played a significant role in accelerating economic development. A 1% rise in the quality of national leadership improved the GDP by approximately 0.046%, indicating that nations with good leadership are more likely to promote economic development. This aligns with Samarasinghe (2018), who argued that leaders with

[2] One may argue the robustness of the findings if the series are cross-sectionally dependent. To address this argument, the Westerlund (2007) panel cointegration approach was employed with cross-sectional dependency. This study did not find any new cointegration evidence related to the findings of Pedroni's cointegration test shown in Table 10.3. In this case, the variables under review were cointegrated and robust either with or without the cross-sectional dependency. To conserve space, the full results are not reported in this section but are available upon request.

Table 9.4 Results of long-run relationships between GDP, capital, population growth, financial development, ICT, and leadership quality

Variables	Model 1	Model 2	Model 3	Model 4
$\ln GFC_{it}$	0.290*** (0.000)	0.143*** (0.000)	0.139*** (0.000)	0.141*** (0.000)
$\ln(n + g + \delta)_{it}$	–0.699*** (0.000)	–0.552*** (0.000)	–0.505*** (0.000)	–0.541*** (0.000)
$\ln FD_{it}$	–	0.064*** (0.000)	0.084*** (0.000)	0.066*** (0.000)
$\ln ICT_{it}$	–	0.072*** (0.000)	0.072*** (0.000)	0.074*** (0.000)
$\ln QL_{it}$	–	0.046*** (0.000)	0.055*** (0.000)	0.046*** (0.000)
$\ln(FD \times QL)_{it}$	–	–	0.021*** (0.000)	–
$\ln(FD \times ICT)_{it}$	–	–	–	0.003*** (0.000)
Marginal effects: $\partial \ln GDP/\partial \ln FD$				
Minimum	–	–	–0.042*** [-9.920]	0.046*** [11.185]
Mean	–	–	0.084*** [40.009]	0.066*** [35.569]
Maximum	–	–	0.153*** [41.689]	0.073*** [30.630]

Note The asterisk *** denotes the statistical significance at the 1% level. Figures in the parenthesis (.) are the *p*-values, whereas [.] indicates the *t*-statistics for marginal effects calculated based on the procedure suggested in Brambor et al. (2006)

less corruption and higher accountability contribute to a higher balance in the political environment, which encourages economic development.

The analysis was expanded to examine the role of national leadership and telecommunication technology in enhancing the positive effect of financial sector growth on economic development. To assess the moderating effect, the interaction terms of $\ln(FD \times QL)_{it}$ and $\ln(FD \times ICT)_{it}$ were accommodated into Model 3 and Model 4. The findings in Model 3 showed a positive and significant coefficient of the interaction term between financial growth and national leadership at the 1% level, indicating that national leadership is crucial in enhancing the contribution of finance to the development of emerging economies. In Model 4, the estimated outcomes for the interaction term between financial growth and ICT had a positive relationship with GDP and were statistically significant at the 1% level. This aligns with findings by Andrianaivo and Kpodar (2011) and Sassi and Goaied (2013), who highlighted that advanced ICT infrastructure significantly improved financial deepening and output increase. The results also agreed with Nguyen et al. (2022) and Shamim (2007), who stated that progress in ICT development reinforced the influence of finance on development.

Given this, initiatives aimed at improving information technology diffusion across all economic sectors, particularly in financial sector development, will enhance efficiency and accessibility to credit for investments. Consequently, economic growth can be significantly expanded through the synergistic development of both ICT and the financial sector.

After verifying the existence of facilitative mechanisms, we evaluated the marginal impacts of financial sector growth on output development at different levels of national leadership and ICT. Furthermore, the marginal effects were tested based on the standard error formula suggested by Brambor et al. (2006).[3] The results showed significant marginal impacts at all levels of national leadership, but positive impacts were only observed at the mean and maximum levels. Specifically, at the mean level of national leadership, a 1% increase in financial sector development resulted in a 0.084% growth in output, while this effect rose to 0.153% at the maximum level. This indicates that countries with moderate or superior national leadership quality are more likely to witness reinforced financial growth leading to output growth. This is because effective leaderships foster a conducive environment for the financial sector to thrive, thereby amplifying its impact on overall economic growth. Moreover, strong national leadership can facilitate the channelling of financial resources towards productive and sustainable investments, further accelerating the transition towards sustainable economic growth.

Likewise, the marginal impacts calculated for the interaction between financial growth and ICT were also favourable and statistically significant across different levels of ICT. To illustrate, a 1% increase in financial growth at the minimum, mean, and maximum levels of ICT would correspondingly boost economic development by approximately 0.046%, 0.066%, and 0.073% respectively. These results highlight the influence of the financial sector on output growth in emerging nations, depending heavily on the degree of ICT development. The rationale behind this finding is not difficult to understand. The development of ICT infrastructure enables the financial sector to efficiently translate savings into capital for productive sectors. Moreover, it helps in monitoring borrowers and managing risks. Therefore, it is reasonable to find that improving ICT enhances access to finance and stimulates economic growth.

9.5 Discussion of Implications

The findings presented in the earlier section not only contributed to policymaking in emerging countries but also have significant implications for shaping the future economic development of Asian economies. This is because Asian economies are

[3] The new standard errors were used to examine the effect of the marginal impacts of national leadership standard, and telecommunication technology could be computed using $\hat{\sigma}_{\partial \mathrm{GDP}/\partial \mathrm{X}} = \left[var\left(\hat{\beta}_1\right) + \mathrm{lnQL}_{it}^2 var\left(\hat{\beta}_4\right) + 2(\mathrm{lnQL}_{it})cov\left(\hat{\beta}_1\hat{\beta}_4\right)\right]^{0.5}$ and $\hat{\sigma}_{\partial \mathrm{GDP}/\partial \mathrm{X}} = \left[var\left(\hat{\beta}_1\right) + \mathrm{lnICT}_{it}^2 var\left(\hat{\beta}_4\right) + 2(\mathrm{lnICT}_{it})cov\left(\hat{\beta}_1\hat{\beta}_4\right)\right]^{0.5}$.

experiencing rapid technological advancements and are often led by emerging leaders who can greatly influence economic development. Moreover, Asia holds a significant influence in shaping a new era for global economies, as it represents over half of the global GDP growth (52%), global trade growth (59%), and manufacturing value-added (53%) over the past two decades (Seong 2024). Therefore, leaders and policymakers in developing countries, particularly those in the Asia region, must recognise the interplay between financial sector development, telecommunication technology, and leadership in order to foster long-term sustainable growth. This holistic approach can lead to more effective and efficient utilisation of financial resources for economic development.

Considering the key findings of this study, policymakers must establish an effective financial architecture that can mobilise and distribute savings into productive investments. Additionally, policies and reforms that could unveil the latent prospect of financial institutions as contributing factors to economic development should be accelerated. These included further liberalisation of the financial sector that could develop subsidiary capital markets, complementing the primarily available bank-centred financial architecture in emerging countries. Furthermore, given the significant role of ICT in the finance-growth nexus, measures to encourage ICT implementation should be a priority. These measures should include incentives such as: (a) tax holidays for investments related to ICT; (b) increased investment by the government and private sector in technology to improve and broaden the available ICT infrastructure; (c) reducing tariffs on the import of ICT-related hardware and software; and (d) implementing new ICT protocols, including 5G-based frameworks. In addition, governments should also promote policies to increase internet usage, mobile cellular and telephone line subscriptions, and enhance telecommunications infrastructure to support economic development.

Since national leadership holds a crucial responsibility in boosting economic development, policymakers should enhance the quality of national leadership to sustain and boost economic development. This involves emphasising measures aimed at preserving effective governance, including establishing a Competent, Accountable and Transparent (CAT) administrative framework that is free from red tape and corruption while upholding the rule of law. Last but not least, leaders should enhance their progressive credentials by advocating for participatory and inclusive democracy, promoting freedom of expression, engaging civil society, and ensuring the free flow of information to appeal to investors in the financial sector.

The key summary of the implications of this study is not far-fetched. The future of emerging and Asia's economies hinges on more than just the growth of the financial sector. In addition to financial sector development, the advancement of telecommunications technology and the cultivation of talented leaders are essential for sustainable growth. Emphasising these areas will not only drive economic growth and development but also foster a robust and innovative society prepared for the challenges of tomorrow.

9.6 Future Directions

Asia and emerging economies stand at the threshold of a transformative era where rapid technological advancements, resilient economic growth, and evolving financial systems will redefine their global standing. The trajectory of this transformation hinges on how these economies leverage the digital revolution to unlock new growth frontiers. The financial technology (Fintech) sector, propelled by the widespread adoption of digital banking, blockchain, and peer-to-peer lending, is not merely reshaping financial landscapes but actively knocking down traditional barriers to financial inclusion. This evolution fosters a more agile, efficient, and inclusive financial ecosystem, enabling businesses and individuals to participate more fully in economic activities (Feyen et al. 2023).

Beyond financial technology, sustained investment in cutting-edge innovations—such as artificial intelligence, big data, and the Internet of Things (IoT)—is pivotal in enhancing productivity, optimising resource allocation, and strengthening regional competitiveness. To remain at the frontline of global innovation, Asia must not only embrace these advancements but also cultivate an ecosystem that encourages technological diffusion, fosters digital entrepreneurship, and upskill its workforce to meet the demands of a rapidly evolving economy. By proactively shaping this future, the region can position itself as a global leader in technological innovation and sustainable economic development (Tang and Rosidi 2025). Sustainable finance is also a cornerstone of Asia's future growth, shaping investment strategies that balance economic growth with environmental and social responsibility. The growing integration of environmental, social, and governance (ESG) criteria into investment decisions to expedite the transition process towards sustainable economic development. Therefore, green bonds and other sustainable financial instruments are no longer niche but essential tools for mobilising projects that prioritise environmental sustainability and climate resilience (Kudryavtseva et al. 2024). This shift not only aligns with global efforts to combat climate change but also enhances long-term financial stability by mitigating environmental risks.

Besides sustainable finance, regional integration will be another potential instrument in driving Asia's economic expansion. Indeed, the computational general equilibrium (CGE) model of Estrades et al. (2023) acknowledged that initiatives such as the Regional Comprehensive Economic Partnership (RCEP) are deepening trade and investment ties, fostering a more interconnected and resilient regional economy. By strengthening economic cooperation, Asia can unlock new opportunities for growth, enhance supply chain efficiencies, and boost financial stability amid global uncertainties. However, alongside integration, resilience-building must remain a strategic priority. The region may face persistent challenges, including geopolitical tensions and domestic economic vulnerabilities, which necessitate proactive risk management (Deng 2024). To mitigate these risks, Asia has already drawn lessons from past financial crises that led to stronger financial regulations, credible monetary and fiscal policies, and higher national savings rates that form a strong foundation to shield Asia's economic framework from future shocks.

Furthermore, the investment in human capital is equally important in determining the ability of Asia to sustain future growth and innovation. Education, skills development, and research must take centre stage in policy agendas to cultivate a highly skilled workforce capable of driving technological advancements and adapting to evolving market demands. The development of local expertise will not only accelerate knowledge creation but also strengthen Asia's position as a global hub for innovation (Cheng and Chan 2022). In sum, the future of Asia and emerging economies is defined by transformation and opportunity. Digital innovation, sustainable finance, deeper regional integration, economic resilience, human capital development, inclusive growth, and climate adaptation will collectively shape the region's trajectory. By navigating these strategic priorities effectively, Asia can solidify its leadership in global economic growth while ensuring long-term sustainability and shared prosperity.

9.7 Conclusion

This study examined the role of national leadership and ICT in the finance-growth nexus in emerging nations from 1997 to 2017. While financial sector development is known to drive economic growth, it alone is not enough to sustain growth. Our findings indicate that both ICT and national leadership not only have a positive effect on economic growth, but also play crucial roles in maximising the influence of financial development on economic growth. Given these findings, it is imperative for policymakers in emerging nations to recognise the inter-connected nature of these growth catalysts and to formulate strategies that leverage the synergistic effects of national leadership, ICT, and financial sector development. By doing so, they can harness the full potential of these catalysts to drive and sustain the future growth of emerging nations.

Although this chapter contributes to the economic growth literature by highlighting the role of national leadership, financial development, and technologies in shaping the future of Asia and emerging economies, its focus is mainly on developing economies. Consequently, the analysis is less comprehensive, and the findings may not be easily generalised to advanced economies. Future studies could expand the scope to include both developed and developing countries, offering a more holistic perspective on these dynamics. This broader approach would enhance the robustness and representativeness of the findings, particularly in strengthening the finance-growth nexus through higher-quality national leadership and technological innovation.

Highlights

- Sustained future growth in Asia requires a robust financial architecture.
- Financial development, ICT, and leadership quality are inter-connected growth drivers.
- Leadership quality and ICT enhance the impact of financial development on economic growth.
- Synergistic policies integrating ICT and governance can maximise growth.
- Promoting quality leadership is pivotal in driving the future growth of Asia and emerging economies.

Acknowledgements We thank the two anonymous reviewers for their suggestions to improve the earlier version of this research paper. This research is supported by the Fundamental Research Grant Scheme [FRGS/1/2023/SS06/USM/02/1], Ministry of Higher Education, Malaysia. We are solely responsible for any remaining errors found within this research paper.

References

Adewosi OA, Donga M, Idi A, Abdullahi B (2018) An examination of drivers of financial development: evidence in West African countries. Pak J Humanit Soc Sci 6(1):132–143. https://doi.org/10.52131/pjhss.2018.0601.0038

Ahmed AD (2010) Financial liberalisation, financial development and growth linkages in Sub-Saharan African countries. Stud Econ Financ 27(4):314–339. https://doi.org/10.1108/10867371011085156

Andrianaivo M, Kpodar K (2011) ICT, financial inclusion, and growth: evidence from African Countries. IMF Working Paper 11–73, International Monetary Fund

Ang JB, McKibbin WJ (2007) Financial liberalization, financial sector development and growth: evidence from Malaysia. J Dev Econ 84:215–233. https://doi.org/10.1016/j.jdeveco.2006.11.006

Asghar N, Hussain Z (2014) Financial development, trade openness and economic growth in developing countries: Recent evidence from panel data. Pak Econ Soc Rev 52(2):99–126. http://www.jstor.org/stable/24398853

Asterious D, Spanos K (2022) The mechanisms linking the finance-growth relationship in view of the financial crisis: an empirical investigation of the EU countries. J Econ Stud 49(7):1240–1268. https://doi.org/10.1108/JES-03-2021-0170

Bagehot W (1873) Lombard street: a description of the money market. Armstrong and Co., Scriber

Bahrini R, Qaffas AA (2019) Impact of information and communication technology on economic growth: evidence from developing countries. Economies 7(21):1–13. https://doi.org/10.3390/economies7010021

Brambor T, Clark WR, Golder M (2006) Understanding interaction models: improving empirical analyses. Polit Anal 14(1):63–82. https://doi.org/10.1093/pan/mpi014

Brown CO (2020) Economic leadership and growth. J Monetary Econ 116:298–333. https://doi.org/10.1016/j.jmoneco.2019.11.004

Cheng IH, Chan SJ (2022) Asian perspectives and approaches for developing assistance in promoting human capital. Oxford Res Encyclopedia Educ.https://doi.org/10.1093/acrefore/9780190264093.013.63
Cherif M, Dreger C (2016) Institutional determinants of financial development in MENA countries. Rev Dev Econ 20(3):670–680. https://doi.org/10.1111/rode.12192
Chinn MD, Fairlie RW (2010) ICT use in the developing world: an analysis of differences in computer and internet penetration. Rev Int Econ 18(1):153–167. https://doi.org/10.1111/j.1467-9396.2009.00861.x
Clayton T, Waldron K (2003) E-commerce adoption and business impact, a progress report. Economic Trends 591(2):33–40. https://escoe-website.s3.amazonaws.com/wp-content/uploads/2020/01/01233759/ET-591-E-commerce-adoption-and-business-impact-a-progress-report-T-Clayton-K-Waldron-Feb-2003.pdf. Accessed 4 Feb 2025
Dahmani M, Mabrouki M, Youssef AB (2023) The ICT, financial development, energy consumption and economic growth nexus in MENA countries: dynamic panel CS-ARDL evidence. Appl Econ 55(10):1114–1128. https://doi.org/10.1080/00036846.2022.2096861
De Gregorio J, Guidotti PE (1995) Financial development and economic growth. World Dev 23(3):433–448. https://doi.org/10.1016/0305-750X(94)00132-I
Demetriades P, Law SH (2006) Finance, institutions and economic development. Int J Financ Econ 11(3):245–260. https://doi.org/10.1002/ijfe.296
Deng Y (2024) Do domestic and external geopolitical risks matter for food availability: implications for global food security. Appl Econ 56(60):9211–9218. https://doi.org/10.1080/00036846.2023.2300766
Duarte J, Andrade JS, Bação P (2023) Dynamic effects of financial development on economic activity. Appl Econ 55(29):3329–3346. https://doi.org/10.1080/00036846.2022.2114994
Elmi Z, Ariani F (2010) Governance and financial development in MENA region. Aust J Basic Appl Sci 4(12):6021–6024
Estrades C, Maliszewska M, Osorio-Rodarte I, Pereira MS (2023) Estimating the economic impacts of the regional comprehensive economic partnership. Asia Glob Econ 3(2):100060. https://doi.org/10.1016/j.aglobe.2023.100060
Feyen E, Natarajan H, Saal M (2023) Fintech and the future of finance: market and policy implications. World Bank Group, Washington DC. http://documents.worldbank.org/curated/en/099450005162250110/P17300600228b70070914b0b5edf26e2f9f. Accessed 4 Feb 2025
Girma S, Shortland A (2007) The political economy of financial development. Oxford Econ Pap 60(4):567–596. https://doi.org/10.1093/oep/gpm040
Goldsmith RW (1969) Financial structure and development. Yale University Press, New Haven and London
Gurley JG, Shaw ES (1955) Financial aspects of economic development. Am Econ Rev 45(4):515–538. http://www.jstor.org/stable/1811632
Guru BK, Yadav IS (2019) Financial development and economic growth: panel evidence from BRICS. J Econ Financ Adm Sci 24(47):113–126. https://doi.org/10.1108/JEFAS-12-2017-0125
Haskel J, Heden Y (1999) Computers and the demand for skilled labour: industry- and establishment-level panel evidence for the UK. Econ J 109(454):68–79. https://doi.org/10.1111/1468-0297.00417
Hassan MK, Sanchez B, Yu JS (2011) Financial development and economic growth: new evidence from panel data. Q Rev Econ Financ 51(1):88–104. https://doi.org/10.1016/j.qref.2010.09.001
Im KS, Pesaran MH, Shin Y (2003) Testing for unit roots in heterogeneous panels. J Econom 115(1):53–74. https://doi.org/10.1016/S0304-4076(03)00092-7
Iyoha MA (2008) Leadership, policy making, and economic growth in African countries: the case of Nigeria. In: Brady D, Spence M (eds) Leadership and growth. World Bank Group, Washington, DC, pp 165–193
Jones BF, Olken BA (2005) Do leaders matter? National leadership and growth since World War II. Q J Econ 120(3):835–864. https://doi.org/10.1093/qje/120.3.835

Jorgenson DW, Stiroh KJ (2000) US economic growth at the industry level. Am Econ Rev 90(2):161–167. https://doi.org/10.1257/aer.90.2.161

Jorgenson DW, Stiroh KJ, Gordon RJ (2000) Sichel DE (2000) Raising the speed limit: US economic growth in the information age. Brook Pap Econ Act 1:125–235. https://doi.org/10.1353/eca.2000.0008

Karikari J (2010) Governance, financial liberalization, and financial development in sub-Saharan Africa. African Development Bank's 2010 African Economic Conference, Tunisia, October 27–29, 2010. https://www.afdb.org/fr/documents/document/aec-2010-governance-financial-liberalization-and-financial-development-in-sub-saharan-africa-21778. Accessed 20 Jan 2025

Kilimani N (2007) Financial development and economic growth in Uganda. IUP J Financ Econ 5(1):14–34

Kudryavtseva OV, Kachalov RM, Chemyavskiy SV, Makeeva NN (2024) Ecological and economic effects of green financing. Econ Math Methods 60(4):77–86. https://doi.org/10.31857/S0424738824040073

Kumar RR (2014) Exploring the role of technology, tourism and financial development: an empirical study of Vietnam. Qual Quant 48:2881–2898. https://doi.org/10.1007/s11135-013-9930-5

Kumar RR, Kumar R (2012) Exploring the nexus between information and communication technology, tourism and growth in Fiji. Tour Econ 18(2):359–371. https://doi.org/10.5367/te.2012.0117

Lal K (2007) Information and communication technologies in the context of globalisation: evidence from developing countries. Palgrave Macmillan, London

Law SH, Azman-Saini WNW (2012) Institutional quality, governance, and financial development. Econ Gov 13(3):217–236. https://doi.org/10.1007/s10101-012-0112-z

Law SH, Habibullah MS (2006) Financial development, institutional quality and economic performance in East Asian economies. Rev Appl Econ 2(2):201–216

Lee JW, Brahmasrene T (2014) ICT, CO_2 emissions and economic growth: evidence from a panel of ASEAN. Glob Econ Rev 43(2):93–109. https://doi.org/10.1080/1226508X.2014.917803

Levine R (1997) Financial development and economic growth: Views and agenda. J Econ Lit 35(2):688–726. https://www.jstor.org/stable/2729790

Levine R, Zervos S (1998) Stock markets, banks, and economic growth. Am Econ Rev 88(3):537–558. https://www.jstor.org/stable/116848

Mankiw NG, Romer D, Weil DN (1992) A contribution to the empirics of economic growth. Q J Econ 107(2):407–437. https://doi.org/10.2307/2118477

Mauro P (1995) Corruption and growth. Q J Econ 110(3):681–712. https://doi.org/10.2307/2946696

McKinnon RI (1973) Money and capital in economic development. Brookings Institution Press, Washington DC

Mises L, Batson HE (1953) The theory of money and credit. Yale University Press, United States

Northouse PG (2017) Introduction to leadership: Concepts and practice, 4th edn. Sage Publications, London

Nguyen TP, Dinh TTH, Tran NT, Thuy TDT (2022) U-shaped impacts of institutional quality and ICT penetration on economic growth: evidence from selected emerging countries. Cogent Econ Financ 10(1):2139887. https://doi.org/10.1080/23322039.2022.2139887

Owoye O, Bissessar N (2014) Corruption in African countries: a symptom of leadership and institutions failure. In: Mudacumura GM, Morçöl G (ed) Challenges to democratic governance in developing countries. Springer, Cham, p 227–245 https://doi.org/10.1007/978-3-319-03143-9_15

Pedroni P (1999) Critical values for cointegration tests in heterogeneous panels with multiple regressors. Oxford Bull Econ Stat 61(S1):653–670. https://doi.org/10.1111/1468-0084.0610s1653

Quinn DP, Woolley JT (2001) Democracy and national economic performance: the preference for stability. Am J Polit Sci 45(3):634–657. https://doi.org/10.2307/2669243

Rajan RG, Zingales L (2003) The great reversals: the politics of financial development in the twentieth century. J Financ Econ 69(1):5–50. https://doi.org/10.1016/S0304-405X(03)00125-9

Ram R (1999) Financial development and economic growth: additional evidence. J Dev Stud 35(4):164–174. https://doi.org/10.1080/00220389908422585

Roller LH, Waverman L (2001) Telecommunications infrastructure and economic development: a simultaneous approach. Am Econ Rev 91(4):909–923. https://doi.org/10.1257/aer.91.4.909

Roy S, Ahmed AMM, Abonamah AA (2014) ICT and economic growth: evidence from 12 MENA economies. Int J Cust Relatsh Mark Manag 5(1):16–30. https://doi.org/10.4018/ijcrmm.2014010102

Samarasinghe T (2018) Impact of governance on economic growth. MPRA Paper 89834, University Library of Munich, Germany

Sassi S, Goaied M (2013) Financial development, ICT diffusion and economic growth: lessons from MENA region. Telecommun Policy 37(4–5):252–261. https://doi.org/10.1016/j.telpol.2012.12.004

Schumpeter JA (1961) The theory of economic development: an inquiry into profits, capital, credit, interest, and the business cycle. Oxford University Press, New York

Seong J (2024) Why Asia can lead the world in a new era. https://www.mckinsey.com/featured-insights/future-of-asia/videos/why-asia-can-lead-the-world-into-a-new-era. Accessed 20 Jan 2025

Shamim F (2007) The ICT environment, financial sector and economic growth: a cross-country analysis. J Econ Stud 34(4):352–370. https://doi.org/10.1108/01443580710817452

Shaw ES (1973) Financial deepening in economic development. Oxford University Press, Oxford

Singha K, Singh MA (2022) Political stability and its effect on economy: some lessons from Sikkim Himalaya. J Social Econ Dev 24:340–354. https://doi.org/10.1007/s40847-022-00199-9

Solow RM (1956) A Contribution to the theory of economic growth. Q J Econ 70(1):65–94. https://doi.org/10.2307/1884513

Spyromitros E, Panagiotidis M (2022) The impact of corruption on economic growth in developing countries and a comparative analysis of corruption measurement indicators. Cogent Econ Financ 10(1):2129368. https://doi.org/10.1080/23322039.2022.2129368

Stock JH, Watson MW (2002) Macroeconomic forecasting using diffusion index. J Bus Econ Stat 20(2):147–162. https://doi.org/10.1198/073500102317351921

Tan YY, Eng A, Robinson E (2010) Perspectives on growth: a political-economy framework—Lessons from the Singapore experience. In: Brady D, Spence M (eds) Leadership and growth. World Bank Group, Washington, DC, pp 99–134

Tang CF, Lim YS (2022) Reactions of long-run private consumption in Malaysia to crises, economic leadership and telecommunication technology. Malays J Econ Stud 59(2):221–239. https://doi.org/10.22452/MJES.vol59no2.3

Tang CF, Lim YS (2024) The effects of telecommunication technology and leadership quality on private consumption in Malaysia. Int Soc Sci J 74(53):895–909. https://doi.org/10.1111/issj.12486

Tang CF, Rosidi MAI (2025) Investigating the effects of ICT infrastructure on Malaysia's economic growth: Insights from the Solow growth model. Inform Technol Dev 31(1):124–139. https://doi.org/10.1080/02681102.2024.2338527

Tang CF, Salman A, Abosedra S (2020) Dynamic interaction of tourism, finance, and institutions in explaining growth in Asia's little dragon economies. Int J Tour Res 22(1):15–25. https://doi.org/10.1002/jtr.2315

Tang CF, Tan BW (2014) The linkages among energy consumption, economic growth, relative price, foreign direct investment, and financial development in Malaysia. Qual Quant 48(2):781–797. https://doi.org/10.1007/s11135-012-9802-4

Uddin MA, Ali MH, Masih M (2017) Political stability and growth: an application of dynamic GMM and quantile regression. Econ Model 64:610–625. https://doi.org/10.1016/j.econmod.2017.04.028

Westerlund J (2007) Testing for error correction in panel data. Oxford Bull Econ Stat 69(6):709–748. https://doi.org/10.1111/j.1468-0084.2007.00477.x
Zolcsak D (2015) The effect of political leaders on economic growth through institutional change. Rom Econ J 18(58):175–190

Chor Foon Tang earned his Ph.D. in Economics from the University of Malaya in 2014 and is currently an associate professor at the Centre for Policy Research, Universiti Sains Malaysia. He has authored over 120 scholarly works on topics related to macroeconomics and growth catalysts across diverse platforms. His influential research earned numerous citations and recognition from Stanford University as one of the Top 2% of scientists worldwide since 2020. As a recognition for his outstanding contributions, he received the Malaysia Scopus Research Excellence Award 2024, conferred by the Ministry of Higher Education Malaysia and Elsevier.

Mannir Salisu earned his Bachelor of Science in Economics from Umaru Musa Yar'adua University, Katsina, Nigeria, in 2011 and his Master of Social Science (Economics) from Universiti Sains Malaysia in 2021. He is currently a lecturer in the Department of Banking and Finance at Hassan Usman Katsina Polytechnic, Katsina State, Nigeria. His research primarily focuses on development economics.

Part III
The Future of Sustainability

Chapter 10
Shaping Ocean Literacy Achievement as a Tool for the Future Asian Economies Toward Sustainable Seafood Consumption: The State of the Literature

Clusters of Future Studies: Past & Futures; Environmental Futures

You-How Go and Kwang-Jing Yii

Abstract Environmental sustainability of the ocean provides essential resources for economic development and food security. In that sense, ocean literacy is required to respond to making effective marine management for sustaining the phase of life living on our planet in light of the decline of food security. However, access to accurate and compelling ocean education remains low. In this chapter, we delve into findings from prominent studies. This undertaking allows us to highlight the key empirical challenges in the field of ocean economics and offers insights as follows: First, we describe an evolution of ocean literacy over the period from 2004 to 2024. Second, we provide a discussion about individuals' ocean literacy toward the support of sustainable seafood consumption behaviour. Third, we discuss the influence of environmental ocean on seafood consumption that the literature has uncovered so far. Forth, we provide several recommendations to directly link the research gaps to the avenues for further research. To promote public awareness of marine environmental threats and conservation efforts, this review is expected to leverage on the widespread interest in ocean literacy that becomes the substantial global attention.

Keywords Ocean literacy · Seafood consumption · Ocean sustainability · Economic sustainability · Marine management · Fisheries

---

Y.-H. Go (✉)
Teh Hong Piow Faculty of Business and Finance, Universiti Tunku Abdul Rahman (Kampar Campus), Kampar, Malaysia
e-mail: goyh@utar.edu.my

K.-J. Yii
Faculty of Business, Design and Arts, Swinburne University of Technology (Sarawak Campus), Kuching, Malaysia
e-mail: KYii@swinburne.edu.my

E. Lau et al. (eds.), *Future of Economy in Asia*,
https://doi.org/10.1007/978-3-031-95500-6_10

10.1 Introduction

Human well-being is inextricably connected to the health of the ocean for achieving a sustainable future in terms of food, energy and minerals. Seafood is urgently needed as one of the key elements for protecting the health and well-being of billions of people worldwide. As fundamental sources of global food security, ocean resources are used and marked to be the better diet for humans and the environment. However, growing demand and rapid trade development for seafood are driving the overexploitation of important marine resources that lead to climate change and pollution. These problems could pose threats to the long-term potential impacts on the ability of marine ecosystems and the health of ecological processes.

This chapter contributes to the understanding of ocean literacy in sustaining seafood consumption in three ways. First, this review provides ongoing debate on providing environmental literacy and food consumption for market actors to act proactively. Second, this review aids practitioners and regulators in designing strategies and policies that harness the good of marine management in economic sectors. Finally, this review also provides a critical baseline effort for academia and researchers toward an improved understanding of challenges in seafood consumption preferences and patterns as future research efforts, thereby supporting the broader vision for sustainable development that aims to "transform our world".

To reflect the essence of ocean literacy, individuals must adopt knowledge and technologies associated with marine management to urgent care of the environment for sustainable development (National Marine Education Association 2020). For improved public ocean literacy, there is not only a need for scientific knowledge to convey a personal connection with the ocean, but it also involves consideration of the relationship between humans and the ocean. Indeed, the goals of ocean literacy set out by the United Nations (UN) and other global initiatives are essential prerequisites in the environmental management of marine ecosystems for achieving sustainability goals (Fletcher and Potts 2007; Jefferson et al. 2015). These sustainability goals are embedded in communities. As a result, the approval of a stand-alone United Nations' Sustainable Development Goal (UN SDG) 14 "*Conserve and sustainably use the oceans, seas and marine resources for sustainable development*" has been a major achievement for the global ocean community. Owing to the importance of ocean resources with relation to food security, seafood consumption is considered one of the key elements in advancing national nutrition agendas (von Schuckmann et al. 2020). At the same time, ocean-literate individuals tend to appreciate the health benefits of seafood consumption. In particular, the focus of ocean literacy in an economy makes seafood consumption one of the most critical areas that regulators, industry players, consultants, and manufacturers have to elaborate guidelines for assessing and managing risks associated with harmful dinoflagellate toxins in seafood (McKinley et al. 2023).

Many agencies engage with experts to work on ocean-related issues. Decision-making on these related issues requires the process of learning about how humans and the ocean interact. Through ocean education as one of the important components of ocean literacy, this process of learning could result wide-ranging knowledge of the fundamental principles to underpin marine biota. To optimise strategies for conserving marine biodiversity as outlined in the UN SDG 14, the process of ocean literacy acquisition has gained a lot of public attention for sustaining the human well-being of the entire natural world. This concept is a mechanism for change that enables the market actors to adapt themselves to the environment they are executed in. Several associations in some countries often seek to prevent pollution practices, which can arise from individuals' problems when the individuals violate regulations protecting the environment, including the ocean. The United Kingdom, Australia, and the United States are the countries with the most international collaborative associations to improve the quality of the ocean. For example, the European Marine Science Educators Association (EMSEA), Canadian Network for Ocean Education (CaNOE), the International Pacific Marine Educators Network (IPMEN), and the Australian Association for Environmental Education (AAEE). These associations have been established to further make recommendations for promoting and enhancing ocean literacy in society. Thus, the growing awareness of natural environmental conditions for the ocean is increasingly seen to be critical for effective marine management, which is a means of enhancing seafood self-sufficiency.

10.1.1 Research Questions

Connecting ocean literacy to seafood consumption behaviour remains a challenging task. In this chapter, we conduct a review of the literature on ocean literacy toward sustainable seafood consumption which has received limited focus to date. Therefore, we address five questions as follows:

(i) What do the long-term trends imply for our understanding of ocean literacy?
(ii) What are the connections of ocean literacy to sustainable seafood consumption?
(iii) Where do we stand?
(iv) How do we extend the literature?
(v) What are the ways forward to improve ocean literacy and sustain seafood consumption in the future?

These questions are needed for the following reasons. First, the critical role in retaining an understanding of the ocean amongst communities needs to be devoted to a sustainable environment since the incidence of marine toxins and the associated seafood poisoning episodes happen in several countries. It follows that the role of ocean literacy is largely ignored in global environmental discussions around food security. Second, a comprehensive insight into the literature should support regulators and practitioners in proposing appropriate strategies for managing and preventing the incidence of marine toxins and polluted seafood. Third, knowledge of the pertinent determinants of the incidence of toxins in areas of the ocean that are most susceptible to contamination should be identified and suitable controls should be introduced. This knowledge provides insights into the need of promoting sustainable food consumption behaviour to prevent serious harmful effects on human health. Therefore, conducting a systematic literature review on ocean literacy toward sustainable seafood consumption is worthwhile, especially in the light of nuclear accidents to marine biota and seafood markets.

10.1.2 Ocean Literacy Programmes in Asia

There has been a lack of attention on ocean literacy in Asia since this continent accounts for more than two-thirds of global fish consumption, with China and other Southeast Asian countries leading in seafood production and consumption. In Malaysia and the Philippines, annual seafood consumption is reported to exceed 50 kg per person, which is much higher than the global average of 20 kg per person (Food and Agriculture Organization 2020). Additionally, extensive fishing industries and aquaculture operations in Asia lead to a high level of seafood consumption. For example, China is the world's largest producer and consumer of seafood. The growth of aquaculture in countries like Thailand and Vietnam has significantly increased the availability and consumption of seafood, catering to both domestic and international markets. Finally, geographic factors greatly influence seafood consumption in Asia. The extensive coastline in numerous island countries in Southeast Asia with diverse aquatic environments offers abundant marine resources to provide a wide variety of seafood and contribute to high seafood consumption in the specific regions.

In Asia, there were several successful ocean literacy initiatives to promote sustainable seafood consumption. For example, the Oki Islands Global Geopark in Japan has implemented educational programmes that increase ocean literacy among local communities and visitors by promoting sustainable marine resource use, awareness of marine biodiversity, and responsible seafood consumption (Oki Islands UNESCO Global Geopark 2022). These initiatives integrate traditional ecological knowledge with scientific insights, fostering a sense of stewardship and leading to more sustainable seafood practices. Moreover, Japan's Nakanoshima (Ama-cho) island was facing severe depopulation. It undertook radical reforms in 2007 to save its high school and created the Miryokuka project (Bedford 2023). This project transformed the school into a center for promoting sustainable living. The research involving observations and semi-structured interviews with stakeholders revealed that the school's place-based pedagogy significantly contributed to reversing the population decline and rejuvenating the island community.

With the use of a traffic light system, consumers were guided about sustainable seafood options by classifying seafood into three categories: green for recommended choices, yellow for occasional consumption, and red for avoidance. The goal of the guide was released on World Oceans Day in tandem with the Save Our Seafood (S.O.S.). A campaign by WWF-Malaysia and the Malaysian Nature Society (MNS) encouraged sustainable seafood choices and increased public awareness of the sharp decline in Malaysian fish stocks. This public awareness campaign has been conducted to distribute a pocket-sized bilingual guide in both English and Malay languages. Its result has been a significant increase in consumer awareness and a shift in purchasing behaviours toward more sustainably produced seafood consumption (World Wide Fund for Nature 2010).

10.2 Review Methodology

We attempt to answer the above questions through a concrete methodology used in the literature review and suggest planning for future research. To achieve our objective, we follow the Scientific Procedures and Rationales for Systematic Literature Reviews (SPAR-4-SLR) protocol developed by Paul et al. (2021). As shown in Fig. 10.1, this protocol consists of three stages, namely “Assembling”, “Arranging”, and “Assessing”.

For the first stage “Assembling”, we focus on the domain of the critical review related to the ocean environment and seafood consumption behaviour. We primarily source journal articles, as they are considered more scientific and have undergone robust peer review processes. For the acquisition, we use Google Scholar, Web of Science (WoS), and Scopus to search for articles from a wide range of leading journals. Targeting well-recognised journals in the field, we search for articles published between 2004 and 2024 by using keywords such as “ocean literacy”, “seafood consumption”, “ocean environmental sustainability”, “economic sustainability”, “marine management”, and “fisheries”. This search yields a total of 220 articles.

For the second stage “Arranging”, we organise the literature using the 5W1H framework (what, why, where, when, who, and how) developed by Lim (2020). In the sub-stage of purification, we conduct an initial screening based on article titles, selecting those with highly relevant titles and then reading the abstracts to ensure relevance. This process results in the exclusion of 155 irrelevant articles, remaining 65 relevant articles.

For the last stage “Assessing”, we evaluate the selected articles using domain-based reviews based on the structured themes (Lim 2016). These themes aim to assess the connection between individuals’ ocean literacy and their sustainable behaviour in seafood consumption within the review domain. Ultimately, we identify the gaps in the current literature and propose avenues for further research.

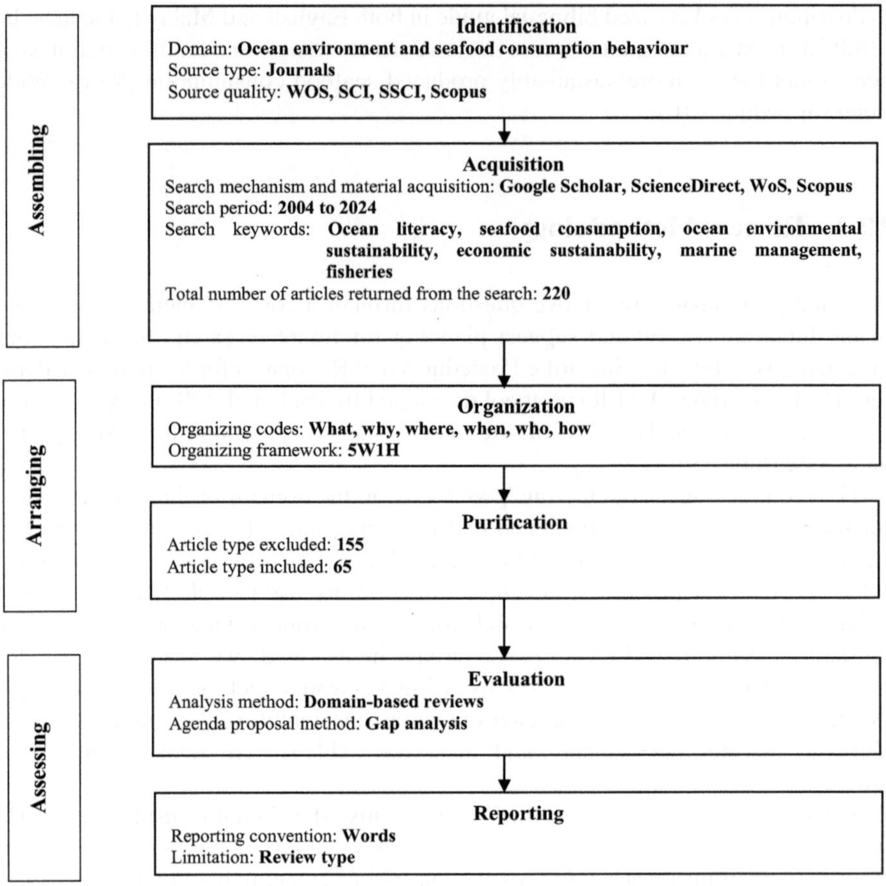

Fig. 10.1 Stages of conducting a critical literature review on ocean literacy toward sustainable seafood consumption

10.3 Findings

10.3.1 Presenting the State of the Literature on Ocean Literacy: A Twenty-First Century Guide

Ocean literacy has evolved over the past 20 yearsand echoed the increasing attention of the government towards Sustainable Development Goals (SDGs). In different countries and cultures, building on definitions of ocean literacy requires different views on the human-ocean relationship. This section underlines the drivers of long-term trends in the frameworks of ocean literacy over time with different audiences in different cultural contexts.

According to theoretical literature in 2004, a group of ocean scientists from at least seven organisations in the United States recognised a lack of ocean-related subjects in formal education that provided explanation of how consumption impacts the ocean environment from the time consumers procure sea food-related products. This group of scientists has taken an initiative action to coin ocean literacy. Through their collaborative and bottom-up process of the Ocean Literacy Campaign, they have developed a framework of ocean sciences in an attempt to improve environmental education that is aimed at teachers and educators. With the application of this original education-based framing of ocean literacy in formal education, individuals were expected to understand the relevance of the ocean to humankind. This understanding could make them communicate about the ocean and make responsible decisions concerning the ocean resources in a meaningful way (Cava et al. 2005a; Ocean Literacy Initiative 2005; Santoro et al. 2017). In the interim, newspapers and the internet were used as information sources for improving ocean knowledge among citizens, while information transmission through television and radio did not provide better effects (Steel et al. 2005). In November 2005, ocean literacy was grounded more firmly into the following seven principles.

- The Earth has one big ocean with many features;
- The ocean and life in the ocean shape the features of Earth;
- The ocean is a major influence on weather and climate;
- The ocean made the Earth habitable;
- The ocean supports a great diversity of life and ecosystems;
- The ocean and humans are inextricably interconnected; and
- The ocean is largely unexplored.

The incorporation of these principles into education facilitates the understanding and comprehension of the importance of marine and coastal environments for the younger generation in building the ocean-literate society in the future. For those who understand these principles, they could participate in effective decision-making to communicate about the ocean and its resources in a meaningful way (Fauville et al. 2019). Therefore, these principles were recognised as the key concepts central to raising a community of ocean stewards. The Ocean Literacy Campaign has further considered these principles for developing a comprehensive report to drive education settings in schools by reflecting the need for ocean literacy (Cava et al. 2005b).

Several studies were carried out to aid the development of ocean literacy. For instance, Kearns and Collins (2012) stated that the relationship between humans and the ocean should consider people's feelings and perceptions in a specific ocean context. In the Canada region with strong connections with the sea, Guest et al. (2015) found that those students who aged 12–18 years old actively engage in ocean-related activities with better knowledge were likely interested in ocean-related careers. In China, Umuhire and Fang (2016) have placed more emphasis on taking an interdisciplinary perspective to define ocean literacy. They found that ocean environmental awareness is vital in ensuring effective use of ocean resources for sustainable coastal development for students to participate in ocean related actions.

According to Gough (2017), ocean literacy contained key principles and concepts that engage learners of all ages. In the context of Ireland. Brennan et al. (2019) proposed six dimensions to the original education-based framing of ocean literacy in Galway. These dimensions included: (i) awareness, (ii) knowledge, (iii) attitude, (iv) communication, (v) behaviour, and (vi) activism. Stoll-Kleemann (2019) encapsulated the human-ocean relationship to further explore ocean literacy in a depth manner. By doing so, they considered extending the behaviour change model by Kollmuss and Agyeman (2002). Public knowledge about the ocean and coastal issues is typically low to moderate, resulting in an ocean-illiterate society (Steel et al. 2005; Boubonari et al. 2013; Perry et al. 2014; Guest et al. 2015; Heck et al. 2016; Umuhire and Fang 2016; Costa and Caldeira 2018). In this regard, they provided the most comprehensive insight on levels of ocean literacy into internal factors (emotions and values), external factors (socio-cultural and politico-economic), internal drivers (social norms and moral involvement), and external drivers (economic incentives).

In recent years, several scholars have defined ocean literacy from an interdisciplinary perspective. For instance, Uehara et al. (2020) revealed that ocean literacy could be an effective and feasible management measure for sustainable socio-ecological production landscapes in Satoumi through the cultivation of relational values. In the United Kingdom, Winks et al. (2020) demonstrated residential marine courses were determined by the cognitive characteristics of 16–18-year-old students rather than their behavioural or affective characteristics. Kelly et al. (2021) outlined the potential aspects for positive change towards a sustainable future. From their outline, they categorised these aspects into education, cultural connections, technological developments and knowledge exchange, and science-policy interconnections to develop an ocean literacy toolkit. According to the toolkit, ocean literacy activities in another step towards Sustainable 2030 are based on a broad range of stakeholders and benefit a wide variety of communities, societies, and environments. As encapsulated different worldviews and linguistic communities, MacNeil et al. (2021) argued that ocean literacy was still inadequate to capture the scope of individuals' experiences with the ocean.

In Brazilian curricula, Pazoto et al. (2022) found that regional guidelines addressed more ocean literacy content than other countries. However, the concepts of ocean literacy that could be used for dealing with environmental risks in coastal zones were poorly represented in Brazilian curricula. Aurélio et al. (2022) found that visitation to the local fish market and hands-on activities of fish biology could increase the learning motivation and encourage active citizenship regarding ocean conservancy. McKinley et al. (2023) presented an updated framework for the United Nations Ocean decade and beyond to highlight multiple dimensions regarding the evolution of ocean literacy. Based on their several highlighted dimensions, they proposed 11 recommendations. By employing the analytic hierarchy process method, Liu et al. (2023) connected the natural sciences on ocean-related issues in their proposed and integrated framework of ocean literacy with social sciences, thereby providing empirical support for future marine policy-making.

In Brazil's ocean and coastal governance system, Diederichsen et al. (2024) applied the Decalogue for Coastal Management. Their analysis revealed that the country has initiated to deal with issues of the Ocean Decade Challenges with strong policy and management tools. McKinley (2024) stated that the role of ocean literacy is important in the Ocean constitution to achieve a balance between the protection of the ocean and the use of marine resources for sustainable ocean futures. In the Macaronesia region of the North Atlantic Ocean, Costa et al. (2024) attempted to explore the perception of school teachers on ocean literacy during the 2021–2022 school year. They found that most of their teaching on awareness of ocean literacy did not influence the pedagogical practice, meaning they did not integrate ocean-related themes into school-related activities. In the Republic of Korea, Choi et al. (2024) aimed to assess the association between people's ocean literacy and climate change mitigation behaviours. Based on 2,000 respondents in November 2022, their findings revealed that those respondents who have a better understanding of the ocean were prompt to follow a range of recommendations for climate and action.

10.3.2 Connecting Ocean Literacy to Sustainable Seafood Consumption

This section reviews the recent empirical literature on whether individuals could sustain their food consumption through their environmental literacy. The way to discover the pulse of their seafood consumption behaviour is to set out with ocean literacy. To manage ocean resources, ocean literacy is widely known to provide individuals with the tools required to evaluate their potential choices and decision-making in their food consumption. The influence of ocean literacy on seafood consumption behaviour could be achieved through the following channels, as argued in existing studies.

10.3.2.1 Individuals' Preferences

In seeking a balance between individuals' personal need for tasty seafood (delicious option) and healthy seafood (nutritious option), ocean literacy is required to raise their concerns about environmental sustainability and drive better choices of seafood toward responsibly sourced species. Broadening ocean literacy could mobilise the ocean-literate society to make informed and responsible choices in seafood consumption (Mogias et al. 2015; Cretella et al. 2023). For example, those individuals who are health conscious and prioritise seafood aligned with their dietary needs. Communicating about ocean health, ocean conservation has the potential to preserve sea creatures and ensure the safety for producing seafood products in a way that best supports their health. For instance, the understanding of seafood's health effects on neurocognition development is important for those who are in pregnancies and childhood (Hibbeln et al. 2019).

10.3.2.2 Individuals' Demographic Characteristics

In Australia in 2009, the least educated residents with low incomes in the Brisbane metropolitan area regularly consumed fruit or vegetable juice and less consumed sushi (Miura et al. 2012). In Peninsular Malaysia, adults' fish consumption is associated with their ethnicity, age, marital status, residential area, and years of education (Ahmad et al. 2016). In Italy and Spanish from December 2019 to April 2020, those consumers who are older and with a higher educational level tend to read eco-labels before buying fish products (Galati et al. 2022).

10.3.2.3 Individuals' Socio-Economic Status

The frequency and types of foods consumed depend on socio-economic differences. The growing demand for seafood products has stimulated luxury food consumption patterns from individuals' tendency to hold the upper status of society. For example, in China, the rise of the traditional consumption of luxury seafood shows the status of the society, such as umami, delicious, fresh, like to eat, status, face consciousness, high-quality life, and networking with others (Wang et al. 2018). As a result, consumers are normally placing their motivation more on the symbolic value than the food value. For example, lobster and salmon are the most common species consumed for luxury seafood in Beijing, Shenzhen, and Shanghai (Wang and Somogyi 2020). Due to this social aspect, demand for luxury seafood species for consumers' choice motives may bring threats to the sustainability of ocean resources (Wang and Somogyi 2019). Over the period from 2000 to 2018, China's export structure in line with global demand for food has gradually declined in terms of its competitiveness of aquatic products in the market. Consequently, the country has diversified its market destinations to Japan, the United States, South Korea, Southeast Asia, and the European Union (Miao et al. 2021).

Food consumption is a key in the development of economic sectors in the food industry. For example, surimi is an ideal raw material for analogue seafood products and other innovative formulated foods for better utilization of raw material from the sea. As a result, the demand for surimi contributes to the rapid rise in popularity of its related products (Pigott 1986). The outright benefit of developing the industry of fisheries and aquaculture provides sources of protein, omega-3 fatty acids, vitamins, calcium, zinc, and iron for benefiting heart and brain health through food consumption (Serpetti et al. 2017; Diz et al. 2019). With a rich source of these essential nutrients, some communities are being incentivised to alter their food consumption patterns to seafood preferences. As derived from cultural practices for centuries, staple seafood species make up the diet of that population, thereby making seafood has been an essential component of many Asian cultures. For example, seafood is a mainstay of the diet in coastal countries like the Philippines, Thailand, Korea, and Japan. Seafood is a daily staple of meals in many traditional diets and culinary traditions, encompassing a wide range of species from fish and shellfish to seaweed.

As the human population continues to increase, countries have continued to exploit the ocean's resources for providing significance of nutritional value of food for human health (Visbeck 2018). For example, more than 60% of the world's population resides in Asia. With enormous populations, Asian countries like China and India play a significant role in a natural increase in the consumption of food overall, including seafood as well. However, a combination of chemicals and trash from land sources is blown into the ocean and impacts our ability to eat healthy seafood. Reducing populations of seafood species arising from unsustainable use of ocean resources through human activities could threaten the pillar of availability, especially in the regions that depend on fishing as a source of food (De-la-Torre 2020).

10.3.3 Gaps in the Current State of Literature

The main purpose of this section is to carry out the discussion to deal with research gaps within the empirical literature previously presented. Limited empirical evidence highlights the nexus between literacy and consumption behaviour linked to the ocean, and in particular to the seafood that it could provide. Additionally, effective ocean literacy leads to sustainable behaviour in seafood consumption that remains less clearly articulated in the literature. Nonetheless, the understanding of ocean-related topics is crucial to living and acting sustainably. However, the literature suffers from the following gaps which still limit the research regarding the ocean phenomenon.

10.3.3.1 Marine Biodiversity

For educational outreach, ocean literacy is still in its infancy. As a result, the content of ocean-related topics has not been adequately addressed and delivered through formal education curricula, which pointed out a lack of ocean-related subjects. Existing studies focused on the contribution of ocean literacy to the public's support for marine management, but failed to expand the concept to broader issues like threatened species, especially for unique marine creatures. The existing marine creatures could keep the balance in the ecological systems by enhancing ocean health.

Demand for food not only increases, but increasing population growth and urbanization appear to trigger a rapid change in food consumption patterns by demanding luxury seafood resources. This demand may reduce the population of target marine species (Sadovy de Mitcheson et al. 2019; Ben-Hasan et al. 2021). In the current era, considerable declines in biological diversity have been measured globally. Biodiversity loss is one of the drivers that change the environment rapidly (Leddin 2024; Fermepin et al. 2024). The ocean is susceptible to biodiversity loss as it contains complex ecosystems (Hoegh-Guldberg and Bruno 2010). However, the literature has failed to pay sufficient attention to biodiversity that has a greater impact on the ocean. For example, the influence of individuals' biodiversity literacy on their seafood consumption and preference, and consequently on marine defaunation in the global ocean.

10.3.3.2 Geographical Specificity in Southeast Asia

Looking at the selected studies allows further insights into the ocean environment on different continents. In scientific publications, the term ocean literacy has been quantitatively dominated by the United States, followed by the United Kingdom and Canada. In terms of the number of international collaborative publications, the United Kingdom and Italy presented the leading positions (Costa and Caldeira 2018). In our samples, most studies placed their emphasis on oceans in several continents, such as North America (Kollmuss and Agyeman 2002; Steel et al. 2005; Perry et al. 2014; Guest et al. 2015; Heck et al. 2016; MacNeil et al. 2021), South America (Pazoto et al. 2022; Diederichsen et al. 2024; McKinley 2024), Europe (Boubonari et al. 2013; Mogias et al. 2015; Brennan et al. 2019; Winks et al. 2020; Calado et al. 2021; Aurélio et al. 2022; Cretella et al. 2023; Costa et al. 2024), and Zealandia (Kearns and Collins 2012).

Asia is the largest continent with a vast and diverse geographical spread that connects with the Arctic Ocean to the north, the Pacific Ocean to the west, and the Indian Ocean to the south. There are several studies on ocean literacy in Asia. Most of them placed the emphasis on three countries in East Asia. For example, China (Umuhire and Fang 2016; Liu et al. 2023), Japan (Uehara et al. 2020), and the Republic of Korea (Choi et al. 2024). In China, the ways of consuming the live reef fish for food trade with diverse social could cause devastating environmental consequences for source countries in Southeast Asia (Fabinyi et al. 2012). Among the oceans in Asia, the Indian Ocean and Pacific Ocean provide the intricate oceanic connections of countries across the globe. Those countries are bordered by the Indian Ocean to the south and west, and as well as the Pacific Ocean to the west that contribute to domestic seafood supply because of their complex geographical locations with lowlands coastal areas and long coastlines. However, sea level projection from coastal tide gauges in Southeast Asian regions is expected to rise to 34.87 cm in 2100 (Ahmad Affandi et al. 2024). To the best of our knowledge, ocean literacy initiatives are still under-researched and under-taught in Southeast Asia, particularly island countries.

10.3.3.3 Innovative Methodologies

In the food sector, the role of ocean literacy could enlighten the innovative techniques and tools for producers and developers (Calado et al. 2021). To reduce the adverse impacts from uncontrolled consumers of added-value sustainable seafood, consumers should access information as a key driver for them to select environmentally sustainable sources of seafood (Roheim et al. 2018). The perceptions of consumers' behaviour are assessed through consumers' early literacy education involving formal and informal settings in school activities. All activities in schools aim to promote students' awareness of environmental issues and responsibilities. The innovation of formal and informal activities in schools has led to significant development in educational provision and practice. However, innovative education remains to be questionable whether it can influence individuals' ocean literacy toward more sustainable behaviour in their seafood consumption. In fact, innovative ways of defining education structures and learning approaches that are beyond the education system often comes with an unclear status of classification, depending on the conditions of education institutions.

10.4 Future Directions

There is a general belief that the twenty-first century will be the "greatest threat to global public health" (Atwoli et al. 2021). Ocean diversity is one of the critical aspects for ensuring that our shared future is as healthy as possible. If no effort and incrementally protection of ocean diversity, many severe uncertainties would remain. To promote a healthy and resilient ocean, commitments have added problems related to environmental and food concerns to development debates at local, national, regional, and global levels. Success of approaches in resolving the problems at global, regional, and national levels remains limited, making seafood not sufficiently protected. In addressing these problems, appropriate individual behaviour is a key driver toward more effective ocean literacy (Fletcher and Potts 2007). Individuals only change their behaviour through the use of information for self-development (Bandura 2016). However, individuals' consumption behaviour is extremely difficult to change, making the solution from the focus of their action on ocean challenges is still waning. What is more notable is the emphasis on calls for systematic behaviour change (Leakey 2022).

Based on the review's findings, the work can be further extended and improved by answering potential research questions as follows: Does biodiversity literacy contribute to ocean literacy and sustainable seafood consumption? How does ocean literacy fit with sustainable seafood consumption behaviour in Southeast Asian countries? What innovations should we expect in education to enhance ocean literacy? What are the effects of innovative education on ocean literacy? How much innovation is needed to promote ocean literacy? To answer these questions, we make the following tentative recommendations to stimulate further research in more determined ways for substantially advancing knowledge.

10.4.1 Exploring the Mediating or Moderating Effect of Biodiversity Literacy in Ocean

Effective utilisation of ocean resources is critical to the overall pursuit of SDGs, which go beyond economic considerations (Winther et al. 2020). As climate change is becoming recognised as the defining issue of the current era, the need for resilient societies becomes even more urgent. Resilient ocean-sustainable societies actively participate in climate change adaptation by enacting policies to protect coastal areas, championing sustainable fisheries management, and advocating for marine ecosystem conservation. They also play a role on a global scale by contributing to larger mitigation efforts, reducing greenhouse gas emissions, and promoting sustainable practices. Achieving this balance requires meticulous care in optimising ocean resource usage to meet societal needs while protecting the delicate balance of marine ecosystems. Navigating the complexities necessitates transformative change, challenging established norms, and encouraging a fundamental shift in how we perceive and interact with ocean resources (Struwig et al. 2024). The inertia in achieving SDGs, particularly in ocean governance, highlights the importance of bold actions to avoid potential resource depletion and environmental degradation, as ocean degradation might lead to a reduction in stewardship behaviour and cause a loss of access to seafood resources (Nash et al. 2022).

Embracing strong ocean stewardship entails recognising the need for lifestyle change and actively promoting sustainable behaviours at various societal levels, with the oceans serving as a focal point for critical adaptation and mitigation efforts. It is a multifaceted endeavour that requires a delicate balance between satisfying societal needs and preserving the health of marine ecosystems. Effective ocean stewardship is critical for transformative lifestyle shifts to recognise resilient and ocean-sustainable societies. Beyond individual choices, this transformation necessitates a thorough reassessment of consumption patterns, emphasising the importance of environmentally conscious behaviour, and the adoption of sustainable practices. Its ethos emphasises the importance of widespread community awareness, and highlighting the complex interplay between human activities and ocean vitality. Also, it emphasises the importance of optimizing the utilization of ocean resources to align with the objectives outlined in SDGs.

To pave the way for a more sustainable future, the achievement of SDGs has proven challenging because of necessitating transformative social change of existing norms and practices. A notable example illustrating the complexities of ocean stewardship is evident in the marine environment of Ireland, where there exists a disconnection between the public's perception of coastal areas and the definitions used by statistical agencies (Hynes et al. 2014). In a nutshell, strong ocean stewardship requires a comprehensive approach that considers societal needs, environmental health, and the importance of meeting global sustainability goals. It requires transformative change, increased awareness and a paradigm shift in how we perceive, use and protect ocean resources. Through these collaborative efforts, societies can work to create a resilient and ocean-sustainable future.

Emphasizing biodiversity education as a part of marine education becomes crucial for garnering public support and cooperation, which are pivotal elements in successful ocean stewardship. Biodiversity education delivered through coordinated public engagement initiatives in marine education plays an important role in educating communities about the far-reaching consequences of individual actions on marine creatures in natural habitats. These initiatives help to develop environmentally conscious communities capable of making informed decisions that are aligned with the overarching goal of ocean sustainability. However, marginalisation of marine education could lead young generations to be unaware of the ocean consequences of their seafood intake. It is not necessary that more knowledge and education of the oceans could equate to greater baseline awareness of the oceans and rationalizing food consumption for particular unique marine creatures. In general, residents in coastal areas are prone to express their feelings and perceptions about the consequences of ocean change (Kearns and Collins 2012). Therefore, they are more aware of the consequences, which are likely to sustain their behaviour in consuming seafood. To achieve sustainability, ocean education must start during the period of childhood (Santoro et al. 2022). In other regions of the world, educators are recommended to choose picture books for improving ocean education (Freitas et al. 2023).

Despite decades of practice through "surface learning", promoting ocean literacy based on knowledge rarely results in any significant change to development plans. This "surface learning" has been proven to be less effective in practice. The replacement of the common "surface learning" with "deep learning" would be helpful as it allows individuals to comprehend the cause and effect of the issue, focusing specifically on the interconnections between humans and the ocean. For example, how their behaviour can affect the ocean, either positively or negatively. Their comprehension should be going along with their experience about fragility, interconnectedness, and awe of nature so that they can become the environmental stewards of the future. This learning not only promotes ocean literacy by providing information about the consequences of unhealthy oceans but could provide concrete ways of preserving an intact marine environment. Generally, individuals who reside in coastal areas prone to aware about ocean related issues.

As looked to the future of ocean literacy, efforts to emphasize marine education should be usually linked to resounding calls for a better understanding of the interaction between human and ocean health. Also, these efforts must be underpinned by marine specialists. For example, before enhancing the concept of ocean literacy, they are necessary to know the readiness and needs of citizens about the ocean. After that, they could embed ocean literacy research in their action plans for an international and interdisciplinary programme in higher educational institutions. A quality of provision within university programmes in delivering ocean-related degrees should be characterised by a clear vision in predominantly natural and physical ocean sciences for assessments. These assessments are expected to generate a wealth of substance to inspire individuals in practicing sustainable consumption on ocean resources for their own and future generation benefit (Tlusty and Thorsen 2017). Lastly, the engagement in efforts from all regional fisheries bodies and regional seas conventions need to be devoted to ocean conservation through the Sustainable Ocean Initiative dialogue conferences.

The basic principle of ocean literacy is in different concepts of environmental ocean. The concept of biodiversity is best used as a focal point to develop the foundation of the relationship between environmental status and human behaviours in sustaining ocean practices. This concept is becoming an important aspect of education of marine biodiversity to enhance individuals' ocean literacy. Thus, we suggest that the current studies are advisable to continue the line of Hoegh-Guldberg and Bruno (2010) and Leddin (2024) to explore how biodiversity literacy could influence ocean literacy and its implications for sustainable seafood consumption. Examining the role of biodiversity literacy that mediates or moderates the relationship between ocean literacy and seafood consumption may provide numerous benefits, such as improved consumption behaviour, responsible environmental behaviour, and support for marine conservation activities.

10.4.2 Exploring the Interaction of Potential Development Toward Sustainable Ocean Economies in Southeast Asian Regions with Global Trends

Supportive seafood-sustainability policies and practices that protect the ocean sustainability are required to engage in economic activities associated with the ocean. However, from the geographical perspective, sustainable seafood production and consumption in Southeast Asian regions are rarely considered in economic development policies. Those individuals recognise democratic channels are prone to put pressure on politicians. In many cases, a lack of political agreements causes governments to slowly react to marine-related issues and perform marine education for achieving sustainable ocean development. Large retailers should have partnerships with environmental non-government organizations (NGOs) with a joint pledge to improve fishery sustainability in the business sector and certify their source seafood products as sustainable. By doing this now and in the future, states and NGOs could have greater coherence in attempting to steer seafood sustainability for maintaining environmental integrity (Tlusty and Thorsen 2017; Tlusty et al. 2019).

Besides that, the flexibility in considering economic and nutrition policies is crucial to meet diverse needs in various types of seafood. This could be done by reinforcing dialogue about the ocean in providing a paramount outcome for synergy across nations, agencies, and institutions in different geographic locations. Encouraging sustainable practices entails a variety of initiatives, including the reduction of single-use plastics, the promotion of environmentally friendly alternatives, the endorsement of sustainable seafood options, and the reduction of carbon footprints. These initiatives such as integrated and adaptive ocean management strategies, sustainable fishing practices, and the establishment of marine protected areas are critical components. Collaboration among governments, industry players, and environmental organisations is critical for facilitating policy implementation that encourages sustainability and penalises harmful practices (Ali et al. 2022). Transparent stakeholder dialogue processes are the necessary transformation toward ocean sustainability by advancing ocean governance in marine regions. These processes could provide inaugural opportunities about what are actions at the climate-ocean interaction should advocate the prioritisation of types of marine species for food purposes in climate conversations (Dobush et al. 2022).

Southeast Asia has been experiencing rapid industrialisation and urbanisation in recent decades, increasing demand for marine resources, but coming with the substantial sea pollution. Looking ahead, ocean literacy in Southeast Asia is a promising and fertile field. There is room for research when considering river and sea to reflect the current state, but also have the potential to shape future regional trajectories in ocean literacy and sustainable seafood consumption. For example, in Southern Vietnam, the Mekong River is facing an increasing pollution crisis caused by plastic bags and bottles. As a result, there are significant of river inputs of chemical elements to the sea. Apart from this issue, time is needed to acquire a better understanding of the interplay between river and sea life for better measures in the development of ocean literacy. After all, there is expected only one true solution to the issues of what exists, without real shortcuts for it.

10.4.3 Connecting Innovative Methodologies to Ocean Literacy

The transformative agenda calls for scientific research and technological advancements in responsible ocean monitoring (Lin and Yang 2020; Aguzzi et al. 2024), waste management (Schmaltz et al. 2020; Prajapati et al. 2021), and sustainable aquaculture practices (Yue and Shen 2022). This not only informs decision-making but also promotes a culture of continuous improvement and adaptability to changing challenges. Now, we turn to innovative methodologies, another suggested way of promoting ocean literacy.

Adopting innovative methodologies are critical to prioritise long-term sustainability over short-term gains. For example, the use of Collective Intelligence is demonstrated to be a methodological tool for a range of stakeholders across marine education, outreach, regulation, and policy to advance and enhance the promotion of ocean literacy among young people, thereby improving ocean literacy and boarding society's awareness of sustainable marine environments (McCauley et al. 2019). In a curriculum that lacks marine science, the fundamental design intention of a prototype iBook is embedded within the context of secondary classrooms through multiple interventions. A participatory design process in which a prototype iBook is embedded increases students' ocean literacy (McHugh et al. 2020). The use of innovative ways through virtual learning could improve marine education programmes within Australia's temperate region at the Great Southern Reef. Such ways not only improve students' ocean literacy, but they allow students to experience real-life examples to build their knowledge connections with the concepts related to the ocean (O'Brien et al. 2023).

In the modern era, the use of technological equipment alone is not sufficient to solve ocean pollution issues (Cordier and Uehara 2019). The solutions are not only technological ones, but it is likely to have greater educational attainment through innovation. Thus, an attempt to connect innovative education to ocean literacy should be made to generate knowledge and awareness toward protecting the ocean environment. Future researchers need to dive deeper into exploring innovative practices as such offering solutions even for complex teaching situations. Alternatively, innovative practices can be simply merged with traditional practices in establishing ocean literacy and see how the outcome on seafood consumption behaviour varies in the short term. Innovative practices in higher education can be used to attract motivated students by providing them with background information on the ocean concerning prior exposure to marine sciences and environments.

10.5 Conclusion

The ocean provides hope for the future as it plays a role in supporting individuals' well-being through food availability and quality on the growth, survival, and reproduction of marine species. Individuals tend to connect their actions with impacts on the ocean by thinking how their behaviour choices that link seafood and human health could impact the ocean. In line with this reasoning, ocean literacy has been proven to provide key terms related to the fundamental ocean concept, making it is an urgent need to reduce the environmental impact of seafood consumption. Therefore, effective ocean literacy is devoted to providing diagnostics of the ocean situation, and ways forward for establishing a direction toward sustainable seafood consumption.

In this chapter, by reviewing the literature, we explore the concept of ocean literacy that emerged in the early 2000s and outline some connections of ocean literacy to sustainable seafood consumption behaviour. Attention in the literature is found to be insufficiently drawn to three aspects, including marine biodiversity, geographical specificity in Southeast Asia, and innovative methodologies. Despite the delicate challenges of global ocean sustainability with the use of resources and environmental conservation, individuals have to be equipped with ocean literacy for their sustainable seafood consumption in general. However, the ocean literacy is particularly difficult to broadly establish accepted definitions with a wide range of domains. To deal with this difficulty, there are three ways that should be put forward for the UN 2030 SDGs and beyond. First, emphasizing biodiversity education as a part of marine education toward ocean stewardship. Second, providing supportive seafood-sustainability policies and practices toward sustainable ocean economies in Southeast Asian regions. Third, incorporating innovative methodologies for the promotion of ocean literacy.

In the interim, we do not only consider individual behaviour for responsible food consumption behaviour but also consider the need for behaviour change at an institutional and systemic scale. In decision-making institutions and businesses, there are more attainable outcomes that everyone must be involved in to deliver "ocean literacy for all" across regions. Furthermore, we should strive to transform the relationship between society and the ocean, which is relevant in local communities and global contexts.

Highlights

- Defining the concept of ocean literacy that emerged in the early 2000s.
- Outlining some connections of ocean literacy to sustainable seafood consumption behaviour in the literature from 2004 to 2024.
- Suggesting to explore marine biodiversity, geographical specificity in Southeast Asia, and innovative methodologies for the future Asian economies in shaping ocean literacy achievement toward sustainable seafood consumption.

References

Aguzzi J, Laurenz T, Flogel S et al (2024) New technologies for monitoring and upscaling marine ecosystem restoration in deep-sea environments. Engineering 34:195–211

Ahmad Affandi ML, Din AHM, Rasib AW (2024) Sea level rise estimation and projection from long-term multi-mission satellite altimetry and tidal data in the Southeast Asia region. Int J Remote Sens 45(23):9033–9063

Ahmad NI, Wan Mahiyuddin WR, Tengku Mohamad TR et al (2016) Fish consumption pattern among adults of different ethnics in Peninsular Malaysia. Food Nutr Res 60(1):1–15

Ali S, Xu H, Yang K et al (2022) Environment management policy implementation for sustainable industrial production under power asymmetry in the graph model. Sustain Prod Consum 29:636–648

Atwoli L, Baqui AH, Benfield T et al (2021) Call for emergency action to limit global temperature increases, restore biodiversity, and protect health: wealthy nations must do much more, much faster. Nutr Rev 79(11):1183–1185

Aurélio L, Sequeira V, França S et al (2022) Bridging the gap between formal and non-formal science education: traditional fish markets as a tool to promote ocean literacy. Appl Environ Educ Commun 21(3):238–253

Bandura A (2016) Moral disengagement: how people do harm and live with themselves. Worth Publishers, NewYork, NY

Bedford T (2023) Revitalizing an island community: education for sustainable living in Nakanoshima, Oki Islands. Japan. J Educ Sustain Dev 17(1):5–18

Ben-Hasan A, de Mitcheson YS, Cisneros-Mata MA et al (2021) China's fish maw demand and its implications for fisheries in source countries. Mar Policy 132:1–11

Boubonari T, Markos A, Kevrekidis T (2013) Greek pre-service teachers' knowledge, attitudes, and environmental behavior toward marine pollution. J Environ Educ 44(4):232–251

Brennan C, Ashley M, Molloy O (2019) A system dynamics approach to increasing ocean literacy. Front Mar Sci 6:1–20
Calado H, Pegorelli C, Vergílio M et al (2021) Expert knowledge-based co-development of scenarios for maritime spatial planning in the Northeast Atlantic. Mar Policy 133:1–9
Cava F, Society NG, Schoedinger S et al (2005a) Science content and standards for ocean literacy: an ocean literacy update
Cava F, Schoedinger S, Strang C et al (2005b) Science content and standards for ocean literacy: a report on ocean literacy. http://www.cosee.net/files/coseeca/OLit04-05FinalReport.pdf
Choi Y, Jang D, Lee MS et al (2024) Association between ocean literacy and climate change mitigation efforts in the Republic of Korea. Mar Policy 165:1–7
Costa AC, Freitas D, Santos AI et al (2024) Where is ocean literacy in oceanic islands' schools? The Azores case. Mar Policy 163:1–12
Costa S, Caldeira R (2018) Bibliometric analysis of ocean literacy: an underrated term in the scientific literature. Mar Policy 87:149–157
Cordier M, Uehara T (2019) How much innovation is needed to protect the ocean from plastic contamination? Sci Total Environ 670:789–799
Cretella A, Scherer C, Holm P (2023) Tasting the ocean: How to increase ocean literacy using seafood heritage with a visceral approach. Mar Policy 149:1–9
De-la-Torre GE (2020) Microplastics: An emerging threat to food security and human health. J Food Sci Technol 57(5):1601–1608
Diederichsen SD, Sardinha GD, Scherer MEG et al (2024) Opportunities to overcome the Ocean Decade Challenges in Brazil's ocean and coastal governance system. Ocean Coast Manag 247:1–9
Diz D, Morgera E, Wilson M (2019) Marine policy special issue: SDG synergies for sustainable fisheries and poverty alleviation. Mar Policy 110:1–2
Dobush BJ, Gallo ND, Guerra M et al (2022) A new way forward for ocean-climate policy as reflected in the UNFCCC Ocean and climate change dialogue submissions. Clim Policy 22(2):254–271
Fabinyi M, Pido M, Harani B et al (2012) Luxury seafood consumption in China and the intensification of coastal livelihoods in Southeast Asia: the live reef fish for food trade in Balabac. Philippines. Asia Pac Viewp 53(2):118–132
Food and Agriculture Organization (2020) The state of world fisheries and aquaculture 2020. Sustainability in action. Rome. https://doi.org/10.4060/ca9229en
Fauville G, Strang C, Cannady MA et al (2019) Development of the international ocean literacy survey: measuring knowledge across the world. Environ Educ Res 25(2):238–263
Fermepin S, Watson JE, Grantham HS et al (2024) Global marine conservation priorities for sustaining marine productivity, preserving biodiversity and addressing climate change. Mar Policy 161:1–8
Fletcher S, Potts J (2007) Ocean citizenship: an emergent geographical concept. Coast Manag 35(4):511–524
Freitas C, Francis P, Bellgrove A et al (2023) Adopting ocean-themed picture books to promote ocean literacy in primary education. Child Lit Educ 55:701–716
Galati A, Miret-Pastor L, Siggia D et al (2022) Determinants affecting consumers' attention to fish eco-labels in purchase decisions: a cross-country study. Br Food J 124(10):2993–3013
Gough A (2017) Educating for the marine environment: challenges for schools and scientists. Mar Pollut Bull 124(2):633–638
Guest H, Lotze HK, Wallace D (2015) Youth and the sea: ocean literacy in Nova Scotia, Canada. Mar Policy 58:98–107
Heck N, Paytan A, Potts DC et al (2016) Coastal residents' literacy about seawater desalination and its impacts on marine ecosystems in California. Mar Policy 68:178–186
Hibbeln JR, Spiller P, Brenna JT et al (2019) Relationships between seafood consumption during pregnancy and childhood and neurocognitive development: Two systematic reviews. Prostaglandins, Leukot Essent Fatty Acids 151:14–36

Hoegh-Guldberg O, Bruno JF (2010) The impact of climate change on the world's marine ecosystems. Science 328(5985):1523–1528
Hynes S, Norton D, Corless R (2014) Investigating societal attitudes towards the marine environment of Ireland. Mar Policy 47:57–65
Jefferson R, McKinley E, Capstick S et al (2015) Understanding audiences: making public perceptions research matter to marine conservation. Ocean Coast Manag 115:61–70
Kelly R, Evans K, Alexander K et al (2021) Connecting to the oceans: supporting ocean literacy and public engagement. Rev Fish Biol Fish 32:123–143
Kearns R, Collins D (2012) Feeling for the coast: the place of emotion in resistance to residential development. Soc Cult Geogr 13(8):937–955
Kollmuss A, Agyeman J (2002) Mind the gap: why do people act environmentally and what are the barriers to pro-environmental behavior? Environ Edu Res 8(3):239–260
Leakey C (2022) Connecting head, heart and hands for impact in the UN Ocean Decade. The marine biologist. https://www.mba.ac.uk/wp-content/uploads/2022/08/2205_MBA-TheMarineBiologist_Issue_23-POP.pdf
Leddin D (2024) The impact of climate change, pollution and biodiversity loss on digestive health and disease. Gastro Hep Adv 3(4):519–534
Lim WM (2016) Understanding the selfie phenomenon: Current insights and future research directions. Eur J Mark 50(9):1773–1788
Lim WM (2020) Challenger marketing. Ind Mark Manag 84:342–345
Lin M, Yang C (2020) Ocean observation technologies: a review. Chinese J Mech Eng 33(1):1–18
Liu S, Zeng W, Li X (2023) Toward an integrated framework of ocean literacy: a Delphi-AHP approach. Mar Policy 157:1–8
MacNeil S, Hoover C, Ostertag J et al (2021) Coming to terms with ocean literacy. Can J Environ Educ 24(1):233–252
McHugh M, McCauley V, Davison K et al (2020) Anchoring ocean literacy: participatory iBook design within secondary science classrooms. Technol Pedagog Educ 29(1):89–107
McKinley E (2024) Ocean literacy for an ocean constitution. Glob Const 13(1):25–28
McCauley V, McHugh P, Davison K et al (2019) Collective intelligence for advancing ocean literacy. Environ Educ Res 25(2):280–291
McKinley E, Burdon D, Shellock RJ (2023) The evolution of ocean literacy: a new framework for the United Nations Ocean Decade and beyond. Mar Pollut Bull 186:1–9
Miao M, Liu H, Chen J (2021) Factors affecting fluctuations in China's aquatic product exports to Japan, the USA, South Korea, Southeast Asia, and the EU. Aquac Int 29(6):2507–2533
Miura K, Giskes K, Turrell G (2012) Socio-economic differences in takeaway food consumption among adults. Public Health Nutr 15(2):218–226
Mogias A, Boubonari T, Markos A et al (2015) Greek pre-service teachers' knowledge of ocean sciences issues and attitudes toward ocean stewardship. J Environ Educ 46(4):251–270
Nash KL, van Putten I, Alexander KA et al (2022) Oceans and society: feedbacks between ocean and human health. Rev Fish Biol Fish 32:161–187
National Marine Education Association (2020) Ocean literacy: the essential principles and fundamental concepts of ocean sciences for learners of all ages. https://www.marineed.org/ocean-literacy/overview
O'Brien M, Freitas C, Venzo P et al (2023) Fostering ocean literacy through informal marine education programs. Mar Pollut Bull 193:1–9
Ocean Literacy Initiative (2005) Ocean literacy: the essential principles and fundamental concepts of ocean science. Technical report, Ocean Literacy Initiative
Oki Islands UNESCO Global Geopark (2022) An environmental education program utilizing the national park and the Geopark. https://www.oki-geopark.jp/en/education/
Paul J, Lim WM, O'Cass A et al (2021) Scientific procedures and rationales for systematic literature reviews (SPAR-4-SLR). Int J Consum Stud 45(4):1–16
Pazoto CE, Silva EP, Duarte MR (2022) Ocean literacy in Brazilian school curricula: an opportunity to improve coastal management and address coastal risks? Ocean Coast Manag 219:1–10

Perry EE, Needham MD, Cramer LA et al (2014) Coastal resident knowledge of new marine reserves in Oregon: the impact of proximity and attachment. Ocean Coast Manag 95:107–116
Pigott GM (1986) Surimi: The "high tech" raw materials from minced fish flesh. Food Rev Int 2(2):213–246
Prajapati P, Varjani S, Singhania RR et al (2021) Critical review on technological advancements for effective waste management of municipal solid waste—Updates and way forward. Environ Technol Innov 23:1–12
Roheim CA, Bush SR, Asche F et al (2018) Evolution and future of the sustainable seafood market. Nat Sustain 1(8):392–398
Sadovy de Mitcheson Y, To AWL, Wong NW et al (2019) Emerging from the murk: threats, challenges and opportunities for the global swim bladder trade. Rev Fish Biol Fish 29:809–835
Santoro F, Selvaggia S, Scowcroft G et al (2017) Ocean literacy for all: a toolkit (Vol 80). UNESCO Publishing
Santoro F, Tereza DM, Ana VM et al (2022) A new blue curriculum-a toolkit for policy-makers. United Nations Educational, Scientific and Cultural Organization, Paris
Schmaltz E, Melvin EC, Diana Z et al (2020) Plastic pollution solutions: emerging technologies to prevent and collect marine plastic pollution. Environ Int 144:1–17
Serpetti N, Baudron AR, Burrows MT et al (2017) Impact of ocean warming on sustainable fisheries management informs the ecosystem approach to fisheries. Sci Rep 7(1):1–15
Steel BS, Smith C, Opsommer L et al (2005) Public ocean literacy in the United States. Ocean Coast Manag 48(2):97–114
Stoll-Kleemann S (2019) Feasible options for behavior change toward more effective ocean literacy: a systematic review. Front Mar Sci 6:2–14
Struwig M, Berg AVD, Hadi N (2024) Challenges in the ocean economy of South Africa. Dev South Afr 41(1):1–15
Tlusty MF, Tyedmers P, Bailey M et al (2019) Reframing the sustainable seafood narrative. Glob Environ Change 59:1–6
Tlusty MF, Thorsen Ø (2017) Claiming seafood is 'sustainable' risks limiting improvements. Fish Fish 18(2):340–346
Uehara T, Sakurai R, Tsuge T (2020) Cultivating relational values and sustaining socio-ecological production landscapes through ocean literacy: a study on Satoumi. Environ Dev Sustain 22:1599–1616
Umuhire ML, Fang Q (2016) Method and application of ocean environmental awareness measurement: lessons learnt from university students of China. Mar Pollut Bull 102(2):289–294
Wang O, Somogyi S (2020) Motives for luxury seafood consumption in first-tier cities in China. Food Qual Prefer 79:1–9
Wang O, Somogyi S (2019) Consumer adoption of sustainable shellfish in China: effects of psychological factors and segmentation. J Clean Prod 206:966–975
Wang O, Somogyi S, Ablett R (2018) General image, perceptions and consumer segments of luxury seafood in China: a case study for lobster. Br Food J 120(5):969–983
Winks L, Ward M, Zilch J et al (2020) Residential marine field-course impacts on ocean literacy. Environ Educ Res 26(7):969–988
Winther JG, Dai M, Rist T et al (2020) Integrated ocean management for a sustainable ocean economy. Nat Ecol Evol 4:1451–1458
World Wide Fund for Nature (2010) WWF-Malaysia and MNS launch first Malaysian sustainable seafood guide. https://www.wwf.org.my/?10940/WWF-Malaysia-and-MNS-Launch-First-Malaysian-Sustainable-Seafood-Guide
Visbeck M (2018) Ocean science research is key for a sustainable future. Nat Commun 9(1):1–4
von Schuckmann K, Holland E, Haugan P et al (2020) Ocean science, data, and services for the UN 2030 sustainable development goals. Mar Policy 121:1–14
Yue K, Shen Y (2022) An overview of disruptive technologies for aquaculture. Aquac Fish 7(2):111–120

You-How Go is an Assistant Professor in Economics at the Teh Hong Piow Faculty of Business and Finance, Universiti Tunku Abdul Rahman, Malaysia. His research areas are informational efficiency, commodity futures, palm oil sustainability, climate change, and ocean sustainability. His research is largely published in mainstream economics, finance, and environment journals. Among them are *Resources Policy*, *Journal of Commodity Markets*, *Agricultural Economics*, *The North American Journal of Economics and Finance*, *The Journal of International Trade & Economic Development*, *Environmental Science and Pollution Research*, *Energy & Environment*, *Journal of Economic Surveys*, *Journal of Futures Markets*, *Journal of the Asia Pacific Economy*, and *Environment, Development and Sustainability*. He is also a member of the Malaysian Finance Association and Malaysian Economic Association.

Kwang-Jing Yii is a Senior Lecturer at the Faculty of Business, Design and Arts, Swinburne University of Technology Sarawak Campus, Malaysia. He has published numerous research papers in high-impact journals such as *Journal of Cleaner Production*, *Land Use Policy*, *Environmental and Ecological Statistics*, *Energy & Environment*, and *International Journal of Islamic and Middle Eastern Finance and Management*. His research interests include green finance, carbon finance, and environmental economics.

Chapter 11
Fuelling the Future: Decoding the Influence of Monetary Policy on Asia's Clean Energy Investment

Clusters of Future Studies: Past and Futures; Environmental Futures

Maggie May-Jean Tang, Kwang-Jing Yii, and Siew-Ling Liew

Abstract Rapid global economic evolution has driven industrialisation and urbanisation, driving up energy demands and fossil fuel consumption. This surge exacerbates environmental concerns, notably climate change, environmental degradation, and carbon emissions. Transitioning to clean energy is vital for achieving sustainable growth and combating climate challenges. Asia presently accounts for half of the world's carbon emissions. To achieve the ambitious target of net-zero emissions by 2050, it is essential to comprehensively examine the driving forces behind the clean energy transition in Asia. According to the International Energy Agency (IEA 2023), to meet the ambitious target, annual investments in clean energy need to scale up significantly from $62.3 billion in 2022 to $138.6 billion in 2030 and further surge to $165.8 billion in 2035. Monetary policy shocks have been a significant driver of uncertainty in financial markets. It influences economic dynamics and investment patterns across various sectors, including clean energy. Studying its impact on clean energy investment could offer valuable insight for policymakers to expedite the energy transition. This study examines the influence of monetary policy alongside urbanisation, green technology innovation, and green growth on clean energy investment. Using the System Generalised Method of Moments (SYS-GMM) approach, the findings reveal that monetary policy, urbanisation, and green growth have significant positive impacts on clean energy investment across Asia. Conversely, green technology innovation does not demonstrate a significant relationship. These results

M. M.-J. Tang (✉) · K.-J. Yii
School of Business, Swinburne University of Technology, Sarawak, Malaysia
e-mail: mtang@swinburne.edu.my

K.-J. Yii
e-mail: kyii@swinburne.edu.my

S.-L. Liew
School of Foundation Studies, Swinburne University of Technology, Sarawak, Malaysia
e-mail: lsliew@swinburne.edu.my

E. Lau et al. (eds.), *Future of Economy in Asia*,
https://doi.org/10.1007/978-3-031-95500-6_11

indicate that favourable monetary policies, urbanisation, and green growth are key pillars in fostering clean energy investments and supporting Asia's pathway to a greener and more sustainable future development.

Keywords Clean energy investment · Monetary policy · Urbanisation · Green growth · Green technology innovation · Asia

11.1 Introduction

Climate warming has emerged as a pressing global issue (Lin and Zhao 2023), with Asia playing a crucial role in this challenge. Asia accounts for over 50% of global carbon emissions, driven primarily by its reliance on coal, natural gas, and other fossil fuels for energy production. Remarkably, 90% of the world's new coal-fired power plants have been located in Asia in the past 20 years, contributing approximately 25% of the global annual greenhouse gas (GHG) emissions (Asian Development Bank 2024). In response to this alarming reality, several Asia countries have committed to achieving net-zero GHG emissions in the coming decades.

Clean energy is fundamental to combat climate change as it can significantly reduce the carbon footprint from the energy sector and minimise greenhouse gas emissions. To sustainably meet Asia's growing energy demands and climate targets, a substantial increase in clean energy investment is essential. The International Energy Agency (IEA) (2023) highlights the urgency of this transition, projecting that annual investments in clean energy must rise from $62.3 billion in 2022 to $138.6 billion during 2026–2030 and further to $165.8 billion in 2031–2035.

Monetary policy shocks have been a significant driver of uncertainty in financial markets. It influences economic dynamics and investment patterns across various sectors, including clean energy. As a capital-intensive industry, clean energy investment is particularly sensitive to fluctuations in monetary policy, which can alter the cost of external financing and shape expectations of future profitability. Additionally, monetary policy-driven changes in economic conditions can directly affect energy demand, thereby influencing the outlook for clean energy firms. A deeper understanding of how monetary policy impacts clean energy investment is crucial for policymakers aiming to accelerate the global energy transition and achieve climate objectives.

Urbanisation is often regarded as both a key driver of economic growth and a contributor to climate change. According to ecological modernisation theory, environmental degradation typically worsens during the early stages of modernisation but shows improvement in later stages as green technologies are adopted. Therefore, analysing the interplay between urbanisation, green technological innovation, and green growth offers a more comprehensive understanding of the factors driving clean energy investment. This integrated approach enriches the research by highlighting the pathways through which urbanisation and green advancements can contribute to a cleaner energy future.

This study seeks to explore the roles of monetary policy, urbanisation, green technological innovations, and green growth in driving clean energy investment across Asia. By addressing gaps in the existing literature from both theoretical and empirical perspectives, it makes a significant contribution to the field. Theoretically, the study employs ecological modernisation theory, integrating the influences of monetary policy and technological advancements into the analytical framework. Empirically, it offers fresh insights into the complex interplay between monetary policy and green initiatives, providing insight of how these factors collectively shape clean energy investment in Asia.

The structure of this paper is organised as follows: Sect. 11.2 offers a concise review of the relevant literature; Sect. 11.3 outlines the research methodology; Sect. 11.4 details the empirical findings and provides an in-depth discussion; and Sect. 11.5 concludes the study, summarising key insights and implications.

11.2 Literature Review

11.2.1 Monetary Policy and Clean Energy Investment

A few attempts have been made to empirically examine the relationship between monetary policy and clean energy investment. Zhang et al. (2023) use FMOLS estimations in 9 Asia economies disclosed that expansionary monetary policy encourages clean energy investment especially in developing countries. The findings also indicated a positive relationship exists between CO_2 emissions and clean energy investment in developed countries while negative effect is found in developing countries. This conclusion in line with the findings of He et al. (2019) who revealed that monetary policy encourages investment efficiency of renewable energy enterprises by influencing investment opportunities. In addition, Fruto-Bencze et al. (2020) determined that the monetary policy is responsive in renewable energy generation and concluded a noteworthy relationship between clean energy share and government spending. Muhafidin (2020) and Nga (2021) both found long-term relationships between monetary policy and renewable energy in Indonesia and Vietnam.

Furthermore, Mughal et al. (2021) using panel NADRDL discovered that contractionary monetary policy lessens the CO_2 emissions, whereas expansionary monetary policy increases CO_2 emissions in the long-run in ASEAN. Chishti et al. (2021) also reported that both expansionary and contractionary policies affect CO_2 emissions. They highlighted that monetary policy is anticipated to reduce the level of CO_2 emission by promoting adoption of clean technologies through investment. These findings further supported by Razmi et al. (2021) who indicated that monetary policy influences clean energy investment in Iran, suggesting that higher money supply increases investment. However, Sun et al. (2022) revealed that expansionary monetary policy negatively affects renewable energy investment in G7 countries.

Other than that, Liu et al. (2020) observed that while monetary policy uncertainty hinders non-renewable energy investment, it has no significant effect on renewable energy in China. In contrast, Sohail et al. (2021) found that monetary policy uncertainty negatively impacts renewable energy consumption in the U.S. Thus, it can be concluded that an effective monetary policy measure is an enabler to clean energy investment, which is necessary for environmental sustainability.

11.2.2 Green Growth and Clean Energy Investment

Recently, the concept of green growth has gained significant attention in most countries. Numerous studies have explored the relationship between green growth, economic performance, and clean energy investment. Song et al. (2019) applied an open economic growth model to China's provincial data found that while short-term R&D investments may not immediately favour green growth, they yield long-term benefits. This encourages clean energy investment alongside higher green economic growth. For BRIC nations, Danish (2020) revealed that environmental technologies significantly advance green growth and clean energy investments, promoting renewable technologies for economic gains.

In addition, Hao et al. (2021), using the CS-ARDL model showed that green growth improves environmental quality in G7 countries which fostering a sustainability-focused investment climate. However, their study did not assess its direct economic impacts. Fernandes et al. (2021) examined OECD data across 32 countries and stated that green growth can enrich the economic performance in which it would trigger more clean energy investments and creates job opportunities. Similarly, Naqvi et al. (2022) employed the Markowitz MVO framework identified green energy development as a pathway to economic and environmental sustainability, which drives clean energy investment.

Furthermore, Cui et al. (2022) employed a nonparametric approach to evaluate green productivity in 36 advanced countries and concluded that governmental investment in R&D for clean energy technologies spurs private-sector involvement. Khan et al. (2023) analysed the aggregate and disaggregate impacts of natural resources on economic performance in G7 countries using quantile regression. Their findings suggest that green growth initially hampers economic development due to the need for clean energy adoption, emphasising the importance of investing in clean energy technologies and infrastructure for sustainable development.

11.2.3 Green Technological Innovation and Clean Energy Investment

Ahmed et al. (2020) used CS-ARDL methods in their research and demonstrated that technological innovation reduces carbon emissions and boosts energy efficiency across 22 emerging economies. Their findings align with evidence from Mensah et al. (2018), Cho and Sohn (2018), and Shahbaz et al. (2020), who show the transformative impact of green technologies on ecological footprints in diverse regions. Jiang et al. (2022) analysed China's ecological footprint and observed that green innovation and renewable energy reduce ecological footprints. This indicated the need for strategic investments in green technologies to enhance productivity and competitiveness. Supporting this, Zhang et al. (2022) applied the NARDL approach and concluded that green investment, renewable energy, and technological innovation positively influence ecological sustainability in both the short and long term.

Furthermore, Fang (2023) highlighted the role of renewable energy investments, green innovation, and industrial structure in mitigating carbon emissions in China. Similarly, Cheng et al. (2021) found that technological innovation curbs carbon emissions and consequently promoting sustainable clean energy investments. Cui et al. (2022) utilised nonparametric methods to assess green productivity in 36 advanced countries and found that technological innovation significantly drives clean energy investment. Enhanced investment in renewable energy not only advances technological frontiers but also reduces environmental pollution and waste. Yan et al. (2020) similarly emphasised the importance of clean energy technologies for fostering green productivity as innovation in this sector mitigates greenhouse gas emissions and promotes environmental sustainability, providing a compelling case for increased investment.

While many studies underline the positive effects of green innovation, Santra (2017) revealed that technological advancements could increase carbon footprints in BRICS countries. Nonetheless, evidence generally points to the critical role of green innovation in achieving ecological sustainability and advancing clean energy investments globally.

11.2.4 Urbanisation and Clean Energy Investment

Lin and Wang (2020) examined the link between natural resource rents, renewable energy, urbanisation, and ecological footprint using panel data. Their findings indicated that renewable energy consumption, resource rents, and urbanisation positively impact environmental quality, creating opportunities for clean energy investments to meet the growing demand for sustainable energy. Similarly, Ahmad et al. (2020) confirmed the Environmental Kuznets Curve (EKC) hypothesis for G7 countries using the CS-ARDL model, showing that urbanisation increases environmental pressures. These findings highlight the urgent need for clean energy investments to mitigate urbanisation's environmental effects.

Cheng et al. (2021) studied urbanisation's impact on environmental degradation in 30 Chinese provinces and found it significantly increased carbon emissions, necessitating innovation in clean energy technologies. Strategic investments in such technologies are essential to ensure sustainable urban development. Supporting this, Bilgili et al. (2016) identified rising CO_2 emissions due to urbanisation in Britain, while Wang et al. (2016) argued that urbanisation fosters clean energy investment by concentrating resources and promoting research and innovation through urban institutions.

Majeed and Tauqir (2020), using dynamic estimation methods, found that urbanisation increased carbon emissions across 156 countries, stressing the need for green finance and clean energy investment for sustainable urbanisation. Hussain et al. (2023) echoed these findings for 54 African countries, revealing that urbanisation and non-renewable energy consumption degrade the environment. They emphasised adopting renewable energy and clean technologies to address environmental challenges.

Additional studies like Çetin et al. (2022) and Ozturk et al. (2022) showed varying relationships between urbanisation, economic growth, and CO_2 emissions across regions. These findings highlighted the importance of tailoring clean energy investment strategies to regional circumstances, as urbanisation plays a key role in shaping environmental quality and sustainability.

11.3 Methodology

The study undertakes a comprehensive, multi-stage examination of the data. Structured into several key segments. The process begins with data collection, followed by the application of empirical estimation models, and the use of econometric techniques to analyse the relationships between the variables.

11.3.1 Source of Data

The study uses annual secondary data spanning from 2000 to 2020, encompassing 16 Asia countries selected based on data availability. These countries include Azerbaijan, Bangladesh, China, Egypt, India, Indonesia, Japan, Jordan, Malaysia, Mongolia, Pakistan, Philippines, South Korea, Sri Lanka, Thailand, and Vietnam. The dependent variable refers to clean energy investment (CEI), while the independent variables are monetary policy (MP), green growth (GG), green technology innovation (GTI), and urbanisation (URB). The operationalised definition and source of data for each variable are explained in Table 11.1.

Table 11.1 Overview of data sources and variable definitions related to clean energy investment

Variables	Description	Source
CEI	Installed solar capacity (Gigawatts)	Our World in Data
MP	Central bank policy rate (%)	World Bank Data
GG	Production-based CO_2 productivity, GDP per unit of energy-related CO_2 emissions (US dollars per kilogram, 2015)	OECD Statistics
GTI	Annual patent applications	Our World in Data
URB	Urban population growth (annual %)	United Nations Development Programme

Definitions of the variables. CEI: clean energy investment; MP: monetary policy; GG: green growth; GTI: green technology innovation; URB: urbanisation

The descriptive statistics in Table 11.2 provide key insights into the mean, standard deviation, and range of the variables used in the study. The mean installed solar capacity (CEI) of 4.734 GW, with a range from minimal to highly developed infrastructure, reflects the region's disparities in renewable energy commitment. The average central bank policy rate (MP) of 6.004% suggests diverse monetary conditions across Asia, likely influencing green investment potential differently among countries. CO_2 productivity (GG), with a mean of 4.869 USD/kg, points to moderate energy efficiency in economic output, while the significant variability in annual patent applications (GTI) highlights uneven innovation efforts across the region, possibly due to disparities in R&D and policy support. An average urbanisation rate (URB) of 2.423% reflects rapid urban growth in many areas, shaping energy demand and sustainability challenges uniquely in Asia. These variations underline the importance of examining the interconnected roles these factors play in influencing green growth and clean energy investment across the diverse economies in Asia.

Table 11.2 Descriptive statistics of the key variables related to clean energy investment

Variables	CEI	MP	GG	GTI	URB
Mean	4.734	6.004	4.869	61,602.960	2.423
Std. Dev	23.120	3.826	2.814	191,064.600	1.377
Minimum	0.000002	0.100	1.170	0.000	−0.203
Maximum	253.964	19.250	16.840	1,393,815.000	11.771
Observation	336	336	336	336	336

Note CEI: clean energy investment; MP: monetary policy; GG: green growth; GTI: green technology innovation; URB: urbanisation

11.3.2 Models

We employ an extended IPAT framework, known as STIRPAT (Stochastic, Impacts, through Regression on Population, Affluence, and Technology), as proposed by Dietz and Rosa (1997). STIRPAT overcomes the drawbacks of IPAT and is appropriate for complex and non-monotonic relationships between environmental variables. The fundamental STIRPAT equation is as Eq. (11.1):

$$I_i = C + \alpha P_i + \beta A_i + \gamma T_i + \varepsilon_i \quad (11.1)$$

where P, A, T represent variables across various cross-sectional elements, while C and ε are the constant and error terms, respectively. In this study, we adapt the STIRPAT model by incorporating green-oriented variables to better capture environmental impacts. Specifically, we redefine the variables to align with sustainable growth factors such as clean energy investment as the environmental impact, green technology innovation under technology, green growth for affluence, and urbanisation replaces population. We assume that these green determinants are influenced by government policy including monetary policy, thus enhancing the model's relevance for environmental and economic policy analysis. The empirical model is expressed as Eq. (11.2) follows:

$$lnCEI_{it} = \beta_0 + \beta_1 MP_{it} + \beta_2 GG_{it} + \beta_3 lnGTI_{it} + \beta_4 lnURB_{it} + \varepsilon_{it} \quad (11.2)$$

where CEI denotes clean energy investment, MP denotes monetary policy, GG denotes green growth, GTI is for green technology innovation, URB denotes urbanisation, and ε represents the error term. The natural logarithm is applied for CEI, GG, and GTI, while the other variables are presented as percentages.

11.3.3 Estimation Procedures

Cross-sectional dependence (CSD) in panel data can result from various factors, such as omitted common effects, unobserved influences, shared shocks, and spatial interactions among cross-sectional units. To address this, the first step in our estimation procedure is to test for CSD in the series using Pesaran's (2004) CD test, where rejection of the null hypothesis of cross-sectional independence indicates CSD in the data. Another important issue in panel data models is slope heterogeneity, which can lead to biased results. To improve the reliability of our findings, we use the Pesaran and Yamagata (P&Y) slope heterogeneity test, which is effective for panels with small sample sizes and extended time periods (Pesaran 2007; Pesaran et al. 2008).

To prevent spurious regression, the subsequent estimation step involves conducting panel unit root tests to determine the order of integration for each variable. Given the potential CSD problem in the series, we opt for second-generation panel unit root tests that account for CSD, namely cross-sectional augmented IPS (CIPS) and Pesaran cross-section augmented Dickey Fuller (PESCADF) tests. CIPS test is particularly suitable for cross-sectionally dependent variables and provides reliable results even in the presence of slope heterogeneity (Pesaran 2007). Meanwhile, PESCADF test takes into consideration CSD in the panel data.

To avoid spurious regression, we begin the estimation process with panel unit root tests to determine the integration order of each variable. Due to the likelihood of cross-sectional dependence (CSD) in the dataset where shocks affecting one cross-sectional unit may influence others, we employ second-generation panel unit root tests that are designed to handle such dependence. Specifically, we use the cross-sectional augmented IPS (CIPS) test and the Pesaran cross-section augmented Dickey-Fuller (PESCADF) test. The CIPS test (Pesaran 2007) is particularly suited for datasets with cross-sectional dependence. It extends the traditional IPS test by including cross-sectional averages of lagged levels and first differences, making it effective in identifying unit roots even when variables are interdependent across cross-sections. Besides, the CIPS test provides robust results in the presence of slope heterogeneity, meaning that it can account for variations in the relationships among cross-sectional units. Similarly, the PESCADF test also adjusts for cross-sectional dependence by augmenting the Dickey-Fuller test with cross-sectional averages. This adjustment allows the PESCADF test to more accurately identify the presence of unit roots in panel data that exhibit cross-sectional dependence, reducing the likelihood of misleading results in the presence of interrelated panel units.

Following this, we employ panel cointegration tests, including the Kao, Pedroni, and Westerlund tests, to assess long-run relationships in the panel data. The Kao (1999) test, as an extension of the Engle-Granger approach, will provide a residual-based test for cointegration under the assumption of homogeneous relationships across cross-sections. However, it assumes a common cointegration vector, which may not apply if heterogeneity exists. In contrast, the Pedroni (2004) test allows for heterogeneous cointegration relationships, accommodating distinct long-run equilibrium relationships across different cross-sectional units. It offers multiple test statistics for both within-dimension (panel) and between-dimension (group mean) cases, increasing its robustness in detecting cointegration, particularly in heterogeneous panels. Moreover, the variance ratio in the Westerlund (2007) test can address cross-sectional dependence, a common characteristic in macroeconomic panels where shocks may affect multiple units simultaneously. This test assesses whether cointegration exists across the entire panel or within subgroups, making it effective in settings with cross-sectional dependence. Westerlund test also includes error correction tests, which can estimate the speed of adjustment toward long-run equilibrium.

To estimate the absolute and conditional convergence of clean energy investment across Asia countries, we utilise a dynamic panel data approach, specifically the System Generalised Method of Moments (SYS-GMM) estimator. This method, developed by Arellano and Bover (1995) and Blundell and Bond (1998), addresses the biases and inconsistencies that arise in traditional panel estimation techniques, such as panel OLS or Within Group estimators, which fail to account for dynamic panel bias and potential endogeneity of regressors. Unlike these traditional methods, System GMM provides consistent and efficient parameter estimates by effectively handling endogenous variables, heteroscedasticity, and autocorrelation issues, thus ensuring more reliable results. Moreover, System GMM's use of lagged instruments for endogenous variables enhances the estimator's efficiency, even in cases where the series exhibit near-random walk behaviour. This approach also includes critical diagnostics for validity, including tests for serial correlation, instrument validity, and the absence of over-identifying restrictions, ensuring the robustness of the estimation (Roodman 2009).

By employing lagged variables, this technique helps minimise biases that could affect the findings' validity. To further validate the analysis, two diagnostic tests are applied. The first assesses the presence of first-order autocorrelation in the residuals using the AR(1) procedure, followed by a second test to detect any second-order autocorrelation with AR(2) approach. These diagnostic tests, as suggested by Arellano and Bond, are critical for identifying and resolving autocorrelation problems within the model.

11.4 Findings

Table 11.3 shows that the Pesaran CD test reveals significant CSD for all variables at the 1% significance level, indicated by CD test values well above the critical threshold and p-values of 0.000. Table 11.4 presents the P&Y slope heterogeneity test results, where both detla tilde and delta tilde adjusted are significant at the 1% level. This indicates the presence of heterogeneity in the model, suggesting that the effect of predictors on the dependent variable (CEI) varied across cross-sectional units.

The panel unit root test results in Table 11.5 provide insights into the stationarity of variables in the dataset by using CIPS and PESCADF tests. These tests are applied to determine whether the variables have unit roots, which indicates if they are stationary at level, first difference, or require further differencing. In the CIPS test, the results indicate that MP and GTI are stationary at level or integrated of order

Table 11.3 Cross-sectional dependence (CSD) test results for the key variables related to clean energy investment

Variable	CD test	p-value
CEI	45.18***	0.000
MP	8.84***	0.000
GG	17.61***	0.000
GTI	7.66***	0.000
URB	12.52***	0.000

Note ***, **, and * denote the level of significance at 1%, 5%, and 10%, respectively. CEI: clean energy investment; MP: monetary policy; GG: green growth; GTI: green technology innovation; URB: urbanisation

Table 11.4 Slope heterogeneity test results for clean energy investment

Dependent variable: CEI	Model
Delta tilde	13.649*** (0.000)
Delta tilde adjusted	16.149*** (0.000)

Note ***, **, and * denote the level of significance at 1%, 5%, and 10%, respectively. () indicates p-value. CEI: clean energy investment

0 under the constant specification. URB also appears stationary at level under the trend and constant specification. For the other variables of CEI and GG, stationarity is only achieved after differencing, indicating they are integrated of order 1. When first-differenced, all variables exhibit significant test statistics across both constant and trend and constant specifications, confirming their stationarity at the first difference. The PESCADF test corroborates these findings. For all variables except for CEI, stationarity is observed at the first difference, as they become significant in the first-difference specifications, affirming their integration order of 1. Meanwhile, CEI is shown to be stationary at level under both constant and trend and constant specifications. Both CIPS and PESCADF tests confirm that none of the variables require second differencing, as they are stationary either at level or first difference. This suggests that the dataset is well-prepared for econometric modelling, with all variables meeting the necessary stationarity conditions.

Table 11.5 Cross-sectionally augmented IPS (CIPS) test and Pesaran cross-sectionally augmented Dickey Fuller test (PESCADF) results for the key variables related to clean energy investment

Cross-sectionally augmented IPS (CIPS) Pesaran (2007)				
Variables	Level		First-difference	
	Constant	Trend and constant	Constant	Trend and constant
CEI	−1.605	−1.996	−3.251***	−3.332***
MP	−2.254**	−2.405	−3.902***	−3.770***
GG	−1.275	−1.635	−3.434***	−3.723***
GTI	−2.622***	−2.822**	−4.353***	−4.652***
URB	−1.865	−3.191***	−4.327***	−4.297***
Pesaran cross-sectional augmented Dickey Fuller (PESCADF)				
CEI	−2.331**	−2.867**	−2.541***	−2.504
MP	−1.259	−1.727	−3.610***	−3.506***
GG	−0.957	−1.655	−2.521***	−2.851**
GTI	−1.513	−1.963	−3.457***	−3.780***
URB	−1.827	−1.703	−3.362***	−3.384***

Note ***, **, and * denote the level of significance at 1%, 5%, and 10%, respectively. CEI: clean energy investment; MP: monetary policy; GG: green growth; GTI: green technology innovation; URB: urbanisation

Table 11.6 Kao, Pedroni, and Westerlund panel cointegration test results for the key variables related to clean energy investment

	Model
Kao (1999)	
Modified Augmented Dickey-Fuller	−2.077** [0.018]
Dickey-Fuller	−1.403* [0.080]
Augmented Dickey-Fuller	−1.332* [0.091]
Pedroni (2004)	
Modified Phillips-Perron	4.205*** [0.000]
Augmented Dickey-Fuller	2.089** [0.018]
Westerlund (2007)	
Variance ratio	2.180** [0.014]

Note ***, ** and * denote significance level at 1%, 5%, and 10%, respectively. Coefficients for Kao and Pedroni tests refer to t-statistics while Westerlund test indicates the Z-value. p-values is reported in []

Table 11.6 presents the panel cointegration test results, providing robust evidence of long-run relationships among the variables, as indicated by the Kao, Pedroni, and Westerlund tests. Beginning with the Kao test, all three variations, such as the Modified Augmented Dickey-Fuller, Dickey-Fuller, and Augmented Dickey-Fuller, are significant. This demonstrates the presence of cointegration among the variables, though the significance level varies. The Pedroni test further substantiates this finding, with both the Modified Phillips-Perron and Augmented Dickey-Fuller tests showing high significance. This points to a strong and consistent long-run cointegration across the variables. Finally, the Westerlund variance ratio test also shows significance at the 5% level, further confirming the presence of long-run cointegration. This implies that, despite short-term fluctuations, the variables are likely to move together over time, supporting the model's suitability for long-term econometric analysis.

Table 11.7 indicates the results of the SYS-GMM estimation, examining the relationship between the independent variables (monetary policy, green growth, green technology innovation, and urbanisation) and clean energy investment in selected Asia countries. To ensure the reliability and precision of the SYS-GMM estimates, significance tests are conducted for AR(1) and AR(2) values. The validity of the SYS-GMM technique is supported by the z-statistics from these tests, with AR(1) significance at the 5% level confirming the presence of first-order autocorrelation in the residuals of clean energy investment. Conversely, the insignificance of AR(2) at the 5% level indicates an absence of second-order or higher autocorrelation, supporting the appropriateness of SYS-GMM for consistency and accuracy in the results.

The results show that past values of the dependent variable (clean energy investment) exert a strong and statistically significant positive influence on current values, with a lagged CEI coefficient of 1.039. Monetary policy exhibits a positive and significant effect, with a coefficient of 0.016, suggesting that a 1% increase in monetary policy leads to a 0.016% rise in clean energy investment. Additionally, the coefficient

Table 11.7 Results of System Generalised Method of Moments (GMM) estimations for dynamic relationship among the key variables related to clean energy investment

Variables	SYS-GMM
lag (CEI)	1.039*** (0.039) [0.000]
MP	0.016*** (0.005) [0.001]
GG	0.031** (0.089) [0.036]
GTI	−0.005 (0.035) [0.877]
URB	0.041*** (0.009) [0.000]
Constant	0.061* (0.227) [0.087]
Arellano-Bond test for AR (1)	z = −2.07, p-value = 0.001
Arellano-Bond test for AR (2)	z = −1.61, p-value = 0.054

Note ***, ** and * denote significance level at 1%, 5%, and 10%, respectively. Standard errors and p-values are reported in () and [], respectively. CEI: clean energy investment; MP: monetary policy; GG: green growth; GTI: green technology innovation; URB: urbanisation

of green growth (0.031) indicates that a 1% increase in green growth corresponds to a 0.031% increase in clean energy investment. Urbanisation also shows a positive and statistically significant coefficient of 0.041. This shows that a 1% increase in urbanisation will rise a 0.041% in clean energy investment. In contrast, green technology innovation shows no significant effect on clean energy investment.

These results indicate that urbanisation, green growth, and favourable monetary policies are key pillars supporting economic resilience in Asia. The limited impact of green technology innovation stresses a need for further investment and integration to realise its full economic potential.

11.5 Discussion of Implications

The SYS-GMM estimation results reveal that lagged clean energy investment (CEI) has a strong and positive influence on the current values, implying its persistence over time. This finding highlights the inertia inherent in clean energy investments, where increased investment in prior periods significantly correlates with sustained or amplified investments in subsequent periods. Such trends may reflect long-term commitments and the cumulative nature of clean energy infrastructure development.

Monetary policy emerges as a significant and positive determinant of clean energy investment, emphasising its critical role in fostering economic stability and growth in Asia. Recently, central banks in countries like India and Indonesia have adopted policies to stimulate investment and spending through interest rate adjustments and liquidity provisions. These efforts align with Asia's broader focus on enhancing local economic development, creating employment opportunities, and driving industrial growth. The findings suggest that even small adjustments to accommodative monetary conditions can yield notable positive impacts, particularly in Asia's predominantly developing economies.

Green growth is also found to have a positive and significant effect on clean energy investment. This result emphasises the efficacy of eco-friendly development policies in fostering sustainable economic outcomes. Countries like China, Japan, and South Korea have demonstrated leadership in prioritising green growth through substantial investments in renewable energy, air quality improvement initiatives, and sustainable practices. The positive relationship between green growth and CEI highlights its multifaceted benefits, including job creation, reduced environmental costs, and enhanced resilience to climate-related disruptions. Given that pollution and resource depletion have historically accompanied Asia's rapid growth, green initiatives offer a means to simultaneously support the economy and address environmental challenges.

Urbanisation, another key variable, exhibits a significant positive relationship with clean energy investment. The rapid pace of urbanisation across Asia, particularly in nations such as China and India, has been a driving force behind economic transformation. Urban growth has facilitated industrialisation, improved access to services, and enhanced labour productivity. The expansion of urban centres has increased demand for housing, infrastructure, and consumer goods, further propelling economic growth. Furthermore, Asia's urban hubs often serve as innovation clusters, promoting sustainable practices and aligning with the region's long-term growth aspirations.

Conversely, green technological innovation does not display a significant direct effect on clean energy investment. While green technology is undeniably pivotal for long-term sustainability, its immediate economic impact remains constrained. This result likely reflects the heterogeneous nature of green technology adoption across Asia. While advanced economies like Japan and South Korea lead in renewable technology development, other nations are still in the early stages of transition. Additionally, the high costs and long timelines associated with research, development, and deployment of green technologies may dilute their measurable impact on CEI in the short term. These findings suggest that further advancements in green technology and its integration into economic systems are necessary to unlock its full potential.

11.6 Future Directions

The findings of this study hold significant implications for policymakers, investors, academics, and other stakeholders engaged in sustainable development and energy transition efforts. Future research could build upon these insights by examining specific policy instruments and mechanisms that directly influence clean energy investment. In particular, exploring the roles of fiscal incentives, tax reforms, subsidies, and public–private partnerships may provide actionable strategies to create a more supportive environment for clean energy initiatives in Asia and beyond.

The study also highlights an insignificant relationship between green technological innovation and clean energy investment in the short term. This presents an opportunity for future research to investigate the long-term impacts of green technology development and deployment. Studies could consider lagged effects and the transformative potential of breakthrough innovations, particularly as technology matures and achieves greater integration into energy systems. Such an approach could uncover the cumulative benefits of sustained investment in research and development.

Additionally, the increasing frequency and severity of climate-related risks are likely to have a profound influence on clean energy investment decisions. Future research could explore how strategies aimed at building climate resilience and enhancing adaptation measures impact investment flows within the clean energy sector. This includes assessing the role of insurance mechanisms, risk-sharing frameworks, and disaster-prevention infrastructure in guiding investment decisions.

By addressing these future directions, subsequent studies can contribute to a deeper understanding of the multifaceted factors driving clean energy investments in Asia. Such insights will be critical in designing effective policies and strategies to accelerate the global transition to a sustainable and resilient energy future.

11.7 Conclusion

Clean energy investment is essential in tackling global climate challenges. While various factors drive clean energy investments, limited research has explored the relationship between monetary policy and clean energy investment within the Asian context. This paper offers a comprehensive analysis of how monetary policy and green initiatives influence clean energy investments in Asia. Using annual data from 16 Asia countries from 2000 to 2020, the study employs the SYS-GMM to estimate the absolute and conditional convergence of clean energy investment across the region. The findings provide critical insights into the linkages between monetary policy, green technological innovations, and economic factors such as green growth and urbanisation in driving clean energy transitions.

The results reveal that monetary policy has significant and positive impacts in fostering clean energy investment. Effective monetary policies, such as integrating climate risks into central bank operations could accelerate the clean energy investment. This includes introducing green bonds, establishing climate risk stress testing for financial institutions, and incentivising private sector investments in clean energy projects. Developing regional green finance platforms can also enhance cross-border collaboration in financing clean energy initiatives. Investors will be more willing to allocate funds to clean energy projects with the support from government policies that creates conducive environment for clean energy investment.

Additionally, the positive relationship between green growth, urbanisation and clean energy investment highlights the potential economic benefits of transitioning towards cleaner energy sources. With rapid urbanisation, Asia should focus on creating smart, energy-efficient cities. Policymakers must prioritise clean energy adoption in urban planning, enforce green building standards, and enhance public transportation systems powered by clean energy. Investments in urban clean energy grids and energy storage systems are essential to meet future demands sustainably.

Moreover, policymakers in Asia should align economic growth strategies with environmental sustainability by offering incentives for green businesses, implementing carbon pricing mechanisms, and phasing out fossil fuel subsidies. Furthermore, strengthening public–private partnerships can drive economic growth while ensuring environmental conservation. Policymakers should play a proactive role in de-risking clean energy investments through measures such as public guarantees, co-financing mechanisms, and regulatory stability. By mitigating investment risks, countries in Asia can attract more private capital in clean energy projects, accelerating the energy transition.

Highlights

- Prior clean energy investment shows persistence, driving future growth.
- Supportive monetary policies significantly boost clean energy investment in Asia.
- Green growth positively impacts investment, promoting sustainable development.
- Urbanisation fuels clean energy demand through economic transformation.
- Green technology innovation requires long-term focus for measurable impact.

References

Ahmad M, Jiang P, Majeed A, Umar M, Khan Z, Muhammad S (2020) The dynamic impact of natural resources, technological innovations and economic growth on ecological footprint: an advanced panel data estimation. Resour Policy 69:101817. https://doi.org/10.1016/j.resourpol.2020.101817

Ahmed Z, Asghar MM, Malik MN, Nawaz K (2020) Moving towards a sustainable environment: the dynamic linkage between natural resources, human capital, urbanisation, economic growth, and ecological footprint in China. Resour Policy 67:101677. https://doi.org/10.1016/j.resourpol.2020.101677

Arellano M, Bover O (1995) Another look at the instrumental variable estimation of error-components models. J Econom 68:29–51

Asian Development Bank (2024) Energy transition mechanism. https://www.adb.org/what-we-do/energy-transition-mechanism-etm. Accessed 21 Sep 2024

Bilgili F, Koçak E, Bulut Ü (2016) The dynamic impact of renewable energy consumption on CO_2 emissions: a revisited environmental kuznets curve approach. Renew Sust Energ Rev 54:838–845
Blundell R, Bond S (1998) Initial conditions and moment restrictions in dynamic panel data models. J Econom 87:115–143
Çetin M, Aslan A, Sarıgül SS (2022) Analysis of the dynamics of environmental degradation for 18 upper middle-income countries: the role of financial development. Environ Sci Pollut Res 29:1–18
Cheng C, Ren X, Dong K, Dong X, Wang Z (2021) How does technological innovation mitigate CO_2 emissions in OECD countries? Heterogeneous analysis using panel quantile regression. J Environ Manag 280. https://doi.org/10.1016/j.jenvman.2020.111818
Chishti MZ, Ahmad M, Rehman A, Khan MK (2021) Mitigating pathways towards sustainable development: assessing the influence of fiscal and monetary policies on carbon emissions in BRICS economies. J Clean Prod 292:126035
Cho JH, Sohn SY (2018) A novel decomposition analysis of green patent applications for the evaluation of R&D Efforts to reduce CO_2 emissions from fossil fuel energy consumption. J Clean Prod 193:290–299
Cui L, Mu Y, Shen Z, Wang W (2022) Energy transition, trade and green productivity in advanced economies. J Clean Prod 361:132288
Danish UR (2020) How do environmental technologies affect green growth? evidence from BRIC economies. Sci Total Environ 712:136504
Dietz T, Rosa EA (1997) Effects of population and affluence on CO_2 emissions. Proc Nati Acad Sci 94(1):175–179. https://doi.org/10.1073/pnas.94.1.175
Fang Z (2023) Assessing the Impact of renewable energy investment, green technology innovation, and industrialisation on sustainable development: a case study of China. Renew Energy 205:772–782
Fernandes CI, Veiga PM, Ferreira JJ, Hughes M (2021) Green growth versus economic growth: do sustainable technology transfer and innovations lead to an imperfect choice? Bus Strat Environ 30(4):2021–2037
Fruto-Bencze D, Avdiu K, Unger S (2020) The effect of trade and monetary policy indicators on the development of renewable energy in Latin America. Crit Perspect Int Bus 16(4):337–359
Hao LN, Umar M, Khan Z, Ali W (2021) Green growth and low carbon emission in G7 countries: how critical the network of environmental taxes, renewable energy and human capital is? Sci Total Environ 752:141853
He L, Wu M, Zhong Z, Xia Y, Yu J, Jiang Y (2019) Can monetary policy affect renewable energy enterprises investment efficiency? a case study of 92 listed enterprises in China. J Renew Sustain Energy 11(1)
Hussain MN, Li Z, Yang S (2023) Heterogeneous effects of urbanisation and environment Kuznet curve hypothesis in Africa. Nat Resour Forum, Blackwell Publishing 47(2):317–333
International Energy Agency (IEA) (2023) Scaling up private finance for clean energy in emerging and developing economies. international energy agency and international finance corporation. https://www.iea.org/reports/scaling-up-private-finance-for-clean-energy-in-emerging-and-developing-economies. Accessed on 30 January 2024.
Jiang D, Shu H, Fan Y, Dong Y, Li H (2022) Influence of renewable energy and natural resources on climate change: the role of green innovation in China. Front in Environ Sci 10:966656
Kao C (1999) Spurious regression and residual-based tests for cointegration in panel data. J Econom 90(1):1–44
Khan Z, Hossain MR, Badeeb RA, Zhang C (2023) Aggregate and disaggregate impact of natural resources on economic performance: role of green growth and human capital. Resour Policy, 80
Lin B, Wang M (2020) How to boost energy productivity in China's industrial sector: an integrated decomposition framework based on multidimensional factors. J Clean Prod 259:120902
Lin B, Zhao H (2023) Tracking policy uncertainty under climate change. Resour Policy 83:103699

Liu R, He L, Liang X, Yang X, Xia Y (2020) Is there any difference in the impact of economic policy uncertainty on the investment of traditional and renewable energy enterprises? a comparative study based on regulatory effects. J Clean Prod 255(3):120102

Majeed TM, Tauqir A (2020) Effects of urbanisation, industrialisation, economic growth, energy consumption and financial development on carbon emissions: an extended STIRPAT model for heterogeneous income groups. Pak J Commer Social Sci 14(3):652–681

Mensah CN, Long X, Boamah KB, Bediako IA, Dauda L, Salman M (2018) The effect of innovation on CO_2 emissions of OCED countries from 1990 to 2014. Environ Sci Pol Resour 25:29678–29698

Mughal N, Kashif M, Arif A, Guerrero JWG, Nabua WC, Niedbala G (2021) Dynamic effects of fiscal and monetary policy instruments on environmental pollution in ASEAN. Environ Sci Pol Resour 28:65116–65126

Muhafidin D (2020) The role of fiscal policy and monetary policy in environmental degradation in Indonesia. Int J Energy Econ Policy 10(3):504–510

Naqvi B, Rizvi SKA, Hasnaoui A, Shao X (2022) Going beyond sustainability: the diversification benefits of green energy financial products. Energy Econ 106111

Nga NTV (2021) Monetary policy, exchange rate, renewable energy and economic growth: an empirical analysis of Vietnam. Accounting 7(6):1315–1324

Ozturk S, Cetin M, Demir H (2022) Income inequality and CO_2 emissions: nonlinear evidence from Turkey. Environ Dev Sustain 24(10):11911–11928

Pedroni P (2004) Panel cointegration: asymptotic and finite sample properties of pooled time series tests with an application to the PPP hypothesis. Econ Theory 20(3):597–625

Pesaran MH (2004) General diagnostic tests for cross section dependence in panels (No. 0435)

Pesaran MH (2007) A simple panel unit root test in the presence of cross-section dependence. J Appl Econom 22(2):265–312

Pesaran MH, Ullah A, Yamagata T (2008) A bias-adjusted LM test of error cross-section independence. Econom J 11(1):105–127

Razmi SF, Moghadam MH, Behname M (2021) Time-varying effects of monetary policy on Iranian renewable energy generation. Renew Energy 177(18):1161–1169

Roodman D (2009) How to do xtabond2: An introduction to difference and system GMM in Stata. Stata J 9(1):86–136

Santra S (2017) The effect of technological innovation on production-based energy and CO_2 emission productivity: evidence from BRICS countries. Afr J Sci Technol Innov Dev 9:503–512

Shahbaz M, Raghutla C, Song M, Zameer H, Jiao Z (2020) Public-private partnerships investment in energy as new determinant of CO_2 emissions: the role of technological innovations in China. Energy Econ 86:104664

Sohail MT, Yu X, Usman A, Majeed MT (2021) Renewable energy and non-renewable energy consumption: assessing the asymmetric role of monetary policy uncertainty in energy consumption. Environ Sci Pol and Res 28(3):31575–31584

Song X, Zhou Y, Jia W (2019) How do economic openness and R&D investment affect green economic growth? Evidence from China. Resour Conserv Recycl 146:405–415

Sun C, Khan A, Liu Y, Lei N (2022) An analysis of the impact of fiscal and monetary policy fluctuations on the disaggregated level renewable energy generation in the G7 countries. Renew Energy 189:1154–1165

Wang S, Li Q, Fang C, Zhou C (2016) The relationship between economic growth, energy consumption, and CO_2 emissions: empirical evidence from China. Sci Total Environ 542:360–371

Westerlund J (2007) Testing for error correction in panel data. Oxf Bull Econ Stat 69(6):709–748

Yan Z, Zou B, Du K, Li K (2020) Do renewable energy technology innovations promote China's green productivity growth? fresh evidence from partially linear functional-coefficient models. Energy Econ 90:104842

Zhang D, Wang Y, Peng X (2023) Carbon emissions and clean energy investment: global evidence. Emerg Mark Financ Tr 59(2):312–323

Zhang H, Shao Y, Han X, Chang HL (2022) A road towards ecological development in China: the nexus between green investment, natural resources, green technology innovation, and economic growth. Resour Policy 77:102746

Maggie May-Jean Tang is a lecturer in the School of Business at Swinburne University of Technology Sarawak Campus. Prior to joining Swinburne Sarawak, she was a business strategist in a company and engaged with several tertiary institutions as a part-time lecturer in teaching business, economics and finance subjects. She holds a Ph.D. in Financial Economics. Her research interests include financial and monetary economics, asymmetrical analysis and environmental economics.

Kwang-Jing Yii is a senior lecturer at School of Business, Swinburne University of Technology Sarawak Campus. Yii holds a Ph.D. in financial economics and has previously held academic positions at several prestigious universities. He is also a member of various professional organisations, including Malaysian Finance Association and Malaysian Economic Association. His research interests include financial economics, energy economics, and behavioural finance.

Siew-Ling Liew joined the School of Foundation Studies, Swinburne University of Technology (Sarawak Campus) (SFS) as a Lecturer in 2023. Prior to joining Swinburne Sarawak, she was engaged with several tertiary institutions as a lecturer in teaching Economics-related subjects and she held leadership positions including programme coordinator and head of research and postgraduate studies. She received her Ph.D. in Economics 2018. Her research interests include economics, economics growth and development.

Chapter 12
Energy-Trade-Growth Nexus in the Asia–Pacific: Implications for the Future Economy

Clusters of Future Studies: Environmental Futures; Technological Trends

Yoke Kee Eng, Grace Hooi Yean Lee, and Audrey Kim Lan Siah

Abstract This chapter delves into the nuanced relationship between energy consumption, trade, and economic growth in 15 Asia–Pacific nations from 1965 to 2014, utilising the Panel Vector Autoregressive (PVAR) approach. The research meticulously unpacks the influence of both aggregated and disaggregated energy sources on trade and output. Leveraging orthogonalised impulse-response functions, the study offers a unique lens to scrutinise output reactions to energy-related shocks. Contrary to prevalent fears, the findings reveal that the Asia–Pacific's ambitious energy conservation objectives will not impede its economic trajectory. Instead, trade emerges as a seminal force propelling economic growth, reaffirming the trade-led growth hypothesis. This signifies the region's urgent need for continued trade liberalisation, especially in the context of mounting protectionist tendencies in the West and evolving trade alliances like the Regional Comprehensive Economic Partnership (RCEP), the ASEAN-China Free Trade Agreement (ACFTA), and One Belt, One Road. Crucially, this chapter underscores that trade liberalisation and energy conservation are not conflicting goals; rather, they can symbiotically further the region's economic ambitions. Offering invaluable insights for decision-makers and academics, the study suggests that the Asia–Pacific can concurrently champion energy efficiency and robust economic growth, with trade poised as the cornerstone of its forthcoming economic ascent.

Y. K. Eng
Department of Economics, Teh Hong Piow Faculty of Business and Finance, Universiti Tunku Abdul Rahman, Kampar, Perak, Malaysia
e-mail: engyk@utar.edu.my

G. H. Y. Lee (✉) · A. K. L. Siah
Department of Economics, School of Business, Monash University, Subang Jaya, Malaysia
e-mail: grace.lee@monash.edu

A. K. L. Siah
e-mail: audrey.siah@monash.edu

E. Lau et al. (eds.), *Future of Economy in Asia*,
https://doi.org/10.1007/978-3-031-95500-6_12

Keywords Energy consumption · Economic growth · Trade openness · Panel vector autoregressive · Sustainability

12.1 Introduction

With the World Trade Organisation (WTO) and the Asia–Pacific Economic Cooperation (APEC) failing to make significant strides in promoting trade liberalisation, Asian nations are reassessing their trade liberalisation plans in light of the changing global landscape. Many Asian countries' economies are highly dependent on trade, so any delay in global trade liberalisation can have serious consequences, making this reevaluation all the more important. The creation of the Regional Comprehensive Economic Partnership (RCEP) in 2012 signifies a significant change in the dynamics of regional trade. It includes all ten ASEAN member states as well as their six main trading partners: South Korea, China, India, Japan, and New Zealand. With the United States pulling out of the Trans-Pacific Partnership (TPP) under Trump's presidency, RCEP, which China is leading, has picked up speed. Joined forces with RCEP, the ASEAN Economic Community (AEC) was formed in 2015, marking yet another watershed moment. The two blocs are responsible for a significant chunk of the world's production and commerce.

Concurrent with the region's fast economic growth, primary energy consumption has skyrocketed across the Asia–Pacific. A thorough understanding of these interdependencies is necessary to inform good policy decisions due to the complex and ever-changing nature of the correlation between energy use, economic production, and trade. Existing work frequently ignores the effects of disaggregated energy consumption patterns in favour of overall consumption, ignoring the importance of this link. This study encourages looking at aggregated and disaggregated energy statistics because different countries have different energy consumption habits and different energy sources have diverse effects on economies.

The world's main energy consumption is projected to increase due in large part to the rising demand for energy, especially in nations like China, India, and the ASEAN. Oil and coal dominate this development, which presents possibilities and threats to the energy and economic policies of the region. As part of a worldwide movement towards sustainable development, some Asian and Pacific nations have launched massive energy efficiency initiatives. But if energy use is directly proportional to GDP, then energy conservation programmes could unintentionally slow economic growth.

This study seeks to expand the existing body of knowledge by exploring the growth-energy-trade nexus within the Asia–Pacific region. Focusing on a panel of 15 countries, this research examines the complex interplay between energy consumption, trade, and economic growth. This research synthesises previous work by analysing the interplay between these factors using the Panel Vector Autoregressive (PVAR) model. It critically evaluates the potential impact of energy conservation policies on trade integration and economic growth, considering the causality directions between these variables. Methodologically, this study distinguishes itself by employing the PVAR model, offering a nuanced approach to understanding individual heterogeneity and endogenous relationships among energy consumption, trade, and economic growth. It provides detailed insights through impulse response functions, allowing for an in-depth examination of the responses to energy-related shocks.

Addressing the gaps in current research, this study utilises disaggregated energy consumption data, offering a clearer understanding of the dynamic relationships between different energy sources, economic growth, and trade. The findings of this study are instrumental in shaping development, trade, energy, and environmental policies in the Asia–Pacific region, emphasizing that trade liberalisation and energy conservation are not mutually exclusive, but rather, can synergistically advance the region's economic goals. The ensuing chapters are structured to provide a comprehensive review of the literature, present the methodology and data, report empirical results, and conclude with policy implications.

12.2 Literature Review

This section provides the theoretical underpinnings behind the dynamic relationships between energy use, economic growth and trade. The empirical literature provides mixed and conflicting evidence with respect to the energy consumption-growth, trade-growth and energy consumption-trade nexus. These contradicting findings are due largely to the use of different countries/regions, econometric methodologies, models, data sources and time periods.

12.2.1 Energy Use and Economic Growth

Our knowledge of the connection between energy use and economic growth has been greatly enhanced by the wealth of studies carried out following the energy crisis of the 1970s. This section critically reviews key contributions in this area, highlighting varying findings and methodologies.

In a comprehensive literature review on the topic, Ozturk (2010) captures a wide range of results and theoretical viewpoints on the relationship between energy use and economic growth. Payne (2010) also examines the link between power demand and GDP, but in a different setting, demonstrating how the findings can vary widely. Li et al. (2022) have developed a panel regression model with a threshold that uses data from 120 countries over the past two decades. Their study shows that renewable energy not only contributes to economic growth but also leads to environmental improvements at a global level. Govindaraju and Tang (2013) compare and contrast research on the relationships between energy consumption, GDP growth, and carbon dioxide emissions in India and China. These studies typically revolve around four competing hypotheses: (1) a one-way Granger causality between energy consumption and GDP; (2) a two-way Granger causation between energy consumption and GDP (the conservation hypothesis); or (3) no Granger causality at all. Kilci (2023)'s study examines the causal relationship between renewable energy consumption and economic growth in selected Euro-Area countries from 1990 to 2020. The study found that in Spain, Italy, and the Netherlands, there is a one-way causality from economic growth to renewable energy consumption. Meanwhile, in Finland, there is a one-way causality from renewable energy consumption to economic growth. However, the study did not identify any causal connection for either the panel or individual countries in Austria, Belgium, Germany, and Portugal, which aligns with the neutrality hypothesis. The policy implications of these causal relationships are substantial, particularly regarding the efficacy of energy conservation measures on economic output.

Ghazouani et al. (2020) contribute to this discourse by examining the trade-energy-growth nexus in Asia–Pacific countries using the bootstrap ARDL approach. The Granger causality analysis reveals a bidirectional relationship between real GDP and trade openness across all selected Asia–Pacific countries, suggesting a complex interdependency that supports both the trade openness-led growth and growth-led trade openness hypotheses. Similarly, Pradhan et al. (2024) find that energy consumption strongly drives economic growth in both G-7 and South Sian countries. In addition, higher per capita GDP encourages greater energy consumption in both regions. This theme of bidirectional causality is echoed by Tran et al. (2024), Wesseh and Lin (2018), who demonstrate feedback causality between energy consumption and economic growth in ASEAN.

Liu and Hao (2018) expand the study's geographical scope to include all 69 Belt and Road nations. The researchers found intricate linkages between renewable and traditional fossil energy usage from 1970 to 2013. They found distinct associations across different national subgroups. They discover different patterns of causation in countries that import energy and those that export it, as well as feedback causalities between the two and economic growth. In their studies conducted in thirty-eight International Energy Agency (IEA) countries, Khan et al. (2021), Duran et al. (2015) provide further evidence that both renewable and non-renewable energy consumption is linked to economic growth.

This critical review underscores the multifaceted nature of the relationship between energy consumption, economic growth, and trade openness. The fact that results vary greatly depending on location and time period shows that this link is complex and affected by many distinct things. The findings of this study hold great significance for future energy policy and planning in the Asia–Pacific region, as it emphasises the intricate connections between commerce, energy consumption, and economic growth.

The intricate relationship between trade, economic growth, and energy use has been extensively explored in economic literature, revealing a spectrum of findings and theoretical perspectives.

12.2.2 Trade and Economic Growth

Trade is widely recognised as a catalyst for economic growth through various mechanisms. Li et al. (2023) and Ryu and Nam (2024) highlight that open economies benefit from technology transfers, which foster competition and innovation in developing nations. Exposure to advanced technologies through international trade significantly enhances innovation in these economies (Wang, 2022). Schmitt and Yu (2001) and Valli and Saccone (2009) emphasise the role of increasing scale economies, while Arshed et al. (2023), Lall (2001) and Navas and Licandro (2011) focus on the enhancement of competitiveness. Additionally, Perla et al. (2021) argue that firms upgrade their productivity through technology adoption to stay competitive, with incentives shaped by the productivity distribution. Highly productive firms benefit from reduced trade costs by expanding exports, while low productivity firms face increased competition domestically. In equilibrium, lower productivity firms frequently upgrade their technology, boosting aggregate growth. The pace of technology adoption generates broader benefits, as other firms may adopt improved technologies. However, aggregate growth rates vary with trade cost reductions, depending on technological improvement costs and wage effects. Despite uneven distribution of gains and losses, the representative consumer benefits from trade openness.

Empirical studies affirming this positive relationship are numerous. Mullings and Mahabir (2018), for instance, observe that Africa's bilateral trade with China significantly contributes to the continent's economic growth, especially in resource-rich and oil-producing countries. This finding echoes Busse et al. (2016). Ibrahim and Abdulmalik (2023) have utilised the fixed effects panel estimation approach to analyse the data collected from 2002 to 2020. Their findings suggest that trade openness plays a vital role in driving the economic growth of the 14 countries within the East African Community and the member countries of the West African Economic and Monetary Union. However, Bresnahan et al. (2016) offer a more nuanced view, finding a positive relationship between trade and manufacturing productivity growth in Tanzania, but not in Ghana and Kenya. Sunde et al. (2023) use the ARDL co-integration method to analyse data from Namibia between 1990 and 2020. The study found a significant negative correlation between economic growth and imports, while trade and exports had a strong positive relationship with economic growth. Ghazouani et al. (2020) highlight the bidirectional causality between economic growth and trade openness, indicating that these two variables reinforce each other. Increased trade openness can drive economic growth by expanding market access, enhancing competitiveness, and promoting technological advancements. Conversely, economic growth tends to result in greater trade openness, as expanding economies require more resources, products, and services, many of which need to be imported. This observation aligns with existing studies, including those by Dahmani et al. (2022) and Banday and Aneja (2023).

Grossman and Helpman (2018) propose that countries with a higher capacity for innovation grow faster due to productivity gains from importing foreign goods.

Saito (2018) underscores the importance of patent protection in promoting economic growth through the importation of a variety of intermediate goods.

12.2.3 Energy Use and Trade

The relationship between trade openness and energy use is multifaceted. Increased trade escalates overall economic output and consequently the demand for production factors, including energy. The requisite transportation in trade also boosts energy demand. Energy and trade can either be interdependent and complementary or exhibit a neutral relationship. Since energy is a crucial input in producing traded goods, drastic reductions in energy consumption due to conservation measures could impair the production capacity for these goods.

Several studies have investigated the relationship between trade openness and energy consumption, yielding mixed results. Odhiambo (2021) finds a unidirectional causal flow from trade openness to energy consumption only when exports are used as a proxy, with no causality found using total trade or imports. Zeren and Akkuş (2020) show that non-renewable energy consumption drives trade openness, while trade openness promotes renewable energy use in top emerging countries. Rahman (2021) identifies a bidirectional causality between international trade and energy consumption in BRICS and ASEAN countries. These studies highlight the complex and varied nature of the relationship, influenced by the proxies and methodologies used.

These studies collectively indicate the complexity of the relationships among trade, economic growth, and energy use, with variations attributable to different regional and country-specific contexts. This critical analysis underscores the need for nuanced policy approaches that consider these intricate interdependencies.

12.3 Methodology

12.3.1 Panel-Data Vector Autoregression Methodology

We adopt the panel-data vector autoregression methodology (PVAR) estimation using Love and Zicchino (2006) strategies based on the work on Holtz-Eakin et al. (1988). The PVAR technique treats all the variables in the system as endogenous and it is estimated using the panel System of Generalised Method of Moments (GMM) estimator. We follow a p-order VAR model as follows:

$$Y_{i,t} = \delta_0 + \delta Y_{i,t-p} + \omega_i + \eta_t + \varepsilon_{i,t} \quad (12.1)$$

where $Y_{i,t}$ is a four-variable vector: Y (real GDP per capita); K (capital per capita); *prim/ oil /coal* (primary energy consumption per capita; oil consumption per capita; coal consumption per capita) and *trade* (structure-adjusted trade openness).

The panel approach has the virtue of addressing the problem of unobserved heterogeneity by including country fixed effects. The inclusion of a fixed effect in the model, ω_i, captures the unobserved country-specific and time-invariant factors. In other words, the factors in the model are allowed to differ at the country-specific level. Similarly, it allows us to control for other time-invariant factors, such as different exchange rate regime, institutional quality and regulations. However, the use of lagged Y on the right-hand side of Eq. (12.1) induces correlation between the fixed effects and the regressors, resulting in biased coefficients if we use the mean-differencing to eliminate fixed effects. We therefore employ the forward mean-differencing procedure which removes only the mean of all the future observations available for each country-year, it is also known as the "Helmert Procedure" (Arellano and Bover 1995). This procedure preserves the orthogonality between the transformed variables and the lagged regressors, allowing the use of lagged regressors as instruments and the coefficients can be estimated by a Systemf GMM. This model also allows for the inclusion of common time effects, η_t, to capture any global and regional shocks that may affect all countries in the system such as regional trade agreements, global financial crisis and global energy crisis.

One of the main benefits of the VAR system is that it allows us to evaluate the reaction of a variable to the innovations in another variable in the model, while holding all other shocks constant. This is accomplished through the impulse response functions. However, to isolate shocks to one of the variables in the system, it is crucial for the variance–covariance matrix of the errors to be orthogonal. We performed a decomposition of the residuals to ensure their orthogonality using the Cholesky decomposition method. For detailed derivations and discussions on impulse response functions related to this process, please refer to Hamilton (2020).

12.3.2 Data

This study aims to assess the linkages between energy consumption, trade and output in the 15 Asia and the pacific countries, namely Australia, Bangladesh, China, Hong Kong, India, Indonesia, Japan, Korea, Malaysia, New Zealand, Pakistan, Philippines, Singapore, Thailand and Vietnam. Data on energy consumption such as Primary energy use per capita (*prim*), Coal consumption per capita (*coal*), and Oil consumption per capita (*loil*) expressed in natural logarithms are sourced from Statistical Review of World Energy; for structure-adjusted trade openness (*trade*), we rely on a methodology used by Chang et al. (2009), which is computed by the residual of a regression of the log trade (percentage of GDP), on the logs of area, population, and dummies for oil exporting. We are interested in investigating the share of the trade volume that is not due to structural country characteristics where the data are collected from World Bank WDI. We use real GDP per capita (constant 2011 US$)

Table 12.1 Summary statistics of output, capital, primary energy use, oil consumption, coal consumption and structure-adjusted trade openness for fifteen countries in Asia and the Pacific[a] (1965–2014)

Variable	Mean	SD	Min	Max
Output per capita (*lyp*), in US$	13,092.19	13,384.53	1046.62	62,416.88
Capital per capita (*lkp*), in USD	44,220.53	53,533.20	1539.42	266,589.20
Primary energy use per capita (*prim*), in Mtoe	1.966	2.418	0.023	13.700
Oil Consumption per capita (*loil*), in Mtoe	1.150	1.895	0.013	12.005
Coal Consumption per capita (*coal*), in Mtoe	0.372	0.564	0.001	2.613
Structure-adjusted Trade openness (*trade*)	−0.004	0.492	−1.362	1.013

SD standard deviation, Mtoe million tonnes oil equivalent
[a]The 15 countries include Australia, Bangladesh, China, Hong Kong, India, Indonesia, Japan, Korea, Malaysia, New Zealand, Pakistan, Philippines, Singapore, Thailand and Vietnam; ** Mtoe denotes million tonnes oil equivalent

as a measure of economic output (*lyp*). Both data on capital stock (*lkp*) at constant 2011 national prices (in mil. 2011US$)/Population (in millions) and real GDP per capita are in natural logarithms and are collected from the Penn World Table 9.0. Our sample covers the period 1965–2014. All of these variables were then averaged over five-year periods to mitigate the business cycles effect as well as the problem of too many instrumental variables given the fact that the number of countries (N) in this study is moderately small ($N = 15$).[1] Table 12.1 displays the summary statistics for the variables. The pairwise correlation matrix for the key variables is presented in Table 12.2.

12.4 Findings

To determine the lag structure for the PVAR model, we adopt the Moment and Model Selection Criteria on Akaike Information criterion (MMSC-AIC) developed by Andrews and Lu (2001). They argue that the MMSC-AIC is particularly effective for smaller sample sizes, outperforming other lag selection methods. Table 12.3 illustrates that the most suitable lag structure for our PVAR model is a first lag, labeled as PVAR (1).

Additionally, the validity of the instrumental variables in our model is supported by the results from Hansen's J Statistics and the over-identifying restriction Sargan

[1] The averages are calculated for 1965–1969, 1970–1974, 1975–1979, 1980–1984, 1985–1989, 1990–1994, 1995–1999, 2000–2004, 2005–2009 and 2010–2014.

Table 12.2 Pairwise correlation matrix of output, capital, primary energy use, oil consumption, coal consumption and structure-adjusted trade openness

	lyp	lkp	Prim	Loil	Coal	Trade
lyp	1.000					
lkp	0.953^{***}	1.000				
prim	0.921^{***}	0.881^{***}	1.000			
loil	0.748^{***}	0.776^{***}	0.914^{***}	1.000		
coal	0.608^{***}	0.511^{***}	0.452^{***}	0.136^{***}	1.000	
trade	0.231^{***}	0.267^{***}	0.214^{***}	0.281^{***}	0.169^{***}	1.000

lyp output per capita, *lkp* capital per capita, *prim* primary energy use per capita, *loil* oil consumption per capita, *coal* coal consumption per capita, *trade* structure-adjusted trade openness. *** $p < 0.01$

	lyp	lkp	Prim	Loil	Coal	Trade
lyp	1.000					
lkp	0.953^{***}	1.000				
prim	0.921^{***}	0.881^{***}	1.000			
loil	0.748^{***}	0.776^{***}	0.914^{***}	1.000		
coal	0.608^{***}	0.511^{***}	0.452^{***}	0.136^{***}	1.000	
trade	0.231^{***}	0.267^{***}	0.214^{***}	0.281^{***}	0.169^{***}	1.000

lyp output per capita, *lkp* capital per capita, *prim* primary energy use per capita, *loil* oil consumption per capita, *coal* coal consumption per capita, *trade* structure-adjusted trade openness. *** $p < 0.01$

Table 12.3 Model lag selection based on MAIC and Sargan test for over-identifying restrictions

Dependent variable: lyp	Model 1: primary energy use (*prim*)	Model 2: oil use (*loil*)	Model 3: coal use (*coal*)
Lag			
1	−20.976	−21.210	−22.596
2	−12.064	−7.880	−8.367
Instrumental lags	Lag 1 to 3	Lag 1 to 3	Lag 1 to 3
Sargan Test	35.690 (0.299)	41.395 (0.124)	37.077 (0.246)

MMSCAIC, n: Andrew and Lu's (2001) Moment and Model Selection Criteria on Akaike Information Criteria (MMSCAIC, n); Sargan Test: Hansen's J Statistics of over-identifying restriction with the null of valid instrumental variables. All variables in logarithms form except prim and coal. *P*-values are in parenthesis

test. Furthermore, the Eigenvalue stability test, depicted in Fig. 12.1, confirms the stability of the PVAR (1) model, as all eigenvalues are located within the unit circle.

As in the standard VAR or PVAR analysis, we focus our discussion on the results from Granger-causality tests, impulse responses and forecast error variance

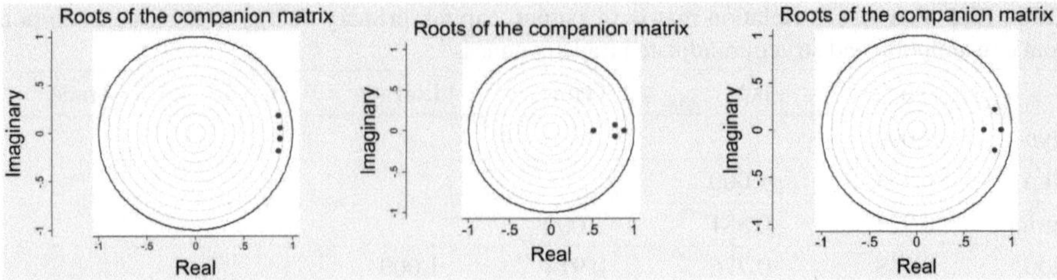

(a) Model 1: *primary energy use* (*prim*)

(b) Model 2: *oil use* (*loil*))

(c) Model 3: *coal use* (*coal)*

Fig. 12.1 Dynamic stability of the PVAR (1) Model for primary energy use, oil use and coal use

decompositions, which are more informative in capturing the dynamics between the variables rather than interpreting the estimated coefficients alone.

12.4.1 PVAR Granger Causality Test Results

Table 12.4 presents the Granger causal relationship between energy, trade and output nexus using the panel Granger causality test. The results show a unidirectional Granger causality running from primary energy consumption and coal consumption to output. The results show a unidirectional Granger causality running from output to oil consumption, confirming the conservation hypothesis that an increase in real GDP causes an increase in oil consumption. In all three models, trade is found to Granger cause output, supporting the trade-led hypothesis. Trade is found to Granger cause primary energy use and oil consumption in the short run. The consumption of primary energy, oil and coal does not have any impact on trade openness.

12.4.2 PVAR and Impulse Response Functions

Table 12.5 reports the PVAR estimation using Love and Zicchino (2006) strategies based on the work on Holtz-Eakin et al. (1988), and the PVAR is estimated using panel system GMM estimator. The results of four-variable vectors: natural logarithms of output per capita *(lyp)*, natural logarithms of capital per capita (*lkp*), energy/oil/coal use per capita (*prim/oil/coal*), and structure-adjusted trade openness (*trade*) indicate that most of the coefficients are significant. Numerous studies have supported the energy–growth nexus (Ajmi et al. 2015; Ang 2007; Ghali and El-Sakka 2004; Shahbaz and Lean 2012) indicating that energy is an essential component in economic growth as it can either directly or indirectly complement to labor and capital in the production process. By decomposing the total primary energy use into its main components of coal and oil, we are able to pinpoint the exact source of energy that is

Table 12.4 Results of panel granger causality test of output, capital, primary energy use, oil consumption and coal consumption and structure-adjusted trade openness

Model 1	*From (X does not Granger cause Y)*			
Primary energy use To	lyp	lkp	Prim	Trade
Lyp	–	13.509**	3.152*	38.882***
Lkp	1.124	–	23.355***	0.738
Prim	0.611	15.206***	–	2.549
Trade	0.079	0.506	0.035	–
Model 2	*From (X does not Granger cause Y)*			
Oil use To	lyp	lkp	Oil	trade
Lyp	–	6.514**	0.060	31.135***
Lkp	0.681	–	6.221**	19.518***
loil	13.329***	18.778***	–	10.444***
Trade	0.002	0.103	0.272	–
Model 3	*From (X does not Granger cause Y)*			
Coal use To	lyp	lkp	Coal	Trade
Lyp	–	0.141	11.662***	32.847***
Lkp	4.805**	–	11.594***	20.644***
Coal	0.601	16.373***	–	2.065
Trade	1.684	4.149**	1.561	–

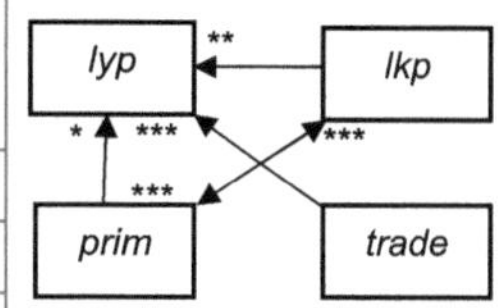

(a) Model 1: primary energy use

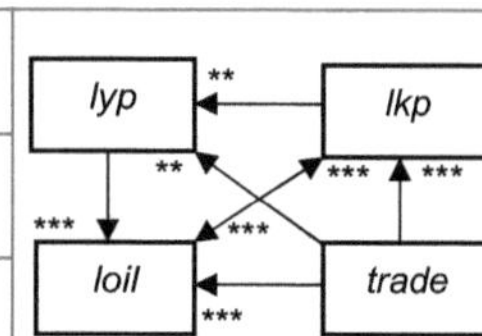

(b) Model 2: oil use

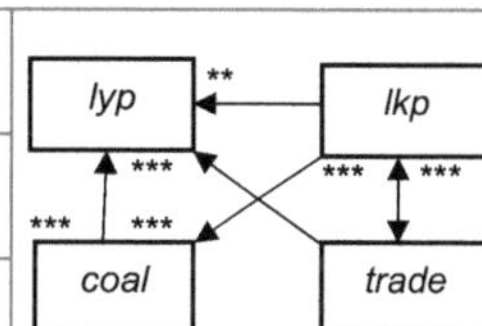

(c) Model 3: coal use

Granger causality in PVAR follows a $\chi 2$ distribution. Null hypothesis: X does not Granger cause Y. *** $p < 0.01$ ** $p < 0.05$ * $p < 0.10$

essential for economic growth. The output shows a positive response to a shock in the coal use, both in the estimated coefficients and in the impulse responses (Fig. 12.4). Oil use, on the other hand, has no significant effect on output and therefore we do not see any impact of total primary energy use on output (Fig. 12.3).

The response of oil consumption to a shock in the output is positive and highly significant in the estimated coefficients and impulse responses, indicating that oil consumption always drops after economic shocks or recession. This is consistent with the conservation hypothesis which implies that policies designed to reduce oil consumption in the region might not adversely affect economic growth. This finding

Table 12.5 Results of PVAR(1) estimation using the system of generalised method of moments (GMM)

Response of	Response to			
Model 1: primary energy use	lyp_{t-1}	lkp_{t-1}	$Prim_{t-1}$	$Trade_{t-1}$
Output per capita (*log*(*yp*))	0.894*** (0.074)	−0.109***(0.027)	0.019 (0.015)	0.284*** (0.047)
Capital per capita (*log*(*kp*))	0.090 (0.063)	0.946***(0.038)	−0.066***(0.014)	0.055 (0.069)
Energy use per capita (*prim*)	−0.050 (0.160)	0.396*** (0.096)	0.814***(0.075)	−0.194* (0.114)
Structure-adjusted trade openness (*trade*)	−0.033 (0.153)	0.039 (0.077)	−0.003 (0.025)	0.798*** (0.067)
Model 2: oil use	lyp_{t-1}	lkp_{t-1}	$loil_{t-1}$	$Trade_{t-1}$
Output per capita (*log*(*yp*))	0.946*** (0.072)	−0.115** (0.049)	−0.019 (0.043)	0.291***(0.039)
Capital per capita (*log*(*kp*)	−0.082 (0.137)	0.797*** (0.064)	0.200** (0.093)	0.223*** (0.061)
Oil use per capita(*log*(*oil*))	0.456*** (0.169)	−0.280*** (0.063)	0.417*** (0.141)	0.327*** (0.085)
Structure-adjusted trade openness (*trade*)	−0.004 (0.238)	−0.017 (0.100)	0.054 (0.162)	0.770*** (0.057)
Model 3:coal use	lyp_{t-1}	lkp_{t-1}	$Coal_{t-1}$	$Trade_{t-1}$
Output per capita (*lyp*)	0.820*** (0.047)	−0.008 (0.023)	0.142*** (0.042)	0.193*** (0.033)
Capital per capita (*lkp*)	0.130* (0.070)	0.817*** (0.025)	−0.245** (0.099)	0.190*** (0.055)
Coal use per capita (*coal*)	−0.061 (0.091)	0.173*** (0.042)	0.717*** (0.134)	−0.061 (0.043)
Structure-adjusted trade openness (*trade*)	0.135 (0.110)	−0.130* (0.072)	−0.093 (0.097)	0.881*** (0.064)

Figures in parentheses are cluster robust standard errors, while the coefficients are estimated by regressing the row variables (response of) on the first lag of the column variables (response to). *** $p < 0.01$ ** $p < 0.05$ * $p < 0.10$

Figures in parentheses are cluster robust standard errors, while the coefficients are estimated by regressing the row variables (response of) on the first lag of the column variables (response to). *** $p < 0.01$ ** $p < 0.05$ * $p < 0.10$

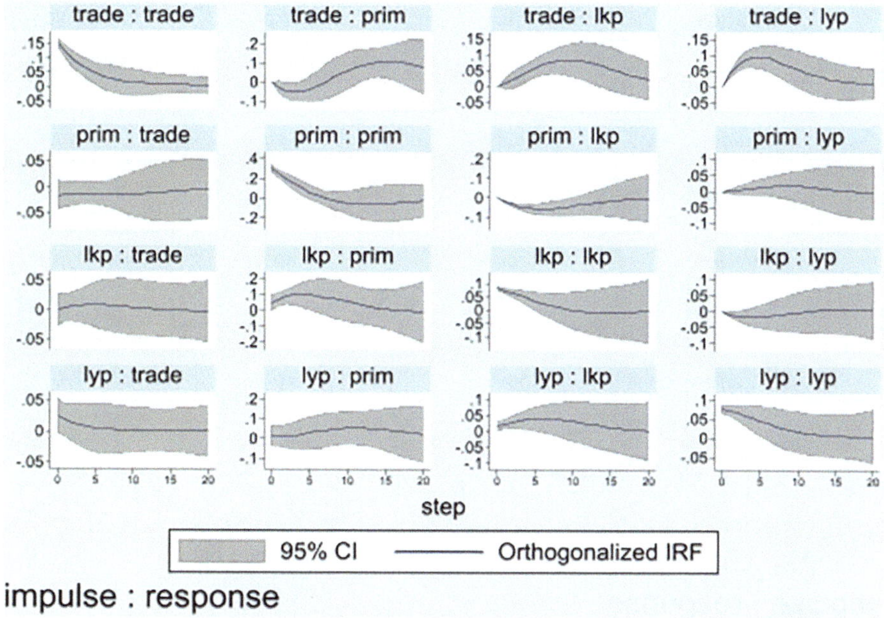

Fig. 12.2 Impulse response functions (Model 1: Total Primary Energy Use). Impulse-response functions of model 1: lypt = f(lkp t–1, prim t–1, trade t–1). Errors are 5% on each side generated by Monte-Carlo with 1000 repetitions

is also confirmed by the variance decomposition result that shows that 24.6% of oil consumption variation is attributed by innovative shocks of output.

Output shows an expected positive response to a shock in trade in all three models examined, both in the estimated coefficients and in the impulse responses (Figs. 12.2, 12.3, and 12.4), implying that increased trade openness stimulates economic growth. While there is no evidence of a statistically significant direct relationship between trade and primary energy and coal consumption, we observe that the response of oil consumption to trade openness is positive in the estimated coefficients and impulse responses (Fig. 12.3). The findings of neutrality effect are consistent with our earlier findings (Sect. 12.4.1), suggesting that primary energy and coal conserving policies are not detrimental to trade liberalisation policies and vice versa. Nevertheless, trade liberalisation policies established in stimulating economic growth will countervail policies aiming at conserving the usage of oil.

12.4.3 Variance Decomposition Results

Even though impulse responses provide information about the effect of changes in one variable on another, they do not show how important shocks on one variable are

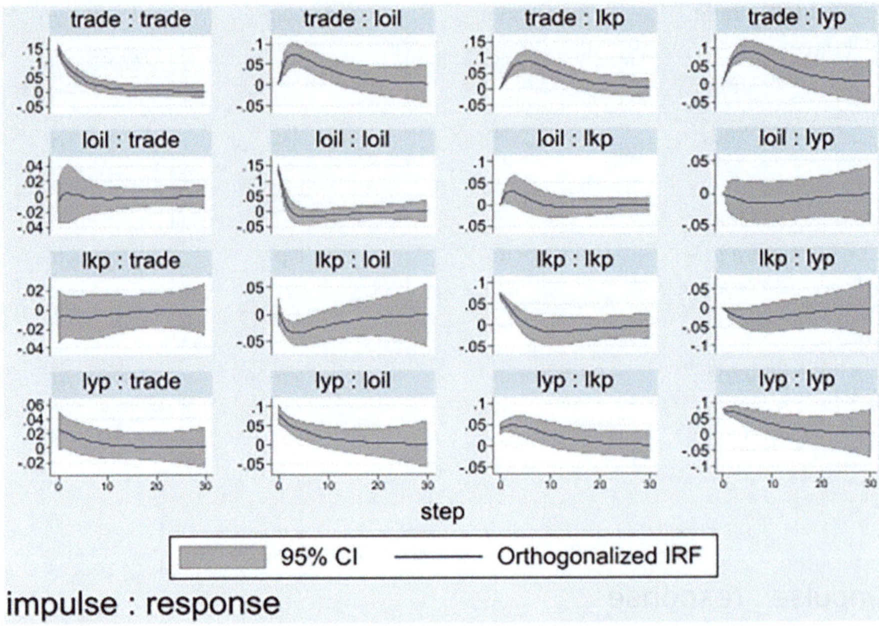

Fig. 12.3 Impulse response functions (Model 2: Oil Use). *Impulse-response functions of model 2: lypt = f(lkpt–1, oil t–1, trade t–1). Errors are 5% on each side generated by Monte-Carlo with 1000 repetitions*

in explaining fluctuations in other variables. We perform variance decomposition that offers essential information on the relative importance of the variation in one variable explained by the shocks received by another variable. The variance decompositions for the different models are presented in Tables 12.6, 12.7, and 12.8. These results report the percentage of variation in each variable that is explained by a shock in another variable, accumulated over time. We report the total effect accumulated over three different periods: short run (the two periods horizon), medium run (the five periods horizon) and long run (the 10 periods horizon).

The variance decompositions for the different models are consistent with the earlier findings using Panel VAR and IRFs. Several interesting results emerge from the analysis. More specifically, the results indicate that primary energy use explains an increasing percentage of output variance, from less than 0.01% in the short run to 1.9% in the long run. This implies that primary use energy conservation policy is unlikely to have any significant impact on output in the short term, its impact in the long run is also rather negligible (explaining 1.9% variation in the output). Coal consumption shocks have the largest impact on output variation, its importance increases as the time horizon increases (from explaining 1.8% of output variation in the short run to 3.9% in the long run).

Variance in primary energy and coal consumption is found to be explained mainly by its own shocks (70.9% for primary energy and 66.3% for coal), followed by shocks

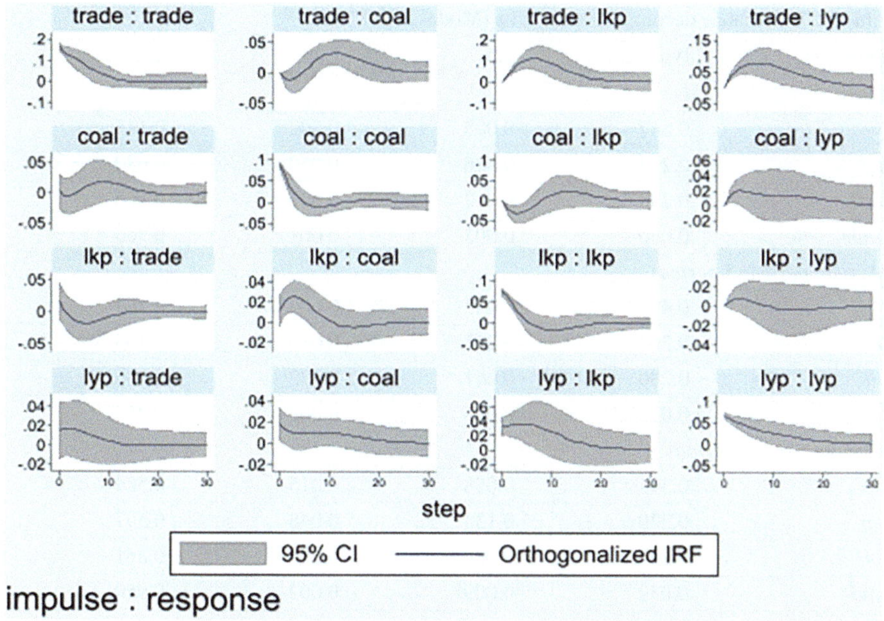

Fig. 12.4 Impulse response functions (Model 3: Coal Use). *Impulse-response functions of model 3: lypt = f(lkp t–1, coal t–1, trade t–1). Errors are 5% on each side generated by Monte-Carlo with 1000 repetitions. All variables in logarithms form except and coal*

Table 12.6 Variance decomposition results (Model 1: Total Primary Energy Use)

Model 1	lyp	lkp	Prim	Trade
Short period: 2 periods				
lyp	0.850	0.005	0.00008	0.145
lkp	0.071	0.893	0.031	0.005
prim	0.002	0.049	0.943	0.006
trade	0.024	0.0003	0.012	0.964
Medium period: 5 periods				
lyp	0.470	0.015	0.004	0.511
lkp	0.122	0.581	0.190	0.107
prim	0.005	0.134	0.831	0.030
trade	0.021	0.003	0.016	0.961
Long period: 10 periods				
lyp	0.312	0.0144	0.019	0.655
lkp	0.130	0.262	0.246	0.362
prim	0.032	0.222	0.709	0.038
trade	0.020	0.007	0.028	0.945

Percentage of variation in the row variable (2(5)10 periods ahead) explained by column variable

Table 12.7 Variance decomposition results (Model 2: Oil Use)

Model 2	lyp	lkp	Loil	Trade
Short period: 2 periods				
lyp	0.847	0.008	0.002	0.143
lkp	0.284	0.586	0.050	0.080
loil	0.295	0.012	0.627	0.066
trade	0.030	0.003	0.001	0.966
Medium period: 5 periods				
lyp	0.494	0.035	0.007	0.464
lkp	0.264	0.251	0.080	0.405
oil	0.280	0.061	0.372	0.286
trade	0.036	0.005	0.002	0.957
Long period: 10 periods				
lyp	0.344	0.058	0.015	0.584
lkp	0.220	0.134	0.048	0.597
loil	0.246	0.089	0.284	0.381
trade	0.039	0.009	0.002	0.950

Percentage of variation in the row variable (2(5) 10 periods ahead) explained by column variable

Table 12.8 Variance decomposition results (Model 3: Coal Use)

Model 3	lyp	lkp	Coal	Trade
Short period: 2 periods				
lyp	0.878	0.002	0.018	0.102
lkp	0.173	0.722	0.033	0.072
coal	0.042	0.043	0.907	0.008
trade	0.010	0.007	0.001	0.982
Medium period: 5 periods				
lyp	0.522	0.006	0.044	0.428
lkp	0.143	0.298	0.083	0.476
coal	0.048	0.138	0.794	0.020
trade	0.015	0.009	0.001	0.975
Long period: 10 periods				
lyp	0.323	0.004	0.039	0.634
lkp	0.107	0.121	0.042	0.729
coal	0.063	0.159	0.663	0.114
trade	0.017	0.019	0.012	0.952

Percentage of variation in the row variable (2(5)10 periods ahead) explained by column variable in the capital use (22.2% and 15.9% respectively). The variance decomposition of the oil consumption model, on the other hand, reveals the importance of output (24.6%) and trade (38.1%) shocks in influencing the variation of oil use. Trade shocks explain an increasing percentage of oil use variance (from 6.6% in the short run to 28.6% in the medium run to 38.1% in the long run). These results again reinforce the conclusion that increased trade increases oil consumption, and trade promoting policies established in stimulating economic growth will countervail oil conservation policy.

12.4.3.1 Sensitivity Analysis

One of the major concerns about the PVAR is that the Cholesky decomposition of the impulse response functions the ordering of the variables in PVAR. To check the sensitivity of our results with respect to different ordering of the variables, we explore two different types of changes in relative ordering among the variables. The first ordering considers *prim/oil/coal*, *lyp*, *lkp* and *trade* from the aggregate demand perspective (see Figs. 12.5). This model assumes that the initial shocks come from the production using energy, rather from the economic growth. The second considers a model from the aggregate supply angle: *trade*, *lyp*, *lkp*, *prim/oil/coal* (as shown in Fig. 12.6). This model assumes that the initial shocks come from the trade shocks, followed by output, capital and energy use. By and large, the results are robust to the order of the variables.

12.5 Discussion of Implications

In examining the nexus between energy consumption, trade, and economic growth in the Asia–Pacific region, this chapter contributes to an area where previous research has yielded inconclusive results, largely due to variations in methodologies, time-frames, and the specific countries or regions analysed. Our findings provide new insights, particularly relevant for policy development in this diverse region.

Our empirical analysis reveals that in the short term, the impact of primary energy and coal consumption on economic growth in the Asia–Pacific region is minimal, accounting for less than 2%. This influence slightly increases in the long term, explaining approximately 6% of economic growth. Additionally, our data does not indicate a significant influence of oil consumption on output variance. These results suggest that policymakers in the region can confidently pursue energy conservation goals without adversely affecting economic growth. This is evident in economies like China, India, Indonesia, Japan, Hong Kong (SAR, China), and Singapore, which have shown marked improvements in energy efficiency, surpassing global averages.

In contrast to the marginal impact of energy shocks on output, trade shocks play a substantial role in economic growth. In the short run, they account for 10.2 to 14.5% of economic growth, and this increases to between 58.4 and 65.8% in the long

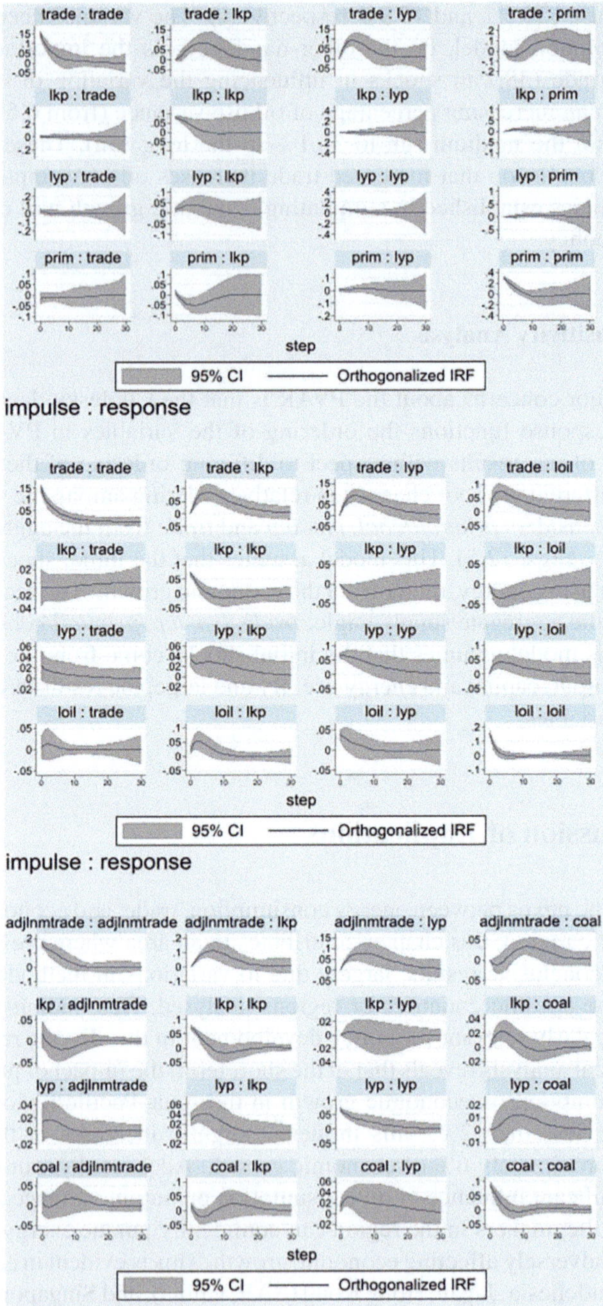

Fig. 12.5 Impulse response functions (First Alternative Ordering: *prim*/*oil*/*coal*, *lyp*, *lkp* and *trade*)

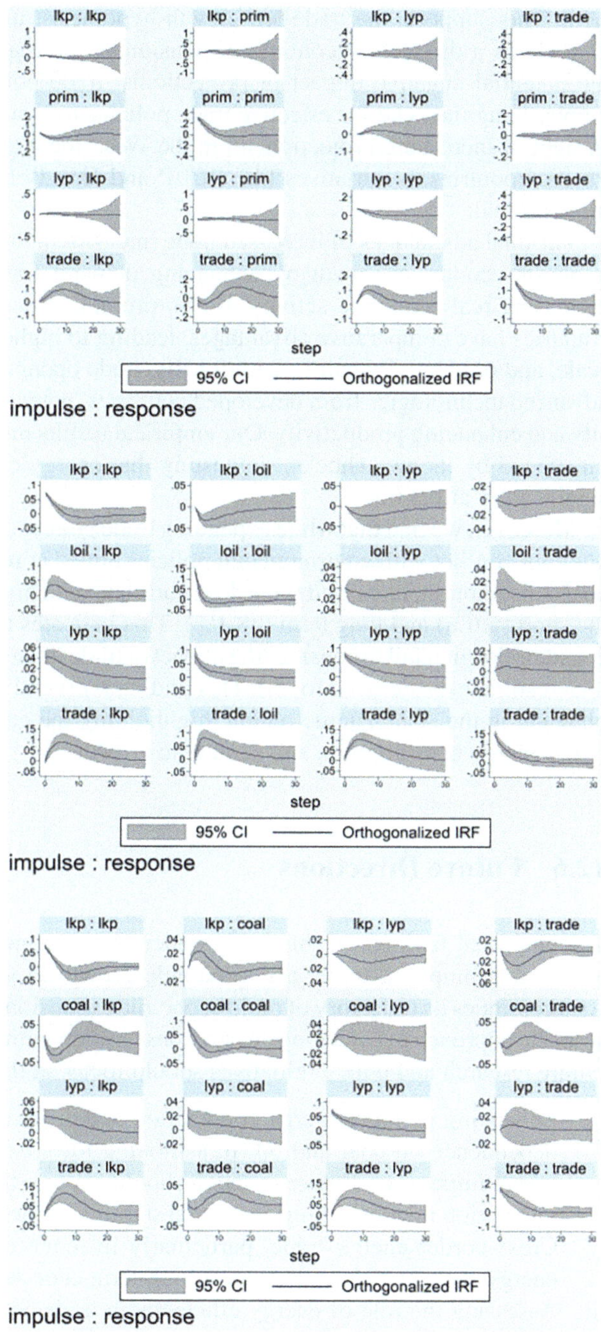

Fig. 12.6 Impulse response functions (Second Alternative Ordering: *trade*, *lyp*, *lkp*, *prim*/*oil*/*coal*)

term. This supports the trade-led growth hypothesis, underscoring the importance of trade as a driver for economic expansion in the region. The findings highlight the potential negative impact of protectionist trade policies on economic growth, emphasizing the need for effective trade policies to foster growth and employment. In light of increasing protectionism in the West, the strategic engagement of Asia–Pacific countries in initiatives like RCEP and One Belt, One Road becomes even more critical.

The dual advantages of increased trade intensity in the region are notable. Firstly, it boosts economic growth by enhancing demand for the country's outputs and increasing real economic activity. It also promotes specialization in sectors where countries have comparative advantages, leading to higher skill levels, economies of scale, and productivity gains. Additionally, trade openness facilitates the transfer of advanced technologies from developed countries, potentially reducing energy intensity and enhancing productivity. Our empirical results indicate that trade is primarily influenced by its own shocks, suggesting that energy conservation policies do not significantly affect trade.

Moreover, we find that while expanding trade policy to stimulate economic growth does not conflict with environmental policies aimed at reducing primary energy and coal consumption, increased trade does correlate with higher oil consumption in both the short (6.6%) and long term (38.1%). This indicates that projections of future oil consumption that fail to consider the impact of trade intensity in the region are likely to underestimate the demand for oil. This finding is crucial for future energy policy and planning in the Asia–Pacific region, highlighting the complex relationships among trade, energy consumption, and economic growth.

12.6 Future Directions

The empirical findings of this study underscore the intricate relationship between energy consumption, trade, and economic growth in the Asia–Pacific region. As Asia continues its trajectory of rapid economic expansion, policymakers must adopt a strategic approach to balancing energy conservation with sustained trade-led growth. Future research and policy initiatives should focus on the following key areas:

i. Enhancing regional energy cooperation and diversification: Given the region's dependence on coal and oil, transitioning towards a more sustainable energy mix is imperative. Future research should explore the role of renewable energy integration in maintaining economic stability while reducing carbon emissions. Cross-border energy trade, particularly in renewable sources, could mitigate energy security risks and support long-term economic growth.
ii. Assessing the role of energy efficiency in trade competitiveness: The findings indicate that primary energy conservation does not adversely impact economic growth, but increased trade openness is associated with higher oil consumption. Future studies should investigate how improvements in energy efficiency can

enhance trade competitiveness while minimizing environmental costs. Comparative analyses between Asia–Pacific economies and other global trade hubs may provide insights into best practices.

iii. Trade policy adaptation in response to evolving global dynamics: With rising protectionist tendencies in the West, Asia–Pacific nations must strengthen intra-regional trade networks. Further research should examine how trade agreements such as RCEP and ACFTA can be optimized to support sustainable economic growth without exacerbating energy demand. Additionally, the long-term effects of these agreements on energy-intensive industries warrant deeper exploration.

iv. Long-term economic impacts of energy policy transitions: While short-term economic fluctuations due to energy policy shifts appear minimal, the long-term trajectory remains uncertain. Future research should utilise extended datasets and advanced econometric models to analyze the cumulative impact of shifting from fossil fuels to renewable energy on trade and growth. Scenario-based policy simulations could help governments navigate economic transitions more effectively.

v. The interplay between digitalisation, trade, and energy consumption: Emerging digital economies and advancements in automation are reshaping global trade patterns. The impact of digital trade and e-commerce on energy demand remains an underexplored area. Future studies should assess whether increased digitalisation contributes to energy savings or intensifies demand, particularly in high-growth sectors such as logistics and manufacturing.

vi. Sectoral-level analysis of trade-energy-growth linkages: While this study provides aggregate insights, future research should adopt a sectoral approach to determine which industries are most sensitive to energy price fluctuations and trade shocks. Understanding these nuances will enable more targeted policy interventions.

By addressing these critical areas, future research can further refine Asia's policy landscape, ensuring that trade liberalisation and energy sustainability evolve in tandem. The Asia–Pacific region's ability to proactively manage these interdependencies will determine its long-term economic resilience in an increasingly uncertain global economy.

12.7 Conclusion

This chapter provides new insights into the nexus between energy consumption, trade, and economic growth in the Asia–Pacific region. Our analysis reveals that primary energy and coal consumption have a minimal impact on economic growth in the short term (less than 2%) and a modest impact in the long term (around 6%), while oil consumption shows no significant influence. In contrast, trade shocks substantially drive economic growth, accounting for 10.2 to 14.5% in the short term and 58.4 to 65.8% in the long term. Increased trade intensity boosts economic activity, promotes specialization, and facilitates advanced technology transfer, although it also correlates with higher oil consumption. These findings highlight the importance of effective trade policies and region-specific strategies to balance economic growth, energy conservation, and environmental goals in the Asia–Pacific region.

Highlights

- Asia–Pacific's energy conservation goals will not hinder its economic growth.
- Trade is a key driver of economic expansion, supporting the trade-led growth hypothesis.
- Highlights the necessity for ongoing trade liberalisation in the region.

References

Ajmi AN et al (2015) On the relationships between CO_2 emissions, energy consumption and income: the importance of time variation. Energy Econ 49:629–638. https://doi.org/10.1016/j.eneco.2015.02.007

Andrews DWK, Lu B (2001) Consistent model and moment selection procedures for GMM estimation with application to dynamic panel data models. J Econom 101(1):123–164. https://doi.org/10.1016/S0304-4076(00)00077-4

Ang JB (2007) CO_2 emissions, energy consumption, and output in France. Energy Policy 35(10):4772–4778. https://doi.org/10.1016/j.enpol.2007.03.032

Arellano M, Bover O (1995) Another look at the instrumental variable estimation of error-components models. J Econ 68(1):29–51. https://doi.org/10.1016/0304-4076(94)01642-D

Arshed N et al (2023) Role of real sector and HDI: Does competitiveness contribute? J Knowl Econ 14(3):3044–3070. https://doi.org/10.1007/s13132-022-00997-w

Banday UJ, Aneja R (2023) Macroeconomic stability, openness, and economic growth nexus: an insight from BRICS countries. J Knowl Econ 15(2):5965–5979. https://doi.org/10.1007/s13132-023-01241-9

Bresnahan L et al (2016) Does freer trade really lead to productivity growth? evidence from Africa. World Dev 86:18–29. https://doi.org/10.1016/j.worlddev.2016.05.007

Busse M et al (2016) China's impact on Africa—the role of trade. FDI and Aid. Kyklos 69(2):228–262. https://doi.org/10.1111/kykl.12110

Chang R et al (2009) Openness can be good for growth: the role of policy complementarities. J Dev Econ 90(1):33–49. https://doi.org/10.1016/j.jdeveco.2008.06.011

Dahmani M et al (2022) ICT, trade openness and economic growth in Tunisia: what is going wrong? Econ Change Restructuring 55(4):2317–2336. https://doi.org/10.1007/s10644-022-09388-2

Duran E et al (2015) Analysis and decomposition of energy consumption in the Chilean industry. Energy Policy 86:552–561. https://doi.org/10.1016/j.enpol.2015.07.033

Ghali K, El-Sakka M (2004) Energy use and output growth in Canada: a multivariate cointegration analysis. Energy Econ 26:225–238. https://doi.org/10.1016/S0140-9883(03)00056-2

Ghazouani T et al (2020) Causal interactions between trade openness, renewable electricity consumption, and economic growth in Asia-Pacific countries: fresh evidence from a bootstrap ARDL approach. Renew Sust Energ Rev 133:110094. https://doi.org/10.1016/j.rser.2020.110094

Govindaraju VGRC, Tang CF (2013) The dynamic links between CO_2 emissions, economic growth and coal consumption in China and India. Appl Energy 104:310–318. https://doi.org/10.1016/j.apenergy.2012.10.042

Grossman GM, Helpman E (2018) Growth, trade, and inequality. Econometrica 86(1):37–83. https://doi.org/10.3982/ECTA14518

Hamilton JD (2020) Time series analysis. Princeton University Press

Holtz-Eakin D et al (1988) Estimating vector autoregressions with panel data. Econometrica 56(6):1371–1395. https://doi.org/10.2307/1913103

Ibrahim A, Abdulmalik MR (2023) Do trade openness and governance matter for economic growth in Africa? A case of EAC and WAEMU countries. Int Econ Econ Policy 20(3):389–412. https://doi.org/10.1007/s10368-023-00565

Khan I et al (2021) The dynamic links among energy transitions, energy consumption, and sustainable economic growth: a novel framework for IEA countries. Energy 222:119935. https://doi.org/10.1016/j.energy.2021.119935

Kilci EN (2023) An investigation on the causality link between renewable energy consumption and economic growth in selected Euro-Area countries. Energy Sources B: Econ Plan Policy 18(1):2250844. https://doi.org/10.1080/15567249.2023.2250844

Lall S (2001) Competitiveness indices and developing countries: an economic evaluation of the global competitiveness report. World Dev 29(9):1501–1525. https://doi.org/10.1016/S0305-750X(01)00051-1

Li R et al (2022) Does renewable energy reduce ecological footprint at the expense of economic growth? An empirical analysis of 120 countries. J Clean Prod 346:131207. https://doi.org/10.1016/j.jclepro.2022.131207

Li X et al (2023) National innovation systems and the achievement of sustainable development goals: effect of knowledge-based dynamic capability. J Innov Knowl 8(1):100310. https://doi.org/10.1016/j.jik.2023.100310

Liu Y, Hao Y (2018) The dynamic links between CO_2 emissions, energy consumption and economic development in the countries along "the Belt and Road." Sci Total Environ 645:674–683. https://doi.org/10.1016/j.scitotenv.2018.07.062

Love I, Zicchino L (2006) Financial development and dynamic investment behavior: evidence from panel VAR. Q Rev Econ Finance 46(2):190–210. https://doi.org/10.1016/j.qref.2005.11.007

Mullings R, Mahabir N (2018) Growth by destination: the role of trade in Africa's Recent growth episode. World Dev 102:243–261. https://doi.org/10.1016/j.worlddev.2017.10.009

Navas A, Licandro O (2011) Trade liberalization, competition and growth 11(1). https://doi.org/10.2202/1935-1690.2087

Odhiambo NM (2021) Trade openness and energy consumption in sub-Saharan African countries: a multivariate panel Granger causality test. Energy Rep 7:7082–7089. https://doi.org/10.1016/j.egyr.2021.09.103

Ozturk I (2010) A literature survey on energy–growth nexus. Energy Policy 38(1):340–349. https://doi.org/10.1016/j.enpol.2009.09.024

Payne J (2010) Survey of the international evidence on the causal relationship between energy consumption and growth. J Econ Stud 37:53–95. https://doi.org/10.1108/01443581011012261

Perla J et al (2021) Equilibrium technology diffusion, trade, and growth. Am Ec Rev 111(1):73–128

Pradhan KC et al (2024) Investigating the relationship between economic growth, energy consumption, and carbon dioxide (CO_2) emissions: a comparative analysis of South Asian nations and G-7 countries. Clean Technol Environ Policy 26(10):3349–3367. https://doi.org/10.1007/s10098-024-02802-5

Ryu D, Nam HJ (2024) Impacts of gender inequality on international trade and innovation. Fin Res Lett 63:105366. https://doi.org/10.1016/j.frl.2024.105366

Saito Y (2018) On the trade, growth, and welfare effects of intellectual property rights protection: trade, growth, and welfare effects of IPR protection. South Econ J 85(1):235–254. https://doi.org/10.1002/soej.12277

Schmitt N, Yu Z (2001) Economies of scale and the volume of intra-industry trade. Econ Lett 74:127–132

Shahbaz M, Lean HH (2012) The dynamics of electricity consumption and economic growth: a revisit study of their causality in Pakistan. Energy 39(1):146–153. https://doi.org/10.1016/j.energy.2012.01.048

Sunde T et al (2023) Testing the impact of exports, imports, and trade openness on economic growth in Namibia: assessment Using the ARDL cointegration method. Economies 11(3):86. https://doi.org/10.3390/economies11030086

Tran T et al (2024) The role of renewable energy in the energy–growth–emission nexus in the ASEAN region. Energy Sustain Soc 14(1):17. https://doi.org/10.1186/s13705-024-00446-3

Valli V, Saccone D (2009) Structural change and economic development in China and India. Europ J Comp Econ 6(1):101–129. https://doi.org/10.2139/ssrn.1486093

Wang X (2022) Capital account liberalization, financial dependence and technological innovation: cross-country evidence. J Bank Finance 145:106642. https://doi.org/10.1016/j.jbankfin.2022.106642

Wesseh JP, Lin B (2018) Energy consumption, fuel substitution, technical change, and economic growth: implications for CO_2 mitigation in Egypt. Energy Policy 117:340–347. https://doi.org/10.1016/j.enpol.2018.03.022

Zeren F, Akkus HT (2020) The relationship between renewable energy consumption and trade openness: new evidence from emerging economies. Renew Energy 147:322–329. https://doi.org/10.1016/j.renene.2019.09.006

Yoke Kee Eng is a Professor in the Faculty of Business & Finance, Universiti Tunku Abdul Rahman, Malaysia. Her research primarily covers the application of econometric analysis to wide-ranging economic issues. Her articles have been published in China Economic Review, Economics Letters, Economic Modelling, International Review of Economics and Finance, Review of Development Economics, Economic Systems, The North American Journal of Economics and Finance, Journal of the Asia Pacific Economy, Energy Policy, and many others.

Grace Hooi Yean Lee is a Professor of Economics and Head of the Department of Economics with Monash University Malaysia. In addition, she is the principal investigator (PI) of the World Values Survey (WVS) for Malaysia (Wave 6, Wave 7 and Wave 8), and is one of the elected executive members of the WVSA (2022–2027). Her principal areas of interest are applied economics, experimental and behavioural economics. She has published numerous articles in top-tier academic journals, including Management Science, Journal of Economic Behavior and Organization, and Public Choice.

Audrey Kim Lan Siah is a Senior Lecturer, Director of Honours/PGDip and Deputy Director of Postgraduate Studies at Monash University Malaysia. Her principal areas of interest are applied economics, and experimental economics. She has supervised Honours and Ph.D. students in these areas. Her articles have been published in Management Science, PLoS ONE, Journal of the Asia Pacific Economy, and many others.

Chapter 13
Green Hydrogen Technology: Analysis of the Economic Preconditions in the Future Development of Asia

Clusters of Future Studies: Environmental Futures; Technological Trends

Juergen Amann, Nicola Cantore, Upalat Korwatanasakul, and Jaime Moll de Alba

Abstract The Paris Agreement's 1.5 °C target drives governments to rethink industrial systems and energy mixes, emphasising decarbonisation technologies like green hydrogen. While green hydrogen is a promising solution, developing countries face challenges due to limited industrial and economic capabilities, which hinder the adoption of green energy technologies. These challenges also complicate balancing industrial growth with sustainability goals. The 2030 Agenda for Sustainable Development urges low- and middle-income nations to develop policies addressing multiple sustainability dimensions, yet many Sustainable Development Goals remain off track with less than five years to 2030. This study introduces a novel methodology to evaluate the industrial capabilities needed for green hydrogen production. Analysing 139 countries from 1995 to 2019 highlights relationships between industrial readiness

---

The original version of the chapter has been revised. A correction to this chapter can be found at https://doi.org/10.1007/978-3-031-95500-6_14

J. Amann
Organisation for Economic Co-Operation and Development (OECD), Paris, France
e-mail: Juergen.AMANN@oecd.org

N. Cantore · J. Moll de Alba (✉)
United Nations Industrial Development Organization (UNIDO), Vienna, Austria
e-mail: J.Moll-de-Alba@unido.org

N. Cantore
e-mail: n.cantore@unido.org; n.cantore@eib.org

N. Cantore
European Investment Bank (EIB), Kirchberg, Luxembourg

U. Korwatanasakul
School of Social Sciences, Waseda University, Shinjuku, Japan
e-mail: korwatanasakul@aoni.waseda.jp

E. Lau et al. (eds.), *Future of Economy in Asia*,
https://doi.org/10.1007/978-3-031-95500-6_13

and green hydrogen production, identifying key bottlenecks and policy implications. The research focuses on Asia, where China leads in green hydrogen production. Expanding the Asian market requires strengthening industrial capacities and enacting robust policies, particularly in lower-income countries. The findings underscore the need for targeted efforts to bridge capability gaps and foster green hydrogen adoption as a critical step toward achieving global decarbonisation and sustainable development goals.

Keywords Green hydrogen · Economic complexity · Industrial preconditions · Revealed comparative advantage (RCA) · Sustainable development · Asia energy transition

13.1 Introduction

There is an almost universal consensus that human beings generate climate change emissions and that countries must undertake all the possible efforts to avoid the most adverse impacts of climate change, especially for developing countries. Current policies and laws for mitigation have expanded over time, but current pledges make it likely that warming will exceed 1.5 °C during the twenty-first century and will hardly be below 2 °C. Moreover, finance flows to support countries in reducing emissions fall short of the levels needed to meet climate goals (IPCC 2023).

According to IEA estimates (2023b), renewable energy represents a key decarbonisation option to reach the net zero emissions target by 2050. In an ideal scenario, almost 90% of global electricity generation in 2050 will come from renewable sources, with solar, photovoltaic, and wind accounting for nearly 70%.

Bottlenecks related to the diffusion of renewable energy are well-known in the current literature. Solar and wind energy can be intermittent depending on the climatic conditions and cannot easily be stored (Cantore et al. 2021). Even though empirical studies observe decreasing costs of renewable energy technologies over time, economic considerations and technical problems may not be cost-competitive compared to fossil fuels (Timilsina 2021). Finally, the policy space for intervention to promote green technologies may vary across countries and time (Cantore et al. 2018).

It is worth noting that there is an increasing trend whereby several leading countries seek to increase hydrogen production and anchor such plans in formal strategies and plans in their transition towards clean energy (Panchenko et al. 2023). Recent literature emphasises that green hydrogen, currently produced only in a few countries and produced only to a limited extent compared to brown hydrogen obtained with fossil fuels, can quickly become an essential option among decarbonisation energy inputs. Increasing consensus is that energy-independent countries must seek a renewable or green hydrogen future soon (Clark and Rifkin 2006).

However, green hydrogen production still struggles to take off. According to the Global Hydrogen Review (IEA 2023a), government targets for low-emission hydrogen production represent only one-fifth of the low-emission hydrogen use in the Net Zero Emissions by 2050 Scenario. Economic history shows that countries need to acquire solid preconditions for breakthrough technologies. Müller and Eichhammer (2023) recently published a ground-breaking paper designing a list of industrial goods and technologies representing hydrogen production conditions. This list of 36 products represents the different value chain tiers to produce green hydrogen, including renewable energy, water electrolysis, hydrogen production, and distribution.

The work of Müller and Eichhammer is embedded in the dynamic product space and economic complexity literature originating from the seminal paper of Hausmann and Hidalgo (2014). Economic complexity refers to the level of sophistication and diversity within an economy based on the range of goods and services it produces and the capabilities needed to manufacture those products. It is a concept that assesses what a country produces and the interconnectedness among those products. Countries with higher economic complexity have economies that produce a wide array of goods, including more technologically advanced and technologically intensive products. Complexity implies a higher level of know-how, skills, infrastructure, and technology required for production. The complexity literature has been further developed in the specific segment of green goods to investigate the capability of countries to produce industrial products useful to reduce climate change emissions or, more in general, pollution (Hamwey et al. 2013).

This paper borrows from the complexity and product space literature, particularly Müller and Eichhammer, to analyse the preconditions for producing green hydrogen. It introduces a composite index explaining the capacity of countries to produce and export the 36 technologies representing the green hydrogen value chain. The present study expands Müller and Eichhammer by (1) measuring the extent to which countries possess the necessary capabilities for production and (2) analysing the export dimension, focusing on the domestic production dimension to conduct a comparison.

Rank 1 and Rank 2 are categories introduced by Müller and Eichhammer to distinguish the most critical and essential technologies to produce green hydrogen from other satellite technologies (Table 13.1). The four Rank 1 components are PV modules, generator sets of a wind turbine, lithium-ion accumulators, and electrolysis stacks, representing the core technologies for renewable energy production, electricity storage, and technologies for water electrolysis. Müller and Eichhammer assign other technologies like pumps for electrolysis or electrical control of photovoltaic modules as Rank 2 because they are necessary but less relevant for the green hydrogen value chain.

We analyse the capacity of countries to produce and export hydrogen technologies by distinguishing between Rank 1 and Rank 2 technologies as two aggregated groups and as a whole group. Overall, the methodological frame derived from the present study is well grounded in the product space and economic complexity literature, revealing production capabilities from existing export data and production performance (Fig. 13.1). To the best of our knowledge, it is innovative because it studies industrial production data and is an original way of interpreting green hydrogen technology export performance without necessarily incorporating complexity or strict and ad hoc interpretations of revealed comparative advantage (RCA).

Table 13.1 List of the 36 Rank 1 and Rank 2 technologies to produce green hydrogen

Technology	Technology sub-group	HS Code	HS Description	Rank
Electrolysis				
	Conditioning, Pumping and Cleaning			
		841,370	Centrifugal pumps, nes	2
		841,430	Compressors for refrigeration	2
		841,490	Pumps, compressors, etc. Parts	2
		841,861	Compression refrigeration equipment	2
		841,869	Freezing equipment, nes	2
		841,899	Freezing equipment, parts	2
		841,950	Heat exchange units, non-domestic	2
		841,989	Machinery for temperature change	2
		841,990	Industrial heating/cooling, parts	2
		842,121	Water filtering machines	2
		842,139	Filtering machinery, gases, nes	2
		842,199	Filter machine, parts	2
	Voltage Transformer/Rectifier			
		850,440	Static converters, nes	2
		850,434	Electric transformers, > 500 kV A	2
	Electrolysis Stack			
		854,330	Apparatus for electro-plating, electrolysis	1
	Ion Exchanger			
		391,400	Ion exchangers based on polymers	2
Battery Storage	*Battery*			
		850,720	Lead-acid electric accumulators, non-vehicle	2
		850,780	Electric accumulators, nes	1
Solar PV	*Support Structure*			
		761,090	Aluminum structures for construction	2
	PV Module			
		854,140	Photosensitive/PV/LED semiconductor device	1
	Static Converter			
		850,440	Static converters, nes	2
	System Controller			
		853,710	Electrical control and distribution boards, < 1 kV	2

(continued)

Table 13.1 (continued)

Technology	Technology sub-group	HS Code	HS Description	Rank
Wind Turbine	*Tower*			
		730,820	Towers, iron/steel	2
	Generating Set			
		850,230	Electric generating sets nes	1
		850,164	AC generators > 750 kV A	2
		848,340	Gearing, ball screws	2
	Electronic Control Equipment			
		853,720	Electrical control and distribution boards, > 1 kV	2
		902,830	Electricity supply, calibration meters	2
		903,020	Cathode-ray oscilloscopes/graphs	2
		903,031	Electrical multimeters	2
		903,039	Voltmeters, non-recording	2
	Bearings			
		848,220	Bearings, tapered roller	2
		848,230	Bearings, spherical roller	2
		848,250	Bearings, cylindrical roller	2
		848,280	Bearings, ball or roller nes	2
	Blades			
		841,290	Hydraulic/pneumatic power engine, parts	2
		850,300	Parts for electric motors and generators	2

List of products identified as potentially relevant for green hydrogen production in a stand-alone configuration using solar photovoltaic and wind power, distinguishing between key components (Rank 1) and other relevant components (Rank 2), classified by the HS 1992 6-digit codes. (Mueller and Eichhammer 2023)

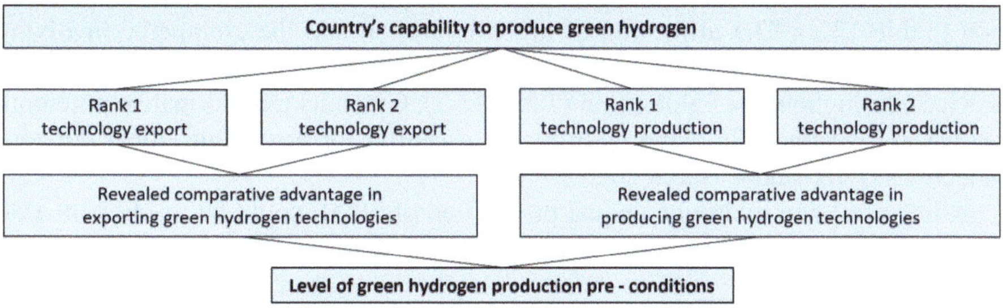

Fig. 13.1 Theoretical framework to measure green hydrogen production preconditions

Table 13.2 Green hydrogen production, conversion and end-use across the energy system (IRENA 2022)

Production	Transformation	Transport	End use
Renewable energy and electrolysis (H2)	**No transformation (H2)** **Transformation** – Synthetic fuels (CO2 from sustainable CO2 capture and H2) – Green ammonia (nitrogen and hydrogen)	Shipping Trucks Pipeline Storage	**Industry (H2)** – Steel – Chemicals – Refineries
			Transport (H2, NH3 and water) – Shipping – Aviation – Cars – Rail – Trucks – Buses
			Heating (H2)
			Power generation (H2)

The term 'synthetic fuels' refers to a range of hydrogen-based fuels produced through chemical processes with a carbon source (carbon monoxide (CO) and CO_2 captured from emission streams, biogenic sources or directly from the air)
H2 green hydrogen used as an energy carrier and chemical input, *CO2* carbon dioxide, *N2* nitrogen, *NH3* ammonia

13.2 Literature Review

13.2.1 The Hydrogen and Green Hydrogen Markets

The global hydrogen trade landscape is characterised by the unique nature of hydrogen as a versatile energy carrier. Unlike fossil fuels, hydrogen is not a primary fuel but is a conduit for energy transfer, comparable to electricity. Its distinct properties[1] make it a crucial enabler in the transition of energy systems, presenting advantages in various applications, including transportation and building heating (Khan and Al-Ghamdi 2023). The hydrogen value chain exhibits complexity, necessitating substantial investment across horizontal and vertical segments for commercialisation (Table 13.2). The utilisation of green hydrogen and the commerce involving hydrogen and its by-products will also influence the trade of necessary goods and services throughout the value chain (Table 13.3). Given its transformative potential, increased investment in clean hydrogen is essential for achieving global net-zero objectives (AIIB and Castalia 2023).

A low-emission hydrogen annual production of 38 Mt could be reached in 2030 if all newly announced production projects were realised (IEA 2023a). However,

[1] Hydrogen must be produced by using another fuel/process and this gives the hydrogen its color name, e.g. grey hydrogen (derived from fossil fuels), blue hydrogen (produced from fossil fuels with carbon capture, utilisation, and storage), and green hydrogen (sourced from renewable energy).

Table 13.3 Overview of tradable goods and services along the green hydrogen and derivatives supply chain (WTO and IRENA 2023)

	Renewable energy production	Hydrogen and hydrogen derivatives production	Transport, storage and reconversion	Use
Equipment (Products)	– Renewable energy generation equipment, e.g., solar panels, wind turbines, etc – Electric equipment	– Electrolysers – Compressors, valves, flow control, metering and related equipment – Ammonia, methanol, direct reduced iron production plants	– Piping and storage systems – Compressors, valves, flow control, metering and related equipment – Hydrogen, ammonia, methanol, hot briquetted iron transport vessels	– Ammonia, methanol, production plants – Direct reduction furnaces for iron production – Fuel cell systems – Hydrogen-ready (industrial) boilers – Hydrogen-ready gas turbines for power generation – Ammonia, methanol and hydrogen storage equipment
Services	– Design – Engineering – Related Construction – Operation and management		– Transportation and storage services – Related construction (e.g., port terminals)	– Wholesale and retail (e.g., hydrogen stations) – Marketing
	Research and development, technical testing and analysis, consulting and training, various professional and business services			

despite the announcement of over 1,000 green hydrogen projects with substantial projected investment through 2030, only a fraction has progressed to the Financial Investment Decision (FID) or operational status, underscoring the early stage of green hydrogen trading (Pepe et al. 2023). The absence of a broad market structure for hydrogen trading results in bilateral transactions, impeding the establishment of comprehensive price indices reflective of supply and demand dynamics. It is worth mentioning that green hydrogen may play a noticeable role in supporting the foreseen growth of the industrial sector in leading economies with significant manufacturing capacity, including the production of steel in India and cement in Vietnam and Indonesia (Cordonnier and Saygin 2022).

Production predominantly relies on natural gas in the global hydrogen trade context, with estimated annual trade flows ranging from USD 150–200 million (WTO and IRENA 2023). However, the trade of derivative commodities like ammonia and methanol holds greater significance, valued at USD 17.5 billion and USD 14.1 billion, respectively, in 2022 (WTO and IRENA 2023). The global hydrogen trade, encompassing all colours but predominantly fossil-fuel-based, reached approximately USD 300 million in 2022, reflecting a 71 per cent increase from 2021 (WTO and IRENA 2023).

While global trade in derivative commodities is substantial, the trade landscape for hydrogen remains relatively small. Except for ethanol and ammonia, the international trade of hydrogen is still at an infant stage, with a few pipelines connecting industrial areas in some European countries and some pilot demonstrations of hydrogen trade by ship (IEA 2023a). The top ten import markets represent nearly 90 per cent of global hydrogen trade, with the United States (US) and the Netherlands as key importers, while Canada and Belgium serve as leading exporters (WTO and IRENA 2023). Bilateral hydrogen trade primarily occurs regionally. On the other hand, the trade landscape for ammonia, as an energy carrier for hydrogen and methanol, is more global than regional. India emerged as the leading ammonia destination market in 2021, followed by the US and Morocco. In contrast, China leads the global methanol market, accounting for 25 per cent of global imports, with other major importers, including India, the Netherlands, the US, the Republic of Korea, and Japan (WTO and IRENA 2023). Methanol primarily originates from natural gas producers such as Trinidad and Tobago, the Kingdom of Saudi Arabia, Oman, the United Arab Emirates, the US, and the Russian Federation. The international trade landscape for hydrogen derivatives and their potential as energy carriers and decarbonisation agents underscores hydrogen-related commodities' evolving dynamics and significance in global and Asian markets.

Within Asia, Singapore and Malaysia stand out as significant hydrogen importers, while Taiwan (the Province of China), Malaysia, Japan, and China emerge as important hydrogen suppliers (WTO and IRENA 2023). However, hydrogen trade remains limited to intra-regional trade, except for the US, a supplier of Singapore and Malaysia. Countries like India, the Republic of Korea, China, Taiwan (the Province of China), and Thailand feature prominently among the top global ammonia importers, while Indonesia and Malaysia serve as significant regional ammonia suppliers (WTO and IRENA 2023). In contrast, countries like China, India, the Republic of Korea, Japan, Taiwan (the Province of China), Indonesia, Thailand, and Singapore feature prominently among global methanol importers. Malaysia is a top methanol supplier in (Southeast) Asia, with Thailand, Indonesia, and Singapore as significant importers (WTO and IRENA 2023).

13.2.2 Policy Landscape in the Hydrogen and Green Hydrogen Sector

The global landscape for hydrogen policy has seen significant developments in recent years, with an increasing number of countries unveiling national strategies to foster the growth of the hydrogen economy. Substantial differences are observable in the role of hydrogen in energy transitions. Asia and Europe emphasised demand, while the Middle East and North Africa focused more on supply (World Energy Council 2021). In 2022, more than 42 countries introduced national hydrogen strategies (Khan and Al-Ghamdi 2023). Most of these national strategies prioritised green hydrogen, emphasising renewable-based hydrogen as the long-term solution. Notably, countries like Chile, Germany, the Netherlands, and Portugal directed their strategies toward developing and deploying green hydrogen technologies. In Asia, Japan, the Republic of Korea, and China emerged as frontrunners in formulating state-level hydrogen strategies. China joined this group in March 2022 by releasing its Medium- and Long-Term Plan for the Development of the Hydrogen Energy Industry (2021–2035), demonstrating a growing commitment to advancing hydrogen technologies and infrastructure (WEF 2023). China's leadership in this field was already proven by the introduction of the first formal green hydrogen standard in 2020 through the Standard and evaluation of low-carbon hydrogen, clean hydrogen and renewable, which may contribute to the definition of a global standard contributing to expanding international trade (Liu et al. 2022).

As countries strive to realise the potential of hydrogen as a clean energy carrier, proactive efforts are required to create the necessary conditions to facilitate hydrogen trade, including the development of infrastructure, market design, regulatory frameworks, and favourable trade policies to support the growth of the hydrogen economy. According to the World Trade Organization (WTO) Environmental Database, WTO members formulated 44 hydrogen-related policies reported between 2009 and 2021 (WTO and IRENA 2023). These policies encompass supportive measures such as grants, loans, or tax concessions, falling under the Agreement on Subsidies and Countervailing Measures (SCM). Ten notifications concerning technical regulations related to hydrogen production and usage were submitted under the Agreement on Technical Barriers to Trade (TBT) (WTO and IRENA 2023). Recent support measures from various countries primarily focus on advancing technology, production, and markets related to renewable energy, hydrogen, and hydrogen fuel cells. These measures include initiatives such as funding programmes, subsidy funds, and investments to promote the development and adoption of low-carbon hydrogen technologies and the advancement of the hydrogen market.

Japan and the Republic of Korea have provided substantial support for hydrogen infrastructure, while others, such as Singapore, India, and Indonesia, have focused on financing local research and development endeavours (RECAP 2022). However, in Asia, the adoption of hydrogen is perceived to be in its early stages due partly to the lack of government policies regarding financial support, the facilitation of developing strategic relationships among stakeholders, and infrastructure development (RECAP 2022).

13.3 Methodology

13.3.1 Data

We utilise the same raw data source as Müller and Eichhammer (2023), namely the *International Trade Data (HS, 92),* which provides trade flows at the 6-digit level of the Harmonised System (HS) 1992 between 1995–2019 and categorises approximately 5,000 commodities (The Growth Lab at Harvard University 2019).[2] The original trade data is taken from UN Comtrade (2023) and is cleaned following the Bustos-Yildirim method. Industrial production data are from the UNIDO INDSTAT dataset whereas data about green hydrogen production are taken from the IEA Hydrogen Production Projects Database.[3] Regarding products, we focus on the *Green H2 Product Space,* which consists of 36 products defined in Table 13.1. The analysis was conducted with 139 countries.

13.3.2 The Revealed Comparative Advantage (RCA) of the Green Hydrogen Product Space

We calculate the *product-based* export index of the revealed comparative advantage (RCA_{cpt}) for each period t and country c, where the export value of a product p is denoted X_{cpt}, as:

$$RCA_{cpt} = \frac{\frac{X_{cpt}}{\sum_p X_{cpt}}}{\frac{\sum_c X_{cpt}}{\sum_c \sum_p X_{cpt}}} \tag{13.1}$$

We calculate the *sector-based* export index of revealed comparative advantage (RCA^s_{ct}) for each period t and country c of all Rank 1 and Rank 2 Green H2 technologies jointly, i.e., p_{all}, such that $p_{all} \in s$, as:

$$RCA^s_{ct} = \sum_{p_{all} \in s} w_{cpt} \times RCA_{cpt} \tag{13.2}$$

with

[2] Data was analysed only until 2019 to mitigate the impact of COVID-19 and subsequent supply chain disruptions.

[3] For trade data see https://atlas.hks.harvard.edu. For industrial production & data see https://stat.unido.org/data/table?dataset=indstat%26revision=3%23data-browser. For hydrogen production data see https://www.iea.org/data-and-statistics/data-product/hydrogen-production-and-infrastructure-projects-database. https://atlas.cid.harvard.edu/about-data

$$\omega_{cpt} = \begin{cases} \frac{X_{cpt}}{\sum_{p\in s} X_{cpt}} & \text{if } p \in s, \\ 0 & \text{otherwise.} \end{cases}$$

where the product-sector concordance is established using the World Integrated Trade Solution (WITS) mapping between the HS 92 and the International Standard Industrial Classification of All Economic Activities (ISIC), Rev.3, nomenclatures,[4] which is performed using the *R* (R Core Team 2024) package *comtradeRggregator* (Amann 2024). We calculate sector-level aggregates at the ISIC Rev.3 II-digit ($s = s_{II}$) as well as IV-digit level ($s = s_{IV}$), respectively.

The RCA is a typical indicator representing the capabilities of countries to produce and export technologies and industrial goods (see, among others, Leromain and Orefice 2014). Our paper assumes that the more countries can produce and export the 36 green hydrogen technologies preconditions, the more they possess the necessary capabilities to produce green hydrogen. However, our index differs from that of Müller and Eichhammer. First, our index does not capture the extent to which countries are producing and exporting the most complex preconditions, being complexity defined a la Hausman and Hidalgo. Second, our index does not capture only the extent to which countries produce goods with RCA >= 1, which is the typical condition assumed by economists to claim that a country produces and exports goods in a competitive way. Conversely, our index also captures and analyses productions and exports with RCA < 1, which may indicate the existence of capabilities of countries to produce and export goods and technologies, even though it is not in a competitive way. Lastly, RCAs are aggregated by technology to represent a synthetic value by country and time.

13.3.3 *Revealed Comparative Advantage (RCA) Over Time (Arithmetic Average)*

To assess the general Green H2 readiness (GH2R) per country and year, we create country-level averages of Rank 1, Rank 2, and all Green H2 technologies. More formally:

$$\overline{RCA}_{ct} = \frac{1}{p} \sum_{p \in P} RCA_{cpt} \tag{13.3}$$

and $P = \{P_1, P_2, P_{all}\}$, with P_1, P_2 (P_{all}) denoting the set of Rank 1, Rank 2, (all) Green H2 technologies, respectively, and $P_1, P_2 \subset P_{all}$.

[4] See: World Integrated Trade Solution (WITS) Product Concordance.

13.3.4 Analysis of Green Hydrogen Industrial Preconditions

We compare the progression of green hydrogen capabilities by economic sector. For this purpose, we compute the country-sector-level average for 1995 to 1999 and 2015 to 2019 to compare the change between average 2015–2019 and average 1995–1999 RCAs by country and ISIC Rev. 3, II-digit industries for Rank 1 and Rank 2 technologies.

More formally, we calculate:

$$\Delta RCA_c^s = \frac{1}{t}\sum_{t=2015}^{2019} RCA_{ct}^s - \frac{1}{t}\sum_{t=1995}^{1999} RCA_{ct}^s \tag{13.4}$$

13.4 Findings

13.4.1 The Revealed Comparative Advantage of the Green Hydrogen Product Space

A preliminary descriptive analysis shows the distribution of Rank 1 and Rank 2 technologies RCAs across countries, as calculated in Eq. (13.1). We are interested in the extent to which countries can produce or export technology. For this, we first check the distribution of product-level RCAs across all four technologies (electrolysis, battery storage, solar PV, wind turbines) for Rank 1 and Rank 2 technologies, which we plot in Fig. 13.2. There, we also partition the RCAs into six categories: RCAs larger than ten are highlighted in dark blue; between 1 and 10 are blue; 0.5 to 1 (light blue); 0.25 to 0. 5 (light red); 0.1 to 0.25 (red) and smaller 0.1 (dark red). According to Fig. 13.2, it is evident that RCAs are distributed according to normal functions across countries. The distribution does not qualitatively change between Rank 1 and Rank 2 technologies, implying that the relevance of the technology in the green hydrogen value chain has no impact on how preconditions for green hydrogen production are distributed across countries.

Next, we visualise the distribution of all Green H2 technologies as described in Eq. (13.3) conditional on the historical World Bank Income group classification[5] in Fig. 13.3. The result shows that high-income and upper-middle-income countries are especially well-equipped in terms of preconditions to produce and export green hydrogen technologies and preconditions. The finding is somewhat predictive of the current global green hydrogen production. According to the existing statistics, high-income countries and China, an upper-middle-income country with an enormous industrial capability potential, are the most important green hydrogen producers

[5] See: World Bank Country and Lending Groups.

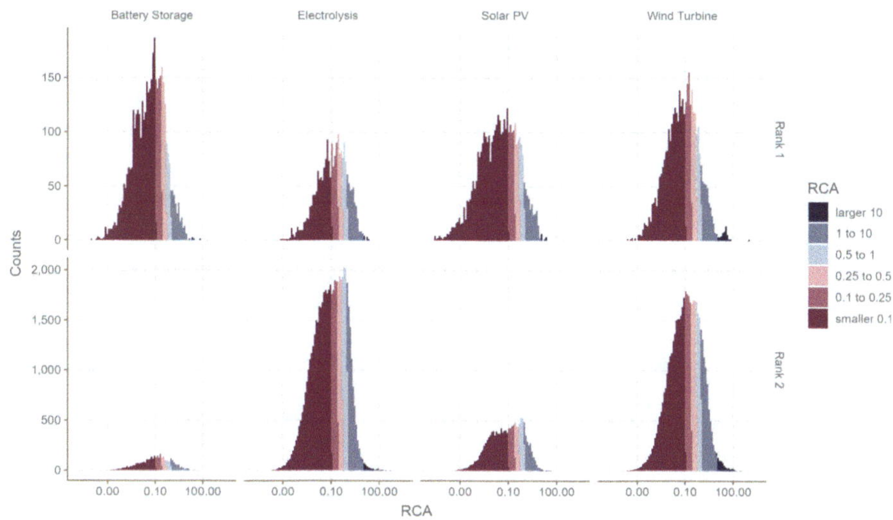

Fig. 13.2 Distribution of the revealed comparative advantage of Rank 1 and Rank 2 technologies at the world level. *RCA* revealed comparative advantage over time (arithmetic average)

(Fig. 13.3). An exception is a lower-middle-income country, Peru, producing green ammonia with investments from a company headquartered in Chile (a high-income country).

However, country income classification cannot be considered the necessary or sufficient condition to trigger a relevant production of green hydrogen. According to the analysis, the US and Japan are performing best regarding RCA of Rank 1 green hydrogen preconditions. As shown in Fig. 13.4, the US is already advancing in the production of green hydrogen, while Japan is lagging behind. In 2017, the Japanese government became the first in the world to formulate a national hydrogen strategy. However, Ohno et al. (2022, page 1) note, '*Japan has performed far below its main goal of increasing uptake of fuel cells and fuel cell vehicles, and the hydrogen stations built across the country have seen little use. Japan's efforts in producing green hydrogen needed to achieve a decarbonised society have fallen behind those of European countries, China, and other nations*'. According to the same report, policy and cost issues have been the main bottlenecks in adopting this new technology. From a cost perspective, European and Chinese companies produce an annual capacity of several hundred megawatts to several gigawatts. In contrast, Japanese companies lack a critical production mass to reduce costs. Halkos et al. (2024) find that Asia is increasing environmental efficiency through management capabilities, optimal resource allocation, and economies of scale. Therefore, the mass production of green hydrogen will play a critical role in increasing the productivity of environmental goods sectors through learning curves in the future. From a policy perspective, the same report states that the Japan Hydrogen Strategy was not strongly tied to the overall country decarbonisation strategy, and this did not give an impulse to the production of green hydrogen.

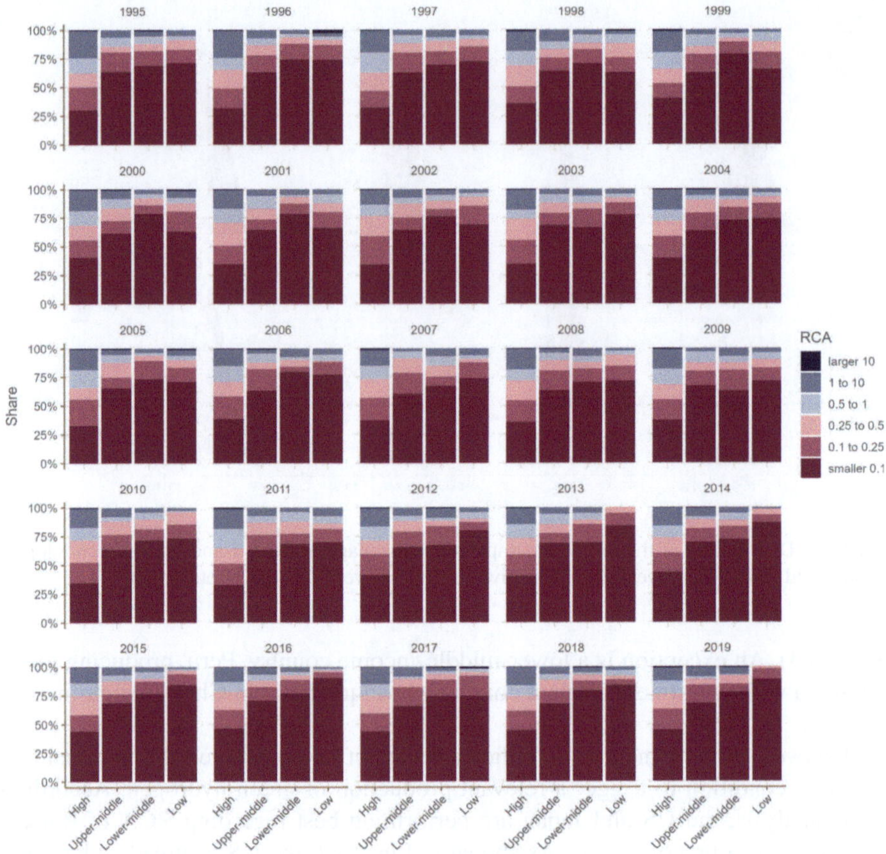

Fig. 13.3 Distribution of all green hydrogen technologies by year and income group. *RCA* revealed comparative advantage over time (arithmetic average)

Fig. 13.4 World producers of green hydrogen, 2023. Production is expressed in thousands of kilotons of hydrogen per year; ISO3 country codes (Klevstrand 2023)

On the other hand, lower-income countries like Egypt, Namibia, and Mauritania are designing projects to start producing green hydrogen. Should these projects materialise in the following years, it would mean that even countries with initial gaps in industrial preconditions could develop this new technology through strong decision-makers' commitment and good policy design. Appropriate imports and technology transfers could mitigate an initial technological gap through these conditions. The International Energy Agency (2023a, page 7) claims, *'only 4% of projects that have been thus far announced are under construction or have taken a final investment decision. Uncertainty about future demand, the lack of infrastructure available to deliver hydrogen to end users and the lack of clarity in regulatory frameworks...'*, meaning that efforts are often still needed to transform projects into concrete facts.

If we consider only the group of Asian Countries, results are quite similar by year and reflect patterns observed at the world level (Fig. 13.5). Low- and lower-middle-income countries show a lower level of RCA in exporting Rank 1 and Rank 2 green hydrogen technology. One difference is that the leading role of China (upper-middle-income country) in green hydrogen production explains the outperformance in terms of RCA of the upper-middle-income country group over the high-income group. China represents about 54% of the current green hydrogen world production[6] and continues to lead on commitment investments in electrolyser projects, followed by Europe. However, low-emission hydrogen only represents about 1% of hydrogen world production, and the world production growth rate is still moderate at 2.5% (IEA Global Hydrogen Review 2024).

13.4.2 Analysis of Green Hydrogen Preconditions with Industrial Data

Next, we consider the capabilities of economies at the sectoral level by aggregating trade data based on their corresponding economic activity according to ISIC Rev. 3 employing official H2-ISIC Rev.3 concordance tables as described in Eq. (13.2). Product-sector concordance is established using the World Integrated Trade Solution (WITS) mapping between the HS 92 and the International Standard Industrial Classification of All Economic Activities (ISIC), Rev.3, nomenclatures.

A caveat of this approach is that HS sectors at VI digits can only be associated with ISIC sectors characterised by a higher level of aggregation (IV digits). This is unavoidable as an ISIC VI digit classification does not exist. Moreover, as we aggregate export HS codes of Rank 1 and Rank 2 technologies at the VI-digit level to their coarse II-digit representation, the transition from the HS VI-digit codes to ISIC

---

[6] Our elaboration from the Hydrogen Production Project database. Considered final green hydrogen products are synthetic methane, hydrogen in molecular form, liquid organic hydrogen carriers, methanol and ammonia. Considered technologies are alkaline electrolysis (ALK), proton exchange membrane electrolysis (PEM), Solid oxide electrolysis cells (SOEC) and 'unknown electrolysis type'.

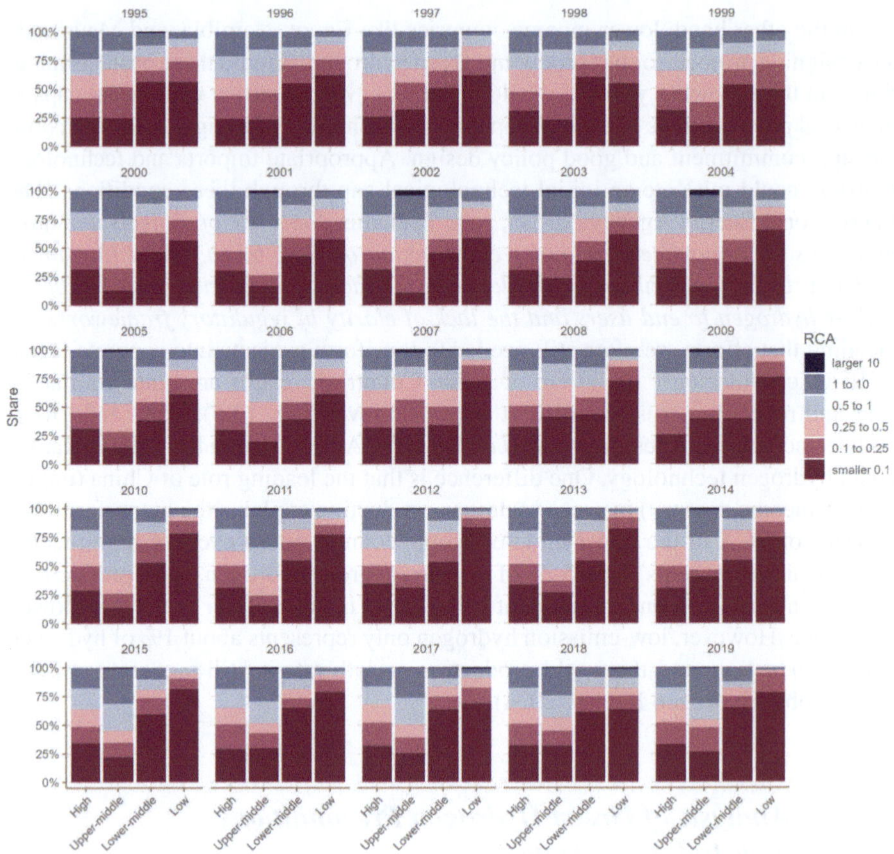

Fig. 13.5 The Asia market—distribution of all green hydrogen technologies for South Asia, East Asia and the Pacific by Income Group and Year. *RCA* revealed comparative advantage over time (arithmetic average)

II-digit codes implies an information loss by aggregation whenever multiple IV-digit sectors get mapped onto their corresponding II-digit aggregate. In our analysis, this is the case Sectors 29—Manufacture of machinery and equipment n.e.c. and 31—Manufacture of electrical machinery and apparatus n.e.c., respectively. All other IV-digit sectors map 1-to-1 onto their II-digit level aggregate, implying no loss of information due to aggregation bias except Sector 28—Manufacture of fabricated metal products that still includes both the solar and the wind energy technology aggregates with reduced possibility to analyse separately the two types of capabilities. We chose the II-digit representation alongside the IV-digit information to provide a broader policy context.

Considering these limitations, we calculate sector-level *RCA* s as defined in Eq. (13.2) and obtain the corresponding ΔRCA s according to Eq. (13.4). In Fig. 13.6, we aggregate the ISIC II-digit level ΔRCA_c^s by income group in the lower panel and show the global aggregate in the top panel. RCA is smaller (larger) than −0.5 (0.5) rounded to −0.5 (0.5) for expositional purposes, and the dashed vertical lines indicate an ΔRCA_c^s of zero.

Figure 13.6 reveals that green hydrogen production's industrial capabilities variations across industries are not necessarily positive across sectors and regions. Positive growth across industries is most frequent across East Asia, the Pacific region, Europe, and Central Asia. Furthermore, notable sector differences emerge. For example, the Middle East and Northern Africa regions have seen an overall positive shift of green hydrogen industrial capabilities for Sector 31—Manufacture of electrical machinery and apparatus and a more consistent negative growth across countries in the regions for Sector 33—Manufacture of medical, precision and optical instruments, watches and clocks. In terms of diversification strategy, this development may be, however,

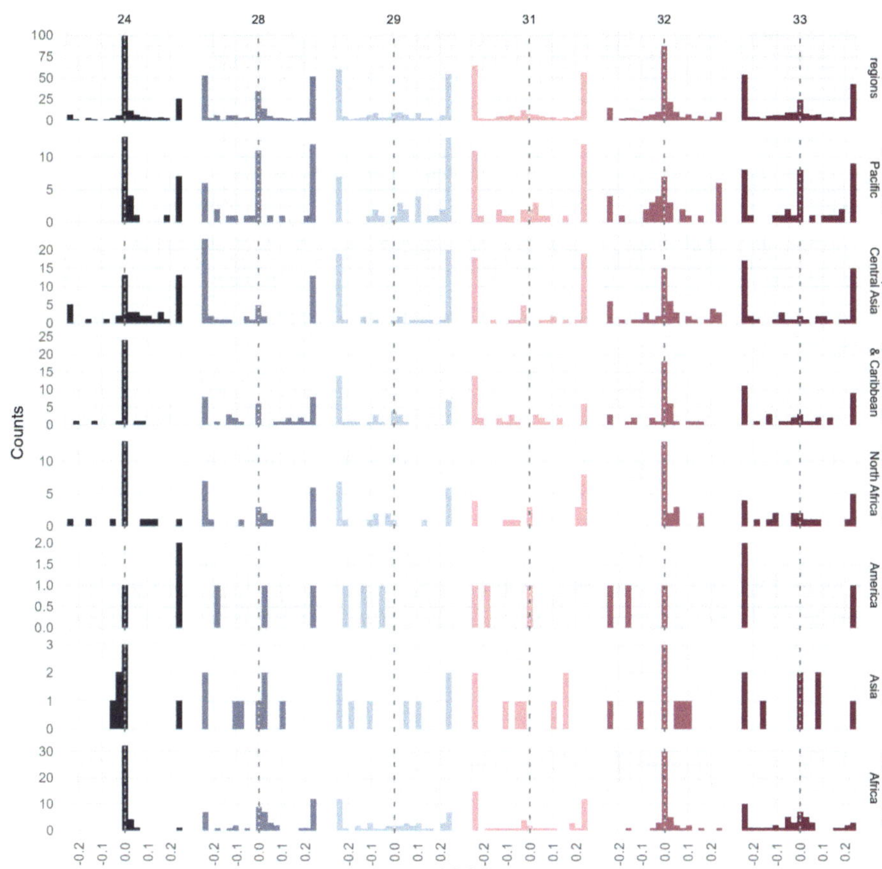

Fig. 13.6 Industrial preconditions and growth—change between average 2015–2019 Revealed Comparative Advantage (RCA) and Average 1995–2000 RCA by Country and 2-digit ISIC Rev. 3 Industry. ΔRCA s smaller (larger) than −2.5 (2.5) rounded to −2.5 (2.5) for expositional purposes; The dashed vertical line corresponds to an ΔRCA of zero

Table 13.4 Share of green hydrogen technology by industrial sector

Technology by ISIC Rev. 3, IV-digit industry	Share
24—Manufacture of chemicals and chemical products	
Electrolysis	100.00
28—Manufacture of fabricated metal products, except machinery and equipment	
Solar PV	51.95
Wind Turbine	48.05
29—Manufacture of machinery and equipment n.e.c	
Electrolysis	68.44
Wind Turbine	31.56
31—Manufacture of electrical machinery and apparatus n.e.c	
Battery Storage	18.91
Electrolysis	25.11
Solar PV	19.77
Wind Turbine	36.21
32—Manufacture of radio, television and communication equipment and apparatus	
Solar PV	100.00
33—Manufacture of medical, precision and optical instruments, watches and clocks	
Wind Turbine	100.00

Rank 1 and Rank 2 technologies combined, classified by the HS 1992 6-digit codes and mapped to ISIC Rev. 3

desirable for the region as a whole as it hints at a broadening and diversification of the green Hydrogen industrial base: while Sector 33 is dominated by wind turbine technology, Sector 31 serves as an enabler of a much wider range of green Hydrogen technologies in addition to wind turbines, including solar PV, electrolysis or batteries (see Table 13.4 for a summary of the green hydrogen technologies attributable to each manufacturing sector; the country-level statistics for Fig. 13.6 are available in Table 13.5, and a complementing IV-digit regional breakdown is provided in Table 13.6).

13.5 Discussion of Implications

These results suggest that the capacity of Asian countries to produce green hydrogen is not only heterogeneous at the country/regional level, as shown by the previous section discussing export data, but also over time, as is evident from industrial data. Policymakers are challenged with country and time-specific conditions to design and implement suitable measures to promote green hydrogen. One–size–fits–all solutions are ineffective as green hydrogen technologies require a very diversified set of technologies and skills for which Asian countries show significantly different levels of

Table 13.5 Change of average 1995–1999 revealed comparative advantage (RCA) and average 2015–2019 RCAs

Country	ISIC Rev. 3, II-digit industries					
	28	*29*	*31*	*32*	*33*	24
ABW	*0.41*	*8.54*	*0.48*	*0.02*	*2.90*	
AFG	*0.00*	*−0.36*	*−0.02*	*0.00*	*0.09*	*−0.04*
AGO	*0.03*	*0.57*	*0.22*	*0.00*	*0.05*	*0.00*
AIA	*0.17*	*5.64*	*0.65*	*0.67*	*0.91*	*24.87*
ALB	*7.15*	*−2.05*	*4.84*	*0.04*	*−0.58*	0.00 ara>
AND	*−0.05*	*0.01*	*−1.80*	*0.22*	*−1.72*	*3.26*
ANS	*0.18*	*2.12*	*0.07*	*−1.22*	*−0.10*	*0.01*
ARE	*0.04*	*0.30*	*0.20*	*0.04*	*−0.14*	*0.08*
ARG	*−0.96*	*−0.11*	*−0.70*	*−0.01*	*−1.88*	*0.00*
ARM	*−0.22*	*−4.32*	*−29.06*	*0.03*	*3.01*	*0.06*
ASM	0.00	*−14.11*	*0.56*	na	0.00	na
ATA	na	na	na	na	*−314.23*	na
ATF	na	*7.50*	*−0.89*	na	na	na
ATG	*1.59*	*−1.74*	*−3.33*	*−0.01*	*−0.53*	na
AUS	*−0.25*	*−0.23*	*−0.61*	*−0.18*	*−0.05*	*0.18*
AUT	*0.37*	*0.24*	*0.09*	*0.05*	*0.11*	*3.06*
AZE	*−0.17*	*−0.80*	*−2.53*	*−0.01*	*−0.02*	*−0.03*
BDI	*0.57*	*−0.13*	*0.18*	*0.00*	*−1.71*	0.00
BEL	*−0.41*	*−0.23*	*−0.04*	*−0.09*	*0.09*	*0.16*
BEN	*−0.23*	*0.02*	*−0.14*	*0.01*	*−0.10*	0.00
BFA	*−0.23*	*−0.32*	*−2.82*	*0.00*	*−2.76*	*0.00*
BGD	*−0.07*	*−0.19*	*−0.04*	*−0.66*	*−0.01*	*0.00*
BGR	*−0.01*	*0.38*	*−3.15*	*0.03*	*4.86*	*−0.08*
BHR	*−0.22*	*−0.08*	*0.20*	*0.01*	*0.07*	*0.01*
BHS	*−0.07*	*0.74*	*−11.88*	*0.12*	*0.14*	0.00
BIH	*2.01*	*4.02*	*7.20*	*0.00*	*−0.50*	*0.01*
BLR	*4.13*	*−7.81*	*0.35*	*0.01*	*−0.46*	*−0.28*
BLZ	*−0.13*	*−0.09*	*0.07*	*0.00*	*−0.29*	0.00
BMU	*0.04*	*−0.14*	*0.00*	*0.00*	*0.24*	0.00
BOL	*−0.10*	*−0.26*	*0.02*	*0.00*	*−0.24*	0.00
BRA	*−3.18*	*−3.51*	*−0.22*	*−0.05*	*−5.30*	*0.08*
BRB	*0.38*	*−0.29*	*−0.13*	*−0.50*	*54.76*	0.00
BRN	*0.31*	*0.11*	*0.00*	*0.00*	*−0.28*	0.00
BTN	*0.03*	*−0.38*	*−0.11*	0.00	*−0.17*	na

(continued)

Table 13.5 (continued)

Country	ISIC Rev. 3, II-digit industries					
	28	*29*	*31*	*32*	*33*	24
BWA	*0.05*	*0.11*	*0.76*	*0.00*	*0.04*	0.00
CAF	*−0.10*	*3.10*	*12.04*	*0.00*	*−0.05*	*0.23*
CAN	*1.02*	*−0.04*	*−0.18*	*−0.17*	*−0.33*	*0.34*
CCK	0.00	*5.98*	*1.34*	*−0.89*	*16.15*	na
CHE	*−1.57*	*−0.89*	*−1.53*	*−0.61*	*−3.27*	*1.25*
CHL	*−0.51*	*−0.15*	*0.08*	*0.02*	*−0.14*	*0.00*
CHN	*1.55*	*0.88*	*−0.33*	*1.95*	*0.32*	*1.04*
CIV	*−1.31*	*−1.03*	*−2.21*	*−0.01*	*−0.02*	*0.00*
CMR	*−0.06*	*0.08*	*−0.04*	*0.01*	*−0.05*	*0.02*
Country	ISIC Rev. 3, II-digit industries (table continued)					
	28	*29*	*31*	*32*	*33*	24
COD	*0.01*	*−0.73*	*−0.03*	*0.00*	*0.02*	0.00
COG	*0.00*	*−0.11*	*0.09*	*0.00*	*0.04*	0.00
COK	0.00	*0.32*	*0.01*	*0.12*	*−36.05*	na
COL	*−0.81*	*−0.19*	*0.38*	*0.00*	*−0.12*	*0.00*
COM	*−0.13*	*0.67*	*−0.04*	*0.00*	*−0.18*	na
CPV	*1.70*	*−0.11*	*−0.25*	*0.62*	*−0.45*	0.00
CRI	*0.12*	*0.03*	*−0.22*	*0.00*	*1.13*	*0.00*
CUB	*0.18*	*−0.76*	*−0.05*	*0.00*	*−0.19*	0.00
CXR	*2.03*	*4.00*	*2.03*	na	*0.00*	na
CYM	*0.01*	*−2.27*	*−1.47*	*0.02*	*−0.03*	na
CYP	*−0.42*	*−2.79*	*−0.27*	*0.09*	*−3.32*	*0.00*
CZE	*0.32*	*0.87*	*−1.08*	*0.18*	*0.58*	*−0.01*
DEU	*−0.97*	*0.20*	*0.46*	*−0.12*	*0.36*	*−4.71*
DJI	*0.36*	*−4.61*	*−4.51*	*0.01*	*0.06*	*−0.06*
DMA	*1.63*	*−20.72*	*−3.48*	*−0.16*	*−0.35*	na
DNK	*7.49*	*5.54*	*10.13*	*−0.38*	*1.03*	*0.15*
DOM	*0.14*	*0.48*	*−1.05*	*−0.13*	*0.35*	*0.00*
DZA	*0.00*	*−0.02*	*0.01*	*−0.01*	*0.01*	0.00
ECU	*0.00*	*−0.22*	*0.05*	*0.00*	*−0.01*	*0.01*
EGY	*2.61*	*−0.77*	*0.22*ara>	*0.01*	*4.79*	*0.82*
ERI	0.00	*−5.76*	*−1.07*	0.00	*−0.52*	0.00
ESH	na	0.00	na	na	na	na
ESP	*1.65*	*0.52*	*2.49*	*−0.19*	*−0.19*	*0.13*
EST	*0.77*	*−1.15*	*2.89*	*−0.07*	*0.09*	*0.85*
ETH	*0.03*	*−0.06*	*0.53*	*0.00*	*0.68*	0.00
FIN	*−0.35*	*1.05*	*0.20*	*−0.01*	*−3.66*	*0.20*

(continued)

Table 13.5 (continued)

Country	ISIC Rev. 3, II-digit industries (table continued)					
	28	*29*	*31*	*32*	*33*	24
FJI	*0.21*	*0.04*	*0.51*	*0.07*	*0.14*	0.00
FLK	*−0.07*	*−5.38*	*−0.19*	0.00	*−0.47*	na
FRA	*−0.49*	*0.46*	*−0.41*	*0.03*	*−0.05*	*1.63*
FRO	*0.00*	*−2.06*	*−0.47*	*0.00*	*0.01*	0.00
FSM	0.00	*0.24*	0.00	0.00	0.00	na
GAB	*0.07*	*−0.04*	*−0.04*	*0.00*	*−0.12*	*0.01*
GBR	*−0.47*	*−0.13*	*−0.37*	*−0.46*	*0.21*	*−0.45*
GEO	*−0.52*	*0.15*	*0.09*	*0.03*	*−0.54*	*0.17*
GHA	*0.04*	*0.22*	*−0.78*	*0.01*	*0.02*	*0.00*
GIB	*−0.23*	*0.04*	*0.15*	*−0.84*	*−0.21*	0.00
GIN	*0.00*	*−2.23*	*0.10*	*0.02*	*0.01*	0.00
GMB	*0.00*	*0.22*	*0.99*	*0.05*	*−0.02*	na
GNB	0.00	*−0.25*	*−0.08*	*0.00*	0.00	na
GNQ	*0.00*	*0.10*	*−0.02*	*0.00*	*0.01*	na
GRC	*−2.00*	*0.77*	*6.57*	*0.01*	*−8.04*	*0.03*
GRD	*0.08*	*0.80*	*−0.70*	*0.04*	*0.99*	na
GRL	*0.02*	*−0.07*	*−0.03*	*0.00*	*−0.35*	0.00
RWA	*0.03*	*−0.32*	*0.18*	*0.05*	*0.50*	*0.03*
SAU	*−0.71*	*0.13*	*−0.08*	*0.00*	*−0.02*	*−0.43*
SDN	*0.00*	*0.00*	*−0.17*	*0.00*	*−0.06*	0.00
SEN	*0.31*	*−0.07*	*−0.93*	*0.01*	*−0.16*	0.00
SGP	*−0.31*	*−0.08*	*0.04*	*0.11*	*−0.10*	*0.30*
SHN	*−10.52*	*29.78*	*0.08*	*0.02*	*−2.88*	na
SLB	0.00	*−1.15*	*0.36*	*0.03*	*−0.04*	0.00
SLE	*0.00*	*−14.77*	*−0.62*	*−0.16*	*−0.74*	0.00
SLV	*−0.46*	*0.65*	*0.03*	*0.00*	*0.07*	*0.00*
SMR	*−0.41*	*5.17*	*2.81*	na	*0.02*	na
SOM	*0.04*	*−0.82*	*−0.12*	*0.01*	*−0.10*	na
SPM	0.00	*2.85*	*0.95*	0.00	*12.40*	na
STP	*6.99*	*−1.39*	*6.20*	*0.03*	*0.54*	na
SUR	*0.00*	*−0.27*	*1.92*	*0.00*	*−0.55*	0.00
SVK	*−6.14*	*−1.67*	*0.32*	*0.05*	*−6.48*	*0.04*
Country	ISIC Rev. 3, II-digit industries (table continued)					
	28	Country	*28*	Country	*28*	Country
SVN	*−0.73*	*−2.14*	*−0.09*	*0.44*	*−74.91*	*0.27*
SWE	*−0.21*	*−1.85*	*−0.76*	*−0.44*	*0.59*	*15.52*
SWZ	*0.05*	*−0.16*	*−0.40*	*−0.01*	*0.16*	0.00

(continued)

Table 13.5 (continued)

Country	ISIC Rev. 3, II-digit industries (table continued)					
	28	Country	*28*	Country	*28*	Country
SYC	*0.21*	*0.15*	*−0.53*	*0.00*	*−0.78*	na
SYR	*0.58*	*0.51*	*−0.01*	*0.00*	*−0.41*	*0.00*
TCA	*9.07*	*2.43*	*−23.65*	0.00	*19.37*	na
TCD	*−0.10*	*0.04*	*−0.30*	*0.01*	*0.00*	0.00
TGO	*−0.43*	*0.14*	*−0.23*	*0.01*	*−0.06*	0.00
THA	*0.62*	*0.82*	*−0.12*	*0.15*	*−0.31*	*0.03*
TJK	*0.01*	*−1.18*	*0.05*	*−0.01*	*139.53*	0.00
TKL	*204.68*	*−9.29*	*85.36*	na	*14.77*	na
TKM	*0.00*	*0.08*	*−0.03*	*0.00*	*9.03*	0.00
TLS	0.00	*−12.12*	*0.07*	na	*−9.64*	na
TON	0.00	*1.88*	*1.97*	0.00	*1.24*	na
TTO	*−0.10*	*−1.25*	*−0.62*	*−0.06*	*−0.13*	*−0.01*
TUN	*−0.96*	*0.48*	*1.47*	*0.16*	*42.82*	*0.00*
TUR	*−0.60*	*0.64*	*0.34*	*0.21*	*0.09*	*0.03*
TUV	*0.43*	*0.02*	*−0.02*	na	0.00	na
CN-TWN	*−0.14*	*0.11*	*−1.75*	*2.42*	*−0.23*	*0.26*
TZA	*0.98*	*−1.67*	*−2.76*	*0.03*	*−0.07*	*0.00*
UGA	*0.34*	*0.11*	*−0.01*	*0.09*	*0.19*	0.00
UKR	*−4.03*	*−0.21*	*−0.59*	*0.00*	*−0.27*	*−2.27*
UMI	*1.05*	*124.68*	*7.78*	na	na	na
URY	*−0.20*	*−0.30*	*0.90*	*0.00*	*−0.24*	*−0.01*
USA	*−0.17*	*−0.22*	*−0.23*	*−0.79*	*−1.43*	*0.75*
UZB	*0.04*	*−0.26*	*−0.26*	*0.00*	*60.20*	0.00
VAT	na	*−1,480*	*11.03*	na	na	na
VCT	*0.16*	*0.00*	*−1.19*	*0.02*	*0.00*	0.00
VEN	*−5.23*	*−0.09*	*−0.08*	*0.00*	*−0.01*	*0.00*
VGB	*0.19*	*−1.53*	*1.11*	*0.10*	*0.13*	na
VNM	*1.28*	*0.20*	*−5.08*	*2.13*	*0.87*	*0.01*
VUT	0.00	*0.09*	*0.02*	*0.24*	*−0.01*	na
WLF	0.00	*83.51*	0.00	0.00	0.00	na
WSM	*3.74*	*2.23*	*1.83*	*0.02*	*0.31*	na
YEM	*0.00*	*−0.11*	*−0.10*	*0.00*	*−0.10*	0.00
ZAF	*−3.55*	*1.57*	*−0.69*	*0.10*	*0.21*	*0.06*
ZMB	*0.02*	*−3.57*	*1.22*	*0.00*	*−4.81*	0.00
ZWE	*−0.26*	*−0.21*	*0.05*	*−0.04*	*−0.03*	0.00

na = not available

By country ISIC Rev. 3, II-digit industries for Rank 1 and Rank 2 technologies; country codes

Table 13.6 Change between average 2015–2019 and 1995–1999 revealed comparative advantages (ΔRCA) by region and ISIC Rev. 3 IV-digit industries

ISIC Rev. 3, IV-digit industry Region	ΔRCA Distribution			
	Mean ΔRCA	Lower quartile	Median	Upper quartile
2413—Plastics in primary forms and of synthetic rubber				
East Asia and Pacific	0.164	0.000	0.013	0.242
Europe and Central Asia	0.500	0.000	0.036	0.216
Latin America and Caribbean	*−0.007*	*−0.002*	0.000	0.000
Middle East and North Africa	0.023	*−0.001*	0.000	0.005
North America	0.365	0.170	0.341	0.547
South Asia	0.194	*−0.015*	*−0.002*	0.001
Sub-Saharan Africa	0.010	0.000	0.000	0.004
2811—Structural metal products				
East Asia and Pacific	1.862	*−0.088*	0.000	0.361
Europe and Central Asia	*−0.241*	*−0.602*	*−0.169*	0.314
Latin America and Caribbean	*0.051*	*−0.160*	0.005	0.185
Middle East and North Africa	0.983	*−0.539*	0.002	0.364
North America	0.295	*−0.069*	0.037	0.530
South Asia	*−1.076*	*−0.550*	*−0.033*	0.028
Sub-Saharan Africa	0.156	*−0.041*	0.025	0.231
2912—Pumps, compressors, taps and valves				
East Asia and Pacific	0.092	*−0.084*	0.003	0.138
Europe and Central Asia	0.066	*−0.461*	*−0.125*	0.488
Latin America and Caribbean	*−0.249*	*−0.085*	*−0.008*	0.193
Middle East and North Africa	*−0.270*	*−0.105*	*−0.041*	0.008
North America	*−0.037*	*−0.139*	*−0.045*	0.061
South Asia	0.095	*−0.170*	0.028	0.208
Sub-Saharan Africa	*−0.295*	*−0.057*	0.015	0.082
2913—Bearings, gears, gearing and driving elements				
East Asia and Pacific	*−0.313*	*−0.002*	0.037	0.353
Europe and Central Asia	*−0.466*	*−0.288*	*−0.023*	0.177
Latin America and Caribbean	*−0.010*	*−0.061*	0.000	0.085
Middle East and North Africa	*−0.078*	*−0.114*	*−0.030*	0.001
North America	0.117	0.081	0.102	0.145
South Asia	*−0.052*	*−0.102*	*−0.025*	0.009
Sub-Saharan Africa	*−0.212*	*−0.107*	*−0.001*	0.036
2919—Other general purpose machinery				
East Asia and Pacific	*−0.306*	*−0.048*	0.035	0.321
Europe and Central Asia	0.186	*−0.055*	0.290	0.661

(continued)

Table 13.6 (continued)

ISIC Rev. 3, IV-digit industry Region	ΔRCA Distribution			
	Mean ΔRCA	Lower quartile	Median	Upper quartile
Latin America and Caribbean	*−0.508*	*−0.321*	*−0.061*	0.139
Middle East and North Africa	*−0.710*	*−0.379*	*−0.010*	0.168
North America	*−0.174*	*−0.209*	*−0.136*	*−0.120*
South Asia	0.103	*−0.076*	0.079	0.336
Sub-Saharan Africa	*−0.093*	*−0.133*	*−0.002*	0.114
3110—Electric motors, generators and transformers				
East Asia and Pacific	*−0.160*	*−0.332*	*−0.020*	0.287
Europe and Central Asia	0.663	*−0.453*	*−0.011*	0.469
Latin America and Caribbean	*−2.077*	*−0.960*	*−0.104*	0.057
Middle East and North Africa	*−0.158*	*−0.236*	0.005	0.358
North America	*−0.222*	*−0.338*	*−0.238*	*−0.113*
South Asia	*−0.001*	*−0.009*	0.046	0.116
Sub-Saharan Africa	0.437	*−0.272*	*−0.021*	0.186
3120—Electricity distribution and control apparatus				
East Asia and Pacific	*−1.046*	*−0.032*	0.010	0.073
Europe and Central Asia	*−0.010*	*−0.187*	0.034	0.517
Latin America and Caribbean	*−0.392*	*−0.090*	*−0.002*	0.123
Middle East and North Africa	0.089	*−0.052*	0.009	0.182
North America	*−0.058*	*−0.136*	*−0.013*	0.043
South Asia	*−0.761*	*−0.161*	*−0.020*	0.023
Sub-Saharan Africa	*−0.090*	*−0.014*	*0.000*	0.012
3140—Accumulators, primary cells and primary batteries				
East Asia and Pacific	0.771	*−0.036*	0.008	0.610
Europe and Central Asia	0.103	*−0.334*	*−0.044*	0.076
Latin America and Caribbean	0.041	*−0.040*	0.008	0.082
Middle East and North Africa	*−0.029*	*−0.024*	*−0.001*	0.056
North America	*−0.018*	*−0.115*	0.051	0.114
South Asia	0.195	0.009	0.024	0.212
Sub-Saharan Africa	0.005	*−0.030*	0.004	0.032
3190—Other electrical equipment n.e.c				
East Asia and Pacific	0.220	*−0.079*	0.000	0.440
Europe and Central Asia	*−0.055*	*−0.064*	0.009	0.210
Latin America and Caribbean	0.124	*−0.002*	0.001	0.016
Middle East and North Africa	0.031	0.000	0.001	0.050

(continued)

Table 13.6 (continued)

ISIC Rev. 3, IV-digit industry Region	ΔRCA Distribution			
	Mean ΔRCA	Lower quartile	Median	Upper quartile
North America	0.118	*−0.117*	0.000	0.294
South Asia	0.030	*−0.007*	0.009	0.051
Sub-Saharan Africa	0.038	0.000	0.000	0.004
3210—Electronic valves and tubes and other electronic components				
East Asia and Pacific	0.053	*−0.047*	0.000	0.091
Europe and Central Asia	*−0.013*	*−0.025*	0.005	0.055
Latin America and Caribbean	*−0.034*	*−0.009*	0.001	0.014
Middle East and North Africa	0.024	0.000	0.006	0.016
North America	*−0.320*	*−0.481*	*−0.168*	*−0.084*
South Asia	*−0.073*	*−0.033*	*−0.001*	0.048
Sub-Saharan Africa	0.021	*−0.002*	0.001	0.013
3312—Instruments and appliances for measuring, checking, testing, navigating and other purposes, except industrial process control equipment				
East Asia and Pacific	0.212	*−0.097*	0.000	0.281
Europe and Central Asia	2.657	*−0.460*	0.007	0.586
Latin America and Caribbean	1.929	*−0.234*	*−0.014*	0.216
Middle East and North Africa	2.986	*−0.145*	*−0.006*	0.204
North America	*−0.509*	*−0.883*	*−0.334*	*−0.047*
South Asia	*−0.033*	*−0.226*	*−0.006*	0.074
Sub-Saharan Africa	*−0.247*	*−0.165*	*−0.018*	0.026

preparedness. Governments need to embed green hydrogen in their overall industrial policy cycles, including the definition of technically sound diagnostics studies, the design of industrial strategies, the adoption of appropriate support instruments and the setting of suitable monitoring and evaluation protocols to adjust the implementation of measures over time. The modern industrial policy cycle needs to overcome traditional boundaries by embracing the principles of inclusive and sustainable industrial development. Policymakers must pursue win–win solutions by reconciling the promotion of manufacturing development, green growth and employment. Modern policies need to strengthen synergies, where possible, and mitigate trade-offs across different dimensions of sustainability when required. The work for the future industrialisation of Asian economies by boosting the penetration of technologies and the diffusion of knowledge and skills can go hand in hand with the creation of the preconditions to produce green hydrogen that can contribute to boosting green growth over time. Findings discussed in the previous sections imply that, whereas China is leading in Asia and the world in green hydrogen production, many other Asian countries are lagging. Lower-income Asian economies lack the appropriate capabilities, whereas countries with higher income levels may lack sufficient policy impulse for breakthrough technology penetration. An important implication is that Asian economies will need a wide array of policies for green hydrogen production: domestic to boost the internal market and international to encourage technology transfers; horizontal to set the right market functioning environment; and vertical to target the specific green hydrogen needs; individual when targeting particular problems pertaining to a specific pillar of sustainability (e.g., green electrification, storage of renewables) or integrated when pursuing synergies across different dimensions of sustainability (e.g., infrastructure creation, promotion of the production of green manufacturing goods). Addressing these challenges in Asian countries will depend on ensuring strong coordination across governments and stakeholders.

13.6 Future Directions

Analysing RCAs based on trade and production data provides insights into countries' technical abilities to produce hydrogen. However, green hydrogen, seen as a game-changer in global energy, faces significant challenges that hinder its adoption and market growth. Table 13.7 summarises the bottlenecks and challenges to the production of green hydrogen derived from the recent literature (Deloitte 2023; Khan and Al-Ghamdi 2023; Qyyum et al. 2021) used as a basis for the following discussions.

One of the primary barriers is the lack of adequate infrastructure and the high associated costs. This encompasses the entire value chain, including production, transportation, and consumption. Additionally, the availability and affordability of green hydrogen hinge on various factors, such as the cost of production and transport, which are influenced by the preferences and roles of different market players, including government entities, large corporations, and small private sector developers. Moreover, policy instruments play a crucial role in accelerating the development of the hydrogen market, emphasising the importance of supportive regulatory frameworks.

Significant technical hurdles persist despite hydrogen's environmental benefits and suitability for power generation and transportation. These include improving process efficiencies, lowering production costs, ensuring cost-effective transmission and distribution, and identifying inexpensive and sustainable feedstocks. Moreover, achieving a stable hydrogen economy requires addressing scientific, technological, social, and economic feasibility challenges. These encompass the entire spectrum of hydrogen production, storage, transmission, and distribution, each posing its unique set of obstacles that must be navigated.

Furthermore, land availability emerges as a potential bottleneck in densely populated countries, where scaling up green hydrogen production may necessitate significant land resources for solar and wind installations. This challenge is compounded by the low energy density per surface area of photovoltaic and wind power, further complicating large-scale deployment. The technical complexities associated with

Table 13.7 Bottlenecks and Challenges to the Production of Green Hydrogen

Area	Bottlenecks and challenges to the production of green hydrogen
Market demand	1. Low national demand for green hydrogen 2. Lack of carbon pricing and business models 3. Uncertainty about future prices and price premiums for green hydrogen 4. Challenges in signing long-term offtake contracts and lack of private sector participation
Technical and infrastructure challenges	1. Dependency on fossil fuels 2. Land availability and water-stressed regions posing challenges for green hydrogen production 3. High production costs of green hydrogen 4. Storage problems associated with green hydrogen 5. Multiplexity and fuzziness in understanding and integrating different hydrogen-related technologies 6. High transportation and storage costs 7. Lack of infrastructure for green hydrogen production, storage, and transportation
Geopolitical and social considerations	1. Slow reaction to the transition from oil and gas legacy positions 2. Safety considerations associated with production, storage, and transport 3. Social acceptance of green hydrogen technologies 4. Lack of collaboration between the public and private sectors
Government policies	1. Lack of or insufficient scope of public support schemes to develop hydrogen markets 2. Lack of integrated hydrogen strategy and a national hydrogen roadmap, particularly among developing countries, resulting in regulatory gaps, risks, and uncertainties affecting development and implementation

hydrogen transportation, characterised by its volatility and flammability, pose significant challenges, particularly regarding safety and affordability. These challenges underscore the multifaceted nature of the barriers hindering the production of green hydrogen and highlight the need for comprehensive strategies to address them effectively. Kannaiyan et al. 2023 point out that safe hydrogen transportation and storage represent another critical aspect of the hydrogen value chain. The refuelling station network needs to be developed for a more extensive penetration of hydrogen as a fuel in the transportation sector. The government may generate a big impulse to use green hydrogen by pushing the decarbonisation of heavy industries.

Compared with the literature covering non-Asian countries (e.g., Pepe, Ansari, and Gehrung 2023 and Bisognin Garlet et al. 2024), common bottlenecks observed among Asian and non-Asian countries are primarily concentrated on technical and infrastructure challenges, including infrastructure readiness, technical hurdles like cost-effective large-scale production and efficient storage, and high production costs. Moreover, there are also common bottlenecks regarding regulatory gaps and a lack of collaboration between the public and private sectors.

13.7 Conclusion

13.7.1 Decarbonisiing Asia Future Economies Through Green Hydrogen

This paper builds upon a taxonomy and methodology developed by Mueller and Eichhammer to analyse the extent to which countries possess preconditions to produce green hydrogen and current patterns of green hydrogen production. Results reveal that high-income and upper-middle-income countries are better positioned to diffuse these technologies as their current capabilities are higher. Results are robust in the context of Asian countries where China is leading the market and is continuing to consolidate its world leadership. In a world market where the expansion of green hydrogen is still moderate, in Asia, high-income countries like Japan will need to disclose their full production potential. In contrast, all countries, particularly lower-income countries, must develop their industrial capacity and reinforce their policy commitment. An industrial capability advantage only sometimes materialises in high-income countries, whereas lower-income countries will need to fill the technology gap with policy interventions aimed at boosting technology transfers, strengthening value chain governance, and shaping a well–functioning market context. The future decade will reveal if higher-income countries can exploit their advantage and if lower-income countries can catch up more rapidly. In the future, countries belonging to both groups may also need to cope with a deterioration of capabilities depending on the national and international circumstances affecting the poly-crisis and turbulent market challenges. Overcoming these crucial challenges will be vital in overcoming economic, technical, and regulatory bottlenecks.

13.7.2 Analysis Limitations and Future Research

Limitations of our analysis lie in using a stylised and non-exhaustive list of technologies necessary to produce green hydrogen and in using data that may not necessarily be consistent for a multi-dimensional analysis. Moreover, such efforts may be hindered by the lack of comprehensive global production data at the granular level required to assess country-level production capabilities. In particular, by mapping export capacity to industrial production capacity, analyses may introduce notable selection bias stemming from the lack of production data, particularly for developing economies and at the IV-digit level. Conversely, resorting to II-digit production data for remedy may, in turn, introduce notable aggregation bias as many IV-digit industries may not be related to green hydrogen production but are contained in the same II-digit sectors as IV-digit industries that do take part in the green hydrogen production network. To address the green hydrogen capabilities more precisely, future research could look at country-level production data and link this to sector-level trade aggregates provided in this study. Case studies will also be pivotal to complement analytical efforts with information from the field.

13.7.3 Disclaimers

13.7.3.1 UNIDO Disclaimer

The views expressed herein are those of the author(s) and do not necessarily reflect the views of the United Nations Industrial Development Organization. (As provided for in Administrative Circular UNIDO/DA/PS/AC.69 of 17. December 1990). This document has been produced without formal United Nations editing. The designations employed and the presentation of the material in this document do not imply the expression of any opinion whatsoever on the part of the Secretariat of the United Nations Industrial Development Organization (UNIDO) concerning the legal status of any country, territory, city or area or its authorities, or concerning the delimitation of its frontiers or boundaries, or its economic system or degree of development. Designations, such as 'developed', 'industrialised', and 'developing', are intended for statistical convenience and do not necessarily express a judgment about the stage reached by a particular country or area in the development process. Mentioning firm names or commercial products does not constitute an endorsement by UNIDO. The opinions, statistical data and estimates contained herein are the responsibility of the author(s) and should not necessarily be considered as reflecting the views of bearing the endorsement of UNIDO.

13.7.3.2 OECD Disclaimer

All opinions, statistical data, and estimates are personal and do not necessarily represent the views of the OECD. This document, as well as any data and map included herein, are without prejudice to the status of or sovereignty over any territory, the delimitation of international frontiers and boundaries, or the name of any territory, city, or area. All designations are intended for statistical convenience only.

13.7.3.3 EIB Disclaimer

The views expressed in this publication are those of the authors and do not necessarily reflect the position of the European Investment Bank.

Highlights

- Green hydrogen is essential for future decarbonisation goals.
- Asian countries are crucial to achieve the necessary green hydrogen production scale.
- Many Asian countries lack the industrial capacity to produce green hydrogen.
- Strengthening industrial capabilities is important for future economic growth and decarbonisation in Asia.
- Asia's market needs policy reform to expand green hydrogen production.

References

AIIB, Castalia (2023) Financing clean hydrogen in Asia and beyond: catalyzing the development of this emerging industry. https://www.aiib.org/en/about-aiib/who-we-are/infrastructure-for-tomorrow/green-infrastructure/pdf/AIIB-Castalia_Financing-Clean-Hydrogen-in-Asia-and-Beyond_Sep-2023.pdf. Accessed 31 May 2024

Amann J (2024) ComtradeRggregator: download and reclassify large chunks of trade data. https://amannj.github.io/comtradeRggregator. Accessed 31 May 2024

Cantore N, Olivari S, Lavopa A, Clara M (2018) Greening the structural change process: ranking sustainability of countries in climate change. Int J Sust Dev World 25(3):195–205. https://doi.org/10.1080/13504509.2017.1402832

Cantore N, Schlör H, Voegele S, Kuckshinrichs W, Haraguchi N, Nussbaumer P, Yan J (2021) Inclusive and sustainable industrial development: measurement approaches for energy transformation. Appl Energy 299:117277. https://doi.org/10.1016/j.apenergy.2021.117277

Clark WW, Rifkin J (2006) A green hydrogen economy. Energy Policy 34(17):2630–2639. https://doi.org/10.1016/j.enpol.2005.06.024

Comtrade, UN (2023). Global trade data. https://Comtradeplus.Un.Org/.2023. Accessed 31 May 2024

Cordonnier J, Saygin D (2022) Green hydrogen opportunities for emerging and developing economies, 205. https://doi.org/10.1787/53ad9f22-en. Accessed 31 May 2024

Deloitte (2023) Green hydrogen: energizing the path to net zero. https://www.deloitte.com/global/en/issues/climate/green-hydrogen.html. Accessed 31 May 2024

Garlet TB, Savian FS, Ribeiro JLD, Siluk JCM (2024) Unlocking Brazil's green hydrogen potential: overcoming barriers and formulating strategies to this promising sector. Int J Hydrogen Energy 49:553–570

Halkos G, Moll de Alba J, Bampatsou C (2024) Determinants of environmental efficiency and sources of productivity change in the manufacturing sector: a comparative analysis between Europe and Asia. Energy 291:130355. https://doi.org/10.1016/j.energy.2024.130355

Hamwey R, Pacini H, Assunção L (2013) Mapping green product spaces of nations. J Environ Development 22(2):155–168. https://doi.org/10.1177/1070496513482837

Hausmann R, Hidalgo CA, Bustos S, Coscia M, Simoes A, Yildirim MA (2014) The atlas of economic complexity: mapping paths to prosperity. The MIT Press.https://doi.org/10.7551/mitpress/9647.001.0001

IEA (2023a) Global hydrogen review 2023. https://doi.org/10.1787/cb2635f6-en. Accessed 31 May 2024

IEA (2023b) Net zero by 2050: A roadmap for the global energy sector. https://iea.blob.core.windows.net/assets/deebef5d-0c34-4539-9d0c-10b13d840027/NetZeroby2050-ARoadmapfortheGlobalEnergySector_CORR.pdf. Accessed 31 May 2024
IEA (2024) Global Hydrogen Review 2024. https://iea.blob.core.windows.net/assets/89c1e382-dc59-46ca-aa47-9f7d41531ab5/GlobalHydrogenReview2024.pdf. Accessed 31 May 2024
IPCC (2023) Summary for policymakers. In: IPCC (ed) Climate change 2023: Synthesis report. Contribution of Working Groups I, II and III to the Sixth Assessment Report of the Intergovernmental Panel on Climate Change. Geneva, IPCC
IRENA (2022) Global hydrogen trade to meet the 1.5°c climate goal: Part II—Technology review of hydrogen carriers. Abu Dhabi: IRENA. https://www.irena.org/-/media/Files/IRENA/Agency/Publication/2022/Apr/IRENA_Global_Trade_Hydrogen_2022.pdf?rev=3d707c37462842ac89246f48add670ba. Accessed 31 May 2024
Kannaiyan K, Lekshmi GS, Ramakrishna S, Kang M, Kumaravel V (2023) Perspectives for the green hydrogen energy-based economy. Energy 284:129358. https://doi.org/10.1016/j.energy.2023.129358
Khan MI, Al-Ghamdi SG (2023) Hydrogen economy for sustainable development in GCC countries: a SWOT analysis considering current situation, challenges, and prospects. Int J Hydrogen Energy 48(28):10315–10344. https://doi.org/10.1016/j.ijhydene.2022.12.033
Klevstrand A (2023) Which ten countries will be the biggest producers of green hydrogen in 2030? https://www.hydrogeninsight.com/production/exclusive-which-ten-countries-will-be-the-biggest-producers-of-green-hydrogen-in-2030-/2-1-1405571. Accessed 31 May 2024
Leromain E, Orefice G (2014) New revealed comparative advantage index: dataset and empirical distribution. Int Econ 139:48–70. https://doi.org/10.1016/j.inteco.2014.03.003
Liu W, Wan Y, Xiong Y, Gao P (2022) Green hydrogen standard in China: standard and evaluation of low-carbon hydrogen, clean hydrogen, and renewable hydrogen. Int J Hydrogen Energy 47(58):24584–24591. https://doi.org/10.1016/j.ijhydene.2021.10.193
Müller VP, Eichhammer W (2023) Economic complexity of green hydrogen production technologies—a trade data-based analysis of country-specific industrial preconditions. Renew Sustain Energy Rev 182:113304. https://doi.org/10.1016/j.rser.2023.113304
Ohno T, Nishida Y, Ishihara T, Hirose A (2022) Re-examining Japan's hydrogen strategy: moving beyond the "hydrogen society" fantasy. https://www.renewable-ei.org/pdfdownload/activities/REI_JapanHydrogenStrategy_EN_202209.pdf. Accessed 31 May 2024
Panchenko VA, Daus Y, Kovalev A, Yudaev I, Litti YV (2023) Prospects for the production of green hydrogen: review of countries with high potential. Int J Hydrogen Energy 48(12):4551–4571. https://doi.org/10.1016/j.ijhydene.2022.10.084
Pepe JM, Ansari D, Gehrung RM (2023) The geopolitics of hydrogen: Technologies, actors, and scenarios until 2040. 13. SWP Research Pape. Berlin: German Institute for International and Security Affairs. https://doi.org/10.18449/2023RP13v02. Accessed 31 May 2024
Qyyum MA, Dickson R, Shah SFA, Niaz H, Khan A, Liu JJ, Lee M (2021) Availability, versatility, and viability of feedstocks for hydrogen production: product space perspective. Renew Sustain Energy Rev 145:110843. https://doi.org/10.1016/j.rser.2021.110843
R Core Team (2024) R: A language and environment for statistical computing'. Vienna, Austria: R Foundation for Statistical Computing. https://www.R-project.org/. Accessed 31 May 2024
RECAP (2022) Perception of the implementation of a hydrogen economy in Asia-pacific: an expert survey. Available via Regional Project Energy Security and Climate Change Asia-Pacific (RECAP) Konrad-Adenauer-Stiftung e.V. https://www.kas.de/en/web/recap/single-title/-/content/perception-of-the-implementation-of-a-hydrogen-economy-in-asia-pacific. Accessed 31 May 2024
The Growth Lab at Harvard University (2019) International trade data (HS, 92). Harvard Dataverse. https://doi.org/10.7910/DVN/T4CHWJ. Accessed 31 May 2024
Timilsina GR (2021) Are renewable energy technologies cost competitive for electricity generation? Renew Energy 180:658–672. https://doi.org/10.1016/j.renene.2021.08.088

WEF (2023) Green hydrogen in China: a roadmap for progress. https://www.weforum.org/publications/green-hydrogen-in-china-a-roadmap-for-progress/. Accessed 31 May 2024
World Energy Council, EPRI (2021) Hydrogen on the horizon: ready, almost set, go?' Innovation insights briefing. London: UK. https://www.worldenergy.org/assets/downloads/Innovation_Insights_Briefing_-_Hydrogen_on_the_Horizon_-_Ready%2C_Almost_Set%2C_Go_-_July_2021.pdf. Accessed 31 May 2024
WTO, IRENA (2023) International trade and green hydrogen: Supporting the global transition to a low-carbon economy. https://www.irena.org/Publications/2023/Dec/International-trade-and-green-hydrogen-Supporting-the-global-transition-to-a-low-carbon-economy. Accessed 31 May 2024

Juergen Amann is an Economist at the OECD Centre for Entrepreneurship, SMEs, Regions and Cities (CFE). His main research interests lie in applied econometrics, environmental economics, industrial development, Input-Output analysis, price setting and structural change. Juergen has a Ph.D. from the University of Nottingham and studied at Duke University on a Fulbright scholarship, the University of Manchester as well as the Vienna University of Economics and Business. He holds graduate degrees in economics and econometrics. Juergen has collaborated with the European Central Bank, the World Bank and the United Nations Industrial Development Organization (UNIDO) on multiple research and policy projects.

Nicola Cantore is a Research and Industrial Policy Officer at the Division of Industrial Policy Research and Statistics at UNIDO, currently seconded at the European Investment Bank in Luxembourg as a senior economist. He is active in policy advice, capacity building and research. He is responsible for the UNIDO work on the country industrial diagnostics strategic studies. Nicola holds a Ph.D. in environmental economics and management at the University of York and a Ph.D. in Economics at the Universita' Cattolica del Sacro Cuore of Milan. He has extensively published articles in peer-reviewed journals, books and outlets concerning economic aspects of industrial development, structural transformation, energy transition and the environment.

Upalat Korwatanasakul is an Associate Professor at the School of Social Sciences, Waseda University. He is an international development economist interested in economic, environmental, and social development topics in Asia and the Pacific, such as climate change, global value chains, and sustainable development. He has worked and had research collaborations with major international organisations and research institutes, such as the Asian Development Bank Institute (ADBI) and specialised agencies and programmes of the United Nations. He holds a Ph.D. in International Studies, majoring in Development Economic Analysis from the Graduate School of Asia-Pacific Studies, Waseda University (Tokyo, Japan).

Jaime Moll de Alba is the Director of the Division of Strategic Programming, Results Monitoring and Quality Assurance at the United Nations Industrial Development Organization (UNIDO). He holds a Ph.D., a Diploma of Advanced Studies, and a Research Sufficiency from Universitat Politècnica de Catalunya (UPC), where he worked as a Visiting Lecturer. He graduated from UPC as an Industrial Engineer specialising in Industrial Organization. His research focuses on sustainable economic development, green industrial performance and structural transformation. Jaime is an Editorial Board Member of Discover Sustainability, Associate Editor of Frontiers Environmental Economics—Resource Economics, and Reviewer Board Member of Economies.

Correction to: Green Hydrogen Technology: Analysis of the Economic Preconditions in the Future Development of Asia

Juergen Amann, Nicola Cantore, Upalat Korwatanasakul, and Jaime Moll de Alba

Correction to: Chapter 13 in: E. Lau et al. (eds.), *Future of Economy in Asia,* https://doi.org/10.1007/978-3-031-95500-6_13

In the original version of this book, the following belated correction have been incorporated: In the original version of chapter 13, the acronym TWN has been updated as CN-TWN. The book and the chapter have been updated with the changes.

The updated version of this chapter can be found at
https://doi.org/10.1007/978-3-031-95500-6_13

E. Lau et al. (eds.), *Future of Economy in Asia*,
https://doi.org/10.1007/978-3-031-95500-6_14

Index

E. Lau et al. (eds.), *Future of Economy in Asia*,
https://doi.org/10.1007/978-3-031-95500-6

MIX
Papier aus verantwortungsvollen Quellen
Paper from responsible sources
FSC® C105338

If you have any concerns about our products,
you can contact us on
ProductSafety@springernature.com

In case Publisher is established outside the EU,
the EU authorized representative is:
Springer Nature Customer Service Center GmbH
Europaplatz 3, 69115 Heidelberg, Germany

Printed by Libri Plureos GmbH
in Hamburg, Germany